Ontario

THE REPORT OF THE

Royal Commission on Electric Power Planning

Chairman: Arthur Porter

February 1980

Published by the Royal Commission on Electric Power Planning
Printed by J.C. Thatcher, Queen's Printer of Ontario

ISBN: The Report (9 volumes): 0-7743-4672-8
ISBN: Volume 1: 0-7743-4663-9

Design and production management: Ken Slater
Photocomposition: Shirley Berch
Text management and photocomposition facilities: Alphatext Limited

Graphics: Acorn Technical Art

Editors: T.C. Fairley & Associates; R.A. Grundy & Associates

For the Commission: Ann Dyer, Editorial Coordinator; DebbieAnne Chown, Terminal Operator

Ontario

Royal Commission
on Electric Power
Planning

416/965-2111

7th Floor
14 Carlton Street
Toronto Ontario
M5B 1K5

29 February 1980

To Her Honour
The Lieutenant-Governor
The Province of Ontario

May It Please Your Honour:

We, the undersigned Commissioners, having been appointed on the 17th day of July 1975 to examine the long-range planning concepts of Ontario Hydro and matters related thereto, and having performed the duties as set out in the Order-in-Council and its subsequent amendments, submit herewith our report for your gracious consideration.

Arthur Porter
Chairman

George A. McCague
Commissioner

Solange Plourde-Gagnon
Commissioner

William W. Stevenson
Commissioner

Royal Commission on Electric Power Planning

ARTHUR PORTER, Chairman
ROBERT E. E. COSTELLO,* Commissioner
GEORGE A. McCAGUE, Commissioner
SOLANGE PLOURDE-GAGNON, Commissioner
WILLIAM W. STEVENSON, Commissioner

ANN DYER, Executive Programme Co-ordinator
FREDERICK R. HUME, Q.C., Legal Counsel
PETER G. MUELLER, Senior Advisor
ROBERT G. ROSEHART,† Scientific Counsellor
ROBERT A. SCOTT, Q.C.,‡ Legal Counsel
RONALD C. SMITH, Executive Director

*Resigned on 9 May 1977 due to ill health
†Resigned on 30 June 1977 to become Dean of University Schools, Lakehead University
‡Lost his life in the air crash on September 4, 1976

Previous publications of the Royal Commission on Electric Power Planning

Shaping the Future. The first report by the Royal Commission on Electric Power Planning. Toronto, 1976

The Meetings in the North. Toronto, 1977

Outreach Guidebook. Toronto, 1976

Issue Paper 1: Nuclear Power in Ontario. Toronto, 1976

Issue Paper 2: The Demand for Electrical Power. Toronto, 1976

Issue Paper 3: Conventional and Alternate Generation Technology. Toronto, 1977

Issue Paper 4: Transmission and Distribution. Toronto, 1977

Issue Paper 5: Land Use. Toronto, 1977

Issue Paper 6: Financial and Economic Factors. Toronto, 1977

Issue Paper 7: The Total Electric Power System. Toronto, 1977

Issue Paper 8: The Decision-Making Framework and Public Participation. Toronto, 1977

Issue Paper 9: An Overview of the Major Issues. Toronto, 1977

A Race Against Time: Interim Report on Nuclear Power in Ontario. Toronto, 1978

Our Energy Options. Toronto, 1978

Report on the Need for Additional Bulk Power Facilities in Southwestern Ontario. Toronto, 1979

Report on the Need for Additional Bulk Power Facilities in Eastern Ontario. Toronto, 1979

List of Volumes

The Report of the Royal Commission on Electric Power Planning is comprised of the following volumes:

Volume 1: Concepts, Conclusions, and Recommendations

Volume 2: The Electric Power System in Ontario

Volume 3: Factors Affecting the Demand for Electricity in Ontario

Volume 4: Energy Supply and Technology for Ontario

Volume 5: Economic Considerations in the Planning of Electric Power in Ontario

Volume 6: Environmental and Health Implications of Electric Energy in Ontario

Volume 7: The Socio-Economic and Land-Use Impacts of Electric Power in Ontario

Volume 8: Decision-Making, Regulation, and Public Participation: A Framework for Electric Power Planning in Ontario for the 1980s

Volume 9: A Bibliography to the Report

VOLUME 1

Concepts, Conclusions, and Recommendations

The Commission

The Commission

Dr. ARTHUR PORTER, Chairman. Professor Emeritus of Industrial Engineering, University of Toronto, formerly Dean of Engineering, University of Saskatchewan, and Professor of Electrical Engineering, University of London. He is a member (former chairman) of the Canadian Environmental Advisory Council and was chairman of Ontario's Committee of Automation and Employment. Dr. Porter was elected a Fellow of the Royal Society of Arts in 1961 and a Fellow of the Royal Society of Canada in 1970.

ROBERT E.E. COSTELLO, Commissioner. Vice-President of Corporate Services, Abitibi Paper Company Limited, Mr. Costello brought to the Commission an extensive industrial background. He was a long-time resident of the North and an engineer by training. Mr. Costello resigned on May 9, 1977 due to ill health. He died in December of that same year.

SOLANGE PLOURDE-GAGNON, Commissioner. Journalist, a former Queen's Park correspondent for *Le Droit* and a commentator for the CBC, Madame Plourde-Gagnon is a mother of three and represented the consumer viewpoint to the Commission.

GEORGE A. McCAGUE, Commissioner. Aside from his own farming operation, Mr. McCague has served on the executive of many farm organizations, including the Ontario Federation of Agriculture, the Farm Products Marketing Board, and the Ontario Milk Commission.

Dr. WILLIAM W. STEVENSON, Commissioner. An economist by training, Dr. Stevenson was a member of the Ontario Energy Board, and brought to the Commission an extensive background and knowledge of the electric power industry.

Chairman's Acknowledgements

An inquiry of such broad scope as the RCEPP which has lasted almost five years, and the associated report, incorporating nine volumes, would not have been possible without the willing help of a large number of people. Indeed, the number is so large that it would be unrealistic to acknowledge their respective contributions individually — instead, albeit with no less sincerity, we do so collectively. The names of many of these people are inscribed not once but many times in the tens of thousands of pages of the Commission's official transcripts. The majority are citizens of Ontario. Additionally, and most important, we had invaluable help in the conduct of the public information hearings, the early phases of the debate stage hearings, and in several seminars and symposia, from eminent Canadians domiciled across the country (from the shores of the Atlantic to the Pacific Ocean) and from scholars and scientists from the United States, the United Kingdom, and West Germany.

To put the planning concepts of Ontario's electric power system into a national framework, members of the Commission and/or staff met with senior members of provincial governments, major utilities, and agencies, and visited several government departments and federal and provincial institutions, notably Hydro-Québec, Manitoba Hydro, the Atomic Energy Control Board, Atomic Energy of Canada Limited, and the National Research Council. The level of co-operation and goodwill during all these visits was exemplary. In thanking all those who participated in the meetings we do so with a deep sense of national pride.

Members of the Commission and staff were fortunate as well to have the opportunity to visit Sweden, Denmark, West Germany, France, the United Kingdom, and the States of California, Arizona, Minnesota, Wisconsin, and New York. Furthermore, three full days of briefing and meetings were held in Washington, D.C. We are grateful to His Excellency the Canadian Ambassador, to the Minister, and to the staff of the Canadian Embassy for the excellent arrangements that were made on our behalf. These visits, which for the most part resulted from suggestions by participants during the hearings, served admirably to provide a global framework for our investigation, especially with respect to energy conservation, alternate energy sources, environmental impacts, and nuclear power. We are extremely grateful for the considerable help and the many kindnesses we received.

We gratefully acknowledge also the excellent co-operation, throughout the inquiry, of several ministries of the Government of Ontario, and of Fisheries and Environment Canada. Many public and special interest groups, industrial associations, public utility commissions, municipalities and townships across the province, and by no means least, individual citizens, presented briefs and participated actively in hearings, workshops, and seminars; their involvement was both central and essential.

We are particularly grateful for the excellent co-operation, during the whole inquiry, of Ontario Hydro; this made the task of the Commission appreciably less arduous. This acknowledgement is coupled with the names of Bob Taylor, Chairman of the Board until June 30, 1979; Bruce Campbell, Counsel for Ontario Hydro; Ian Wilson and John Dobson, who were responsible for liaison between the Corporation and the Commission. The major contributions of Dick Houston, Q.C., Brian Isbister, and Wes James, which were terminated so tragically on September 4, 1976, are recalled with gratitude.

We wish to express our special appreciation to the Staff of the Commission whose names are given in Appendix D, and especially to Ron Smith, Executive Director, Peter Mueller, Senior Advisor, and Ann Dyer, Executive Programme Coordinator, whose many contributions and tireless efforts were vital to the success of the inquiry. During the first 15 months of our work, Robin Scott, Q.C., our first legal counsel, guided us expertly through the intricacies of the Public Inquiries Act and acted as our close advisor and friend; his tragic death was a severe blow to the Commission. We were extremely fortunate when Fred Hume, Q.C., was appointed our legal counsel in September 1976. Through our lengthy (more than 2,000 hours) and vigorous schedule of hearings he provided us with the balanced perspective, and by no means least the sense of humour, which was so welcome and necessary. The Commission is very indebted to Sushil Choudhury, a member of the research team, who came to us from Vancouver on a two-year secondment from B.C. Hydro. We are grateful to Robert Bonner, Q.C., Chairman of the Board of B.C. Hydro, for making this excellent arrangement possible.

We take this opportunity to thank, most warmly, the Provincial Secretary of Resource Development and the Deputy Provincial Secretary for their enthusiastic support of the work of the Commission; this was especially helpful in view of the several major amendments to the original Order-in-Council. It is a pleasure as well to acknowledge the help of the staff of the Provincial Secretariat and of Mrs. Irene

Beattie of the Premier's Office for handling so expeditiously a myriad of administrative and budgetary matters. Throughout our public hearings the court reporters, Angus Stonehouse & Company Ltd., undertook an exacting task, all the more so because of the highly technical evidence, with competence and cheerfulness. We are grateful to them.

Table of Contents

Preface

Like a flock of geese
in the fog they passed
then returned home
this morning
In a flock of small birds
I saw them
they left me a feather
may it lighten as their work is realized — Anon

It is fitting that the Report of the Ontario Royal Commission on Electric Power Planning should begin with the above lines. They were received anonymously on September 6, 1976, at the memorial service held by Grand Council Treaty No. 9 in Timmins, Ontario, to honour those who lost their lives in a tragic accident on September 4, 1976, while participating in the work of the Commission. The contributions of these highly dedicated citizens of Ontario, during the first critical year of the Commission's inquiry, are acknowledged gratefully with a deep sense of pride.

The Ontario Royal Commission on Electric Power Planning was established on July 17, 1975, by Order-in-Council No. 2005B/75. Subsequently, amendments and additions to the Terms of Reference (see Appendix A) were promulgated through Orders-in-Council Nos. 3489/77 dated December 14, 1977, 2065/78 dated July 12, 1978, and 2837/79 dated October 24, 1979. The Commission's mandate, which is interpreted and discussed in Chapter 1, relates essentially to the long-range planning concepts and programmes of the province's electric power utility — Ontario Hydro — for the period 1983-93 and beyond.

The Commission, since its inception, has been particularly educationally oriented. Indeed, the purpose of the Preliminary Public Meetings (October 1975 to April 1976), of the Public Information Hearings (May 1976 to March 1977), and of numerous seminars, symposia, and workshops was to inform and educate the public, the government, and by no means least the Commission, concerning the fundamentals of electric power planning. To complement and supplement the public meetings we published *Shaping the Future*, our first report, which identified the issues; nine issue papers, which structured and explained the issues; and a report on the meetings in the north that addressed the special problems of the people of northern Ontario. Further, to encourage participation in the energy debate by students in schools, colleges, and universities, we initiated the Outreach programme and published a guidebook on energy issues. *Our Energy Options*, a series of seven essays on important electric power planning topics by well-known specialists in the field, was published in September 1978 with the primary object of encouraging energy education in the schools.

The Debate Stage Hearings (May 1977 to May 1979) and the Regional Public Hearings in southwestern Ontario (March 1979) and eastern Ontario (April 1979) gave opportunity for many members of the public to state their views and to debate a multiplicity of contentious issues. Noteworthy features of these hearings were their informality and the Commission's insistence on the complete openness of information. As required by the appropriate Orders-in-Council, we published *A Race Against Time, Interim Report on Nuclear Power in Ontario* in September 1978; the *Report on the Need for Additional Bulk Power Facilities in Southwestern Ontario* in June 1979; and the *Report on the Need for Additional Bulk Power Facilities in Eastern Ontario* in July 1979. In particular, it is gratifying to note that the *Interim Report on Nuclear Power* is being cited in the international literature on nuclear power.[1]

Because of the broad scope of the Terms of Reference, our inquiry has occupied four and one-half exciting, though arduous, years. The time scale was, in fact, extended appreciably by the amendments that required the Commission to report on nuclear power, and on the need in specific regions for additional bulk power facilities, prior to the publication of this Report.

Notwithstanding the technological basis of electric power, we have paid special attention to the social, environmental, health, economic, and political impacts of power planning. During the last 50 years, for example, the increasingly widespread use of electricity has transformed life-styles and industry in Ontario, and indeed in all industrialized countries. Further, of all modern technologies, electricity is probably the most taken for granted, and herein lies a major dilemma. In this dawning age of energy conservation, how can we moderate, voluntarily or through legislation, the use of that which is taken for granted?

The belief that our standard of living and level of employment are tied to an ever-increasing supply of energy is being eroded, but not fast enough. Most people now recognize that the world's primary resources of energy (oil, natural gas, coal, and uranium) are non-renewable. Energy supplies, especially of crude oil, are rapidly running out. It behooves mankind to invent and plan a future, which, while maintaining within reason present living standards, does so without recourse to continuing growth in the utilization of energy. Indeed, we have concluded that zero energy growth per capita by the year 2000, if not before, is a realistic and necessary goal for an industrial society such as exists in Ontario.

Our philosophy for survival, albeit survival with comparatively high living standards, is based essentially on the need to live within environmental constraints (i.e., the Spaceship Earth concept), to ensure adequate food supplies for all people, and to optimize our utilization of energy. Societal values are in a continous state of evolution, and technological developments are clearly a primary cause. Central among these are contemporary developments in energy and communications.

The Commission's Report comprises nine volumes: Volume 1 – "Concepts, Conclusions, and Recommendations" is, in effect, an overview volume; each chapter is reasonably self-contained. However, because of the complex interlinkages between major topics, there are unavoidable, and indeed desirable, redundancies between several chapters of this volume; for example, energy conservation and environmental impact are "horizontal" rather than "vertical" concepts. Volumes 2 to 8 may be thought of as background volumes. They are based on the transcripts of the public hearings, the exhibits, numerous technical memoranda, and research reports; they provide much of the factual evidence and information upon which our conclusions and recommendations are based. Each background volume was prepared by a team of Commission staff and/or outside consultants – the senior authors are identified in each volume. The conclusions and opinions expressed in these volumes are those of the authors and not necessarily those of the Commission. Volume 9 presents bibliographic material. The research and evolution of all volumes of the Report were directed and reviewed for the Commission by Philip A. Lapp and Peter G. Mueller.

Compendium of Recommendations

The Royal Commission on Electric Power Planning, having been empowered and instructed under Order-in-Council 2005B/75 dated the 17th day of July 1975, Order-in-Council 3489/77 dated the 14th day of December 1977, and Order-in-Council 2065/78 dated the 12th day of July 1978 to examine the long-range planning concepts of Ontario Hydro for the period of 1983-1993 and beyond, and matters related thereto, reports its recommendations below, taken from the appropriate chapters in this volume.

Ontario's Electric Power Requirements

3.1 Through the development of demand scenarios based on end-use data, future planning philosophy should be reoriented to emphasize demand management increasingly rather than maintain the focus on supply expansion, as is traditional.

3.2 A comprehensive energy end-use data base for the province should be developed as soon as possible, and Ontario Hydro, in addition to macro-economic or "top down" forecasting models, should develop complementary models based on the detailed building up of electricity demand on an end-use basis. Ontario and federal government ministries and agencies should support Ontario Hydro's efforts to fill the remaining data gaps.

3.3 Ontario Hydro should employ, as a useful analytical device for load-forecasting purposes, the distinction between "captive" and "competitive" end uses of electricity.

3.4 Because of increasing emphasis on end-use forecasting, the role of the public utility commissions in developing load growth patterns should be enhanced to provide opportunities for more input than hitherto by the public.

3.5 As part of a larger objective of planning for an improved annual load shape and higher load factors and as a means of increasing the resiliency of the electric power system and reducing Ontario's dependence on crude oil, Ontario Hydro should give high priority to demonstrating the technical and economic feasibility of new and retrofit hybrid electric/fossil space-heating systems.

3.6 For system planning purposes, Ontario Hydro should base its system expansion plan on a growth range for peak capacity to the year 2000 of 2.5 to 4.0 per cent per annum.

The Technology of Power Generation and Alternative Energy Sources

4.1 During the next decade the Ontario government and Ontario Hydro should actively support the demonstration of fluidized-bed combustion with special reference to its future role in the generation of electric power.

4.2 The Ontario government should support the demonstration of biomass energy projects, including gasification of forest and agricultural residues, testing methanol technologies, evaluating ethanol potential, and generation of biogas.

4.3 During the next decade the Ontario government should continue its programme to demonstrate the suitability of solar space heating and water heating in the Ontario context with special reference to its potential role in energy conservation.

4.4 The Ontario government and Ontario Hydro should make every effort to convert the "mothballed" gas-fired boilers at the R.L. Hearn Generating Station to burn refuse or refuse-derived fuels.

4.5 The Ontario government and Ontario Hydro should assign high priority to the demonstration of industrial co-generation.

4.6 The Ontario government should expand its efforts to put in place a low-temperature hot-water district heating system, to demonstrate its energy efficiency under Ontario conditions, and to test the use of conventional as well as renewable or non-conventional fuels, for the combined generation of heat and electricity.

Nuclear Power

5.1   Ontario Hydro should publish a report as soon as possible on the expected exposure levels resulting from any reactor re-tubing operation, addressing, in particular, the following questions:

- How many workers (Ontario Hydro employees and others) will be subjected to the 5 rem annual dose limit in connection with the re-tubing of a single reactor?

- Will workers be subject to high dose levels on a continuing basis when the re-tubing of the Pickering A and Bruce A reactors begins on a sequential basis?

- A worker could receive an aggregated dose of 50 rems over, say, a 15-year period. Is this medically acceptable? Should these exposures be age-dependent?

- What is the total number of workers required, on a continuing basis, to undertake re-tubing operations? Are that many adequately skilled workers at present available?

- To what extent can the re-tubing operation be undertaken by "remote control", thereby minimizing the aggregated exposures of workers?

- Will workers who may be subjected to higher-than-normal radiation doses, and their unions, be fully informed of the nature of the risk?

5.2   A new division devoted exclusively to nuclear power safety, reporting directly to the Executive Vice-President (Operations) of Ontario Hydro, should be established.

5.3   The new safety division recommended for Ontario Hydro should establish a small emergency task force, available 24 hours a day on an "on call" basis. This force should be one that could be transported expeditiously in an emergency, by road or helicopter or both, to any nuclear generating station in the province.

5.4   A systematic attempt should be made by Ontario Hydro to look for patterns in operating and accident experience available from both CANDU and other reactor systems. These patterns should be fed back into the process of setting design, operating, and safety criteria.

5.5   Operational procedures and especially the reporting systems at CANDU stations should be critically assessed to improve communication.

5.6   The current CANDU control room and indicator design should be reviewed and assessed from a human factors perspective to ensure that the equipment will display clear signals on reactor status to the operator under both normal and accident conditions.

5.7   The educational requirements and training programmes for all nuclear supervisory, operational, and maintenance personnel should be critically reviewed.

5.8   Provision should be made for the continuous updating and monitoring of the performance of all reactor operators and maintenance personnel; there should be much more imaginative use of simulators in this regard.

5.9   The Atomic Energy Control Board should establish a human factors group to ensure that human factors concepts and engineering become central elements in the safe design, construction, operation, and maintenance of Ontario's nuclear stations. Further, human factors concepts should be reflected in the licensing requirements for both nuclear stations and key operating personnel.

5.10   All aspects of contingency planning should be assessed in the light of the experience at Three Mile Island, and a comprehensive plan for each nuclear facility should be made publicly available. The public must be aware of these plans, which must be rehearsed regularly if they are to be credible. Special attention should be paid to preparing in advance for the sensitive and accurate handling of information during an accident.

5.11   Continuing epidemiologic evaluation of Elliot Lake miners and uranium mill workers should be undertaken. The public should be informed of the progress of these studies.

5.12   Ontario should contribute its share to any national programme for uranium mine and mill waste research.

5.13   Measures should be taken to ensure that the costs of long-term tailings monitoring, management, and R&D are reflected in the cost of uranium fuel rather than becoming a general charge to the Ontario taxpayer, not least because most of the uranium is currently being exported (over 90 per cent).

5.14   The future expansion of the nuclear power programme in Ontario, and in particular the uranium mining and milling portion of the fuel cycle, should be contingent on demonstrated progress in research and development with respect to both the short- and the long-term aspects of the low-level uranium tailings waste disposal problem, as judged by the provincial and federal regulatory agencies and the people of Ontario, especially those who would be most directly affected by uranium mining operations. It would be unacceptable to continue to generate these wastes in the absence of clear progress to minimize their impact on future generations.

5.15   All existing and planned Ontario Hydro nuclear stations should be retrofitted or designed for the interim storage on site of their spent fuel for the next 30 years by which time a disposal facility should be available.

5.16   An independent "nuclear waste social advisory committee" should be established to ensure that broad social, political, and ethical issues are addressed. This committee should be chaired by an eminent Canadian social scientist.

5.17   If progress in high-level nuclear waste disposal R&D, in both the technical sense and the social sense, is not satisfactory by at least 1990, as judged by the technical and social advisory committees, the provincial and federal regulatory agencies, and the people of Ontario – especially in those communities that would be directly affected by a nuclear waste disposal facility – a moratorium should be declared on additional nuclear power stations.

5.18   No further development of the 1,250 MW CANDU reactor, even in the concept stage, should be undertaken by Ontario Hydro. Any additional nuclear base-load power stations in the post-Darlington period should be based on 850 MW CANDU reactors. We believe that such standardization will facilitate reactor safety as well as optimizing the average capacity factors of these stations.

5.19   The Ontario government should advise the federal government that Ontario's requirements will be insufficient to ensure an order level of one reactor per year and, therefore, that the maintaining of CANDU as a viable option for the future suggests a need for urgent federal initiatives to fill the order gap. Our estimate of the likely total installed nuclear capacity in Ontario to the year 2000 is in the order of 17,500 MW; this means one additional 3,400 MW four-reactor nuclear station after Darlington, and it could be a high estimate, depending on, for example, actual load growth, success with conservation, co-generation, and potential imports of hydroelectric energy from Manitoba or Quebec. If the industry wishes to survive, it must begin to search for opportunities to diversify.

5.20   Although it is important to keep open the thorium fuel cycle option by engaging in an R&D programme, a firm decision to go ahead with a major demonstration and/or commercial programme should be delayed at least until 1990, and then made only if it is acceptable to the public after appropriate dialogue and study concerning the full implications and impacts of such a project.

5.21   Nuclear power should no longer receive the lion's share of energy R&D funding, and R&D priorities in the nuclear field should be focused primarily on the human factor in reactor safety, on the management and disposal of wastes at the front and back ends of the fuel cycle, and on the decommissioning of nuclear facilities.

5.22   Procedures should be established to ensure fair handling of *bona fide* cases of professional dissent. Procedures should include the following concepts:

  • Concerns should be expressed in writing and considered by a special review group consisting of representatives of management, professional engineering staff, and at least one outside expert.

  • The review group should obtain evidence from the dissenting staff member's colleagues.

  • The review group should assess management's response to the concerns.

5.23   Standard-setting for the nuclear fuel cycle should be done in an open manner, including opportunities for public participation in the process.

5.24   The role of the Atomic Energy Control Board on-site resident inspector should be strengthened and the reports of the inspector should be made public.

5.25   Advisory committees based on the social sciences should be established by the Atomic Energy Control Board.

5.26   Appropriate steps should be taken to guarantee that the Atomic Energy Control Board has adequate human and financial resources. The Atomic Energy Control Board, or its eventual successor, must not become a victim of government spending restraints.

5.27   The Government of Canada should ensure the separation of the promotional and regulatory aspects of nuclear power by drafting appropriate legislation to replace the Atomic Energy Control Act as a matter of the highest priority. This would ensure that the Atomic Energy Control Board and Atomic Energy of Canada Limited would report to separate ministers, reflecting their very different roles, thereby avoiding public confusion and possible conflicts of interest of the sort that have in the past strained public confidence in the regulatory process.

5.28   The Atomic Energy Control Board should expand its membership to include a broad representation of the general public as well as members of the scientific and technical community.

Bulk Power Transmission

6.1   Ontario Hydro should continue to undertake research and explore all alternatives that will permit the upgrading of existing transmission facilities and lead to optimizing the use of existing rights of way. Evidence of this research should routinely form part of Ontario Hydro's submission for approval of the acquisition of a new transmission corridor and/or the siting of a new transmission line.

6.2   Given the advances in converter technology that suggest that high-voltage direct current (HVDC) transmission has now become economically attractive for distances in excess of 650-800 km, Ontario Hydro should carefully re-examine the advantages of HVDC for the proposed east-west interconnection and study its application for the line connecting the proposed Onakawana generating station with load centres in southern Ontario.

6.3   Ontario Hydro should utilize even more imaginative approaches to public involvement in transmission routing. In particular, we believe the utility should leave more of the initiative in the public participation process to affected citizens, permitting those who will be most immediately impacted and involved to select alternate routes and to designate the preferred route; independence will be essential. The chairman of an appropriate citizens study committee should be selected by the citizens. Ontario Hydro should clearly state its criteria for routing, and this information with any other required by the committee should be readily provided by the utility. While the time period for study should be established by the utility, the procedures should be established by the study committee.

6.4   Ontario Hydro should take all possible steps to ensure the safety and convenience of all persons working in the vicinity of extra-high-voltage transmission lines.

6.5   Ontario Hydro should continue to plan the integrated electric power system on the basis of 500 kV and 230 kV transmission lines.

6.6   Ontario Hydro should work with the appropriate farm organizations and the Ministry of Agriculture on the design of an appropriate single-pole and/or lattice tower for use in cultivated fields.

6.7   The farming community with the collaboration of Ontario Hydro should develop, as soon as possible, alternative routes for a second 500 kV transmission line from the Bruce Generating Station that will have minimal and acceptable impact on Class 1 and Class 2 agricultural land. Ontario Hydro should provide the necessary funding.

6.8   In order to facilitate the co-operation of the farming communities, Ontario Hydro should not site a thermal generating station in the vicinity of Goderich or Kincardine, or indeed on the eastern shoreline of Lake Huron south of the Bruce Generating Station, before the year 2000. Ontario Hydro should make a public statement to this effect as soon as possible.

The Total Electric Power System

7.1   Ontario Hydro, working with the municipal electricity utilities, should give high priority to completing the load-management experiments now under way so that the technical problems, cost, and public acceptability of alternate systems can be assessed.

7.2   An in-depth study of the Commission's supply scenarios should be undertaken and the findings should be used as a basis for future planning of the electric power system.

7.3   The studies aimed at strengthening the electricity interchange capability with Quebec should be expedited, and in particular they should be extended to ensure close collaboration between Ontario Hydro and Hydro-Québec in the future planning of their respective systems for the mutual benefit of both provinces.

7.4   Ontario Hydro should co-operate with Manitoba Hydro in studies aimed at strengthening electricity interconnections and the purchase of substantial blocks of hydraulic power from the lower Nelson River; there should be closer collaboration between the two utilities in the future planning of their respective systems for the mutual benefit of the two provinces.

7.5   The interconnections between Ontario Hydro and neighbouring utilities in the United States should be strengthened.

Land Use

8.1   Ontario Hydro and the Ontario government should build on developments already taking place at the Bruce site to test further the concept of a combined energy centre as described in the Ministry of Industry and Tourism's 1976 report.

8.2   Ontario Hydro should accept financial responsibility for the debenture debt load of municipalities in the vicinity of the Bruce Generating Station that is over and above what would have been incurred in the absence of the Ontario Hydro projects.

8.3   Ontario Hydro should not proceed with land-banking programmes for at least the next 10 years.

8.4   Ontario Hydro's planning concepts should reflect the primary objective of conserving Ontario's food lands, particularly in southwestern Ontario.

8.5   The potential of Ontario's forest lands, especially in northern and eastern Ontario, as sources of energy should be the subject of an in-depth feasibility study; and, if the social, environmental, and economic indications are favourable for methanol or ethanol production, a demonstration plant should be built and tested as soon as possible.

8.6   The existing research and development programmes relating to energy plantations, especially the potential of the hybrid poplar in eastern Ontario, with emphasis on abandoned low-quality farmlands, should be expedited.

8.7   On strictly power-systems-planning and economic grounds, the Onakawana lignite deposits should be developed; and an electric power station of 800 MW-1,000 MW capacity should be built at the mine site. However, we recognize that the Royal Commission on the Northern Environment, on social and environmental grounds, with respect to both the power station and the associated transmission corridor, may not support this recommendation, and we believe that their views should have precedence.

Environmental Concerns

9.1   Ontario Hydro should not install sulphur scrubbers at its fossil-fuelled electric power stations as long as the existing policy of utilizing low-sulphur fuels is maintained.

9.2   Ontario Hydro and the Ministry of the Environment should strengthen existing air and water pollution monitoring systems, especially, although not exclusively, in the vicinity of thermal power stations, and environmental impact maps should be prepared for the benefit of the public.

9.3   Interdisciplinary institutes for environmental research in Ontario universities should be involved more actively in the environmental assessment process.

Energy Conservation

10.1   Over a period of 10-20 years, efficiency goals for all energy-intensive industrial processing equipment, machines, and systems should be established by the Ministry of Energy. In setting these goals, efficiency standards already being achieved in several foreign countries, notably Sweden and West Germany, should be taken into account. Efficiency goals should be applied in the first place to the pulp and paper industry, the iron and steel industry, the chemicals industry, the petroleum refining industry, and all heat-treating operations.

10.2   Mandatory heating, insulation, and lighting standards should be enacted for new residential and commercial construction, and these standards should take into account the optimum utilization of passive solar energy measures.

10.3   Progressively stricter efficiency standards for all major energy-consuming appliances, such as water heaters, refrigerators, home furnaces, and air-conditioners, should be put into effect through legislation.

10.4   Direct government loans and other economic incentives should be made available to finance the retrofitting of houses, multi-unit residences, and some commercial buildings with conservation equipment, including insulation and, where appropriate, storm windows and shutters.

Economic and Financial Factors

11.1   In formulating its industrial policy, Ontario should recognize the need for an adequate and competitively priced supply of electricity, but Ontario should not attempt to compete aggressively for power-intensive industry with provinces with large remaining hydraulic resources.

11.2   The Ontario government should continue to support Ontario Hydro's efforts to utilize its surplus generating capacity by undertaking interruptible or firm sales to neighbouring utilities that are both profitable and in the best interests of the people of Ontario. No firm-sale commitments should be made that might jeopardize the generation reserves required to meet Ontario requirements or tie up needed transmission capacity.

11.3   Ontario Hydro should perform system simulations to estimate more accurately the incremental costs of encouraging the substitution of electricity for fossil fuels, especially oil.

The Ministry of Energy should develop comparable cost estimates of alternative means to supply, or save, the same energy at point-of-end-use.

11.4   Time-differentiated electricity rates (seasonal and time-of-day) should be introduced as soon as possible to as many classes of customers as practicable. Seasonal rates should be introduced first, to ensure that the higher long-run costs of supplying low-load-factor space-heating loads are properly recovered. Time-of-day rates should be phased in as day-night electricity supply-cost differentials become significant and obstacles to metering small customers are overcome.

11.5   Means should be sought to ensure that all customers are made aware of the likely future trend in the costs of providing electricity service in each of the rating periods and end uses selected.

11.6   For rate-making purposes, Ontario Hydro should calculate marginal electricity supply costs in each "rating period" on the basis of the current system expansion plan, for comparison with the expected near-future accounting costs proposed by the Ontario Energy Board.

11.7   Ontario Hydro should include, in its tests of time-of-use rates, not only assessments of customer response concerning willingness to change personal energy habits, but also the required technology.

11.8   To encourage the prudent and efficient use of electricity, such features as declining block rates, uncontrolled flat-rate water heaters, and bulk metering of new electrically heated apartment buildings should be modified or eliminated.

11.9   Ontario Hydro should pursue vigorously the potential of the miniaturized solid-state (silicon chip) meter for mass application and include such meters in its current tests of load-management systems and time-of-use rates. A demonstration project involving perhaps 100 residential consumers should be set up during the next few years.

11.10   In analysing the options for increasing the province's capacity for energy self-sufficiency, a systems approach should be adopted in which the incremental costs of conventional electricity generation are compared with the unit costs of conservation or renewable energy technologies, taking into account the load characteristics of each end use.

11.11   Because of institutional and financial obstacles facing decentralized, heavily "front-ended", alternative energy and conservation programmes, and in view of the redeeming social importance of reducing Ontario's oil dependency, provincial loan guarantees, tax and fiscal incentives, and direct financial support should be made available to promote industrial co-generation, heat-loss and building-design standards aimed at optimizing energy-conservation investments, solar water heating, and passive solar systems. The setting up of a mini-utility, backed by the Ontario Energy Corporation, should be considered, to support industrial co-generation initiatives.

Decision-Making

12.1   Ontario Hydro should be encouraged to continue and, where necessary, to expand its public participation programme to ensure that the public is fully involved. Ontario Hydro should adopt joint planning processes whereby real decision-making authority is shared with, and in some cases (see recommendation 6.3) left to the initiative of, citizen representatives.

12.2　Ontario Hydro should ensure that the participants in the utility's participation programme have access to independent expertise whether the expertise is supportive of or opposed to Ontario Hydro's planning concepts.

12.3　In order to enhance the optimum utilization of electricity, both public utility commissions and the Regional Offices of Ontario Hydro should be adequately financed and encouraged to sponsor, in their areas, educational programmes, seminars, and workshops in energy utilization and conservation.

12.4　Ontario Hydro should find practical means to give effect to its commitment to greater openness by commencing to publish a technical-papers series, containing accounts of technical, scientific, and socio-economic research in language understandable to the layman. These publications should be made widely available to libraries across the province.

12.5　A clear statement of the objectives and responsibilities of the utility, especially as they relate to the social objectives as endorsed by government, should be issued by the Ministry of Energy.

12.6　The status of the existing Ontario Energy Board should be enhanced through expanded membership, representing a broad range of interests and disciplines, and the agency should be renamed the Ontario Energy Commission. It should be an authoritative and independent body.

12.7　The chairman of the recommended Ontario Energy Commission should be a person well known to the public and not associated with any of the special interests that should be represented.

12.8　As well as providing a vehicle for the consideration and examination of rate structures for both electricity and natural gas, the Ontario Energy Commission should be responsible for advising the government and people of Ontario on energy policy in general and on electric power planning in particular. The Ontario Energy Commission should be strongly future-oriented and just as strongly people-oriented.

12.9　The Ontario Energy Commission should be provided with a modest increment in staff and consulting budget over and above that of the existing Ontario Energy Board. The designation "Commission" as against "Board" was selected not only to suggest a break from the past but also to provide a broader umbrella to embrace a policy advisory function as well as the traditional regulatory function. The indications are that the additional staff requirements would be small.

12.10　The principle of funding of public interest groups from the public purse should be adopted in connection with energy and environmental hearings in the future. Only in this way will it be possible for disparate views to be aired adequately in public hearings.

The public interest funding programme should be improved in two areas:

- The requirement of adequate accounting practices should be written into contracts between the groups and the funding body.

- Wherever appropriate, an essentially inquisitorial rather than adversarial approach should be adopted in order to reduce the expenses incurred by participating groups.

CHAPTER ONE

The Commission's Mandate

The Royal Commission on Electric Power Planning was established by the Government of Ontario in response to broadly based concerns relating to Ontario Hydro's long-range planning proposals (February 1974),[1] expressed by public interest groups of which the farm groups of southwestern Ontario (with land-use concerns) and environmental groups across the province were the most persistent and articulate. Consequently, very comprehensive terms of reference were necessary to cover the multiplicity of issues that were identified – see Appendix A. Indeed, so complex and interlinked are the concepts and issues associated with energy policy and planning that we are devoting this chapter exclusively to structuring and interpreting them. This is necessary not least because, during the inquiry, there were repeated instances of misunderstandings of the purpose and scope of our task. These misunderstandings were manifest in numerous incorrect descriptions of the primary purpose of the Commission, which was variously stated as being to study and report on the health and safety aspects of nuclear power plants, to consider the land-use implications of the routing of specific bulk power transmission lines, to investigate the siting of power generation facilities, to examine the environmental impacts of generating stations and transmission lines, and to develop a framework for decision-making. In fact, we were required to investigate all of the above, and other issues relating to Ontario's electric power system, and to examine their interdependence.

Fig. 1.1: p. 10   Our Terms of Reference have an intrinsic structure, as shown schematically in Figure 1.1. Although the hearings and investigations relating specifically to the role of nuclear power in Ontario, which culminated in the publication of the *Interim Report on Nuclear Power in Ontario*,[2] and the activities associated with the regional reports on the need for bulk power facilities[3] are shown as independent entities in Figure 1.1, they were in fact handled in the context of the total electric power system.

The Concept of System Planning

All biological, economic, and technological systems consist of a vast number of processes and components that interact in complex ways; each specific system, moreover, interacts with the environment in which it exists. For example, energy systems incorporate decision-making processes, primary sources of energy, transportation, conversions from one form of energy into another, and the end use of the energy. The related environments[4] are population, socio-political institutions, the natural and man-made environments, economic and social factors, and the individual.

The planning of an energy system, and indeed of any man-made system, *ipso facto*, determines the system's evolution, which is based on the concepts of policy formulation, design, construction, operation, and monitoring. This concept of system evolution applies, of course, to Ontario's electric power Fig. 1.2: p. 11   system. The main steps and processes that are involved are shown schematically in Figure 1.2. In the diagram, "environment" should be interpreted in its broadest sense. It is particularly noteworthy, as indicated in Figure 1.2, that the operation of the system, to maintain viability, must be monitored continuously, and, most important and essential, so must the environment. The feedback links (represented by dotted lines) show symbolically the need to take the system's behaviour and performance into account in determining policy and in developing designs for future systems.

Although the Commission's mandate can be interpreted as relating to all the processes identified in the diagram, it is obvious that an inquiry of such magnitude would be impracticable. Consequently, we have limited our concerns to the concepts that relate specifically to policy formulation (especially in the sense of decision-making), planning, and monitoring, with fleeting references to operational issues where appropriate. However, notwithstanding the complexity of the whole, we have endeavoured to maintain a total-system outlook, as well as an environmental outlook, throughout the inquiry. Indeed, we draw special attention to the environmental impacts due to the operation of the system itself. Both can give rise to policy changes. Note that:
- Policy formulation and decision-making must take into account, in an ongoing way, environmental and social changes.
- The magnitude of the above changes must be assessed.
- The assessments must be used to update social, economic, and natural environmental patterns, as required in planning processes.

Clearly, moreover, the values, health, and expectations of all consumers must be regarded as an adjunct to environment. This is why the monitoring process is so central. For instance, recognition of changes in consumers' patterns of energy utilization, in their attitudes to environmental pollution, and in their willingness to conserve energy may have a profound impact on the planning of energy systems. The educational implications are significant. In particular, the electric power system must be responsive to change, and this is only achievable through learning processes – it is in this respect that public input is of great significance. There is an inevitable interaction between culture and technology.

Recognized also, in the Commission's mandate, is the critical importance of time and time-scales in the planning and implementation processes. In a highly pluralistic society, the long delays that separate the development of new knowledge and its application, in all fields of human endeavour, from an associated awareness of the potential social and environmental consequences are undesirable, though unavoidable. Similarly, the long lead times between the commissioning and completion of a thermal generating station complicate the power planning process appreciably, not least because of uncertainties in the electric load projections – the longer the lead time, the greater the uncertainty. No system is absolutely reliable, nor for that matter absolutely safe. Nevertheless, an adequate level of reliability, as well as a capability for responding to environmental changes through design, must be incorporated into the system. Although the complementary concepts of reliability and resilience are not mentioned specifically in our Terms of Reference, they are nevertheless implied. We draw special attention to:

- the relationship between reliability and diversity. In the systems planning sense, this would be manifest in the development of a range of options, and in measures to ensure the responsiveness of the system (e.g. to changing demand patterns) by the incorporation of an appropriate "mix" of generating facilities.
- the fact that all systems with adaptive capability rely on relevant research and development programmes to ensure that planning is forward- rather than backward-looking.

The Commission's mandate to emphasize the centrality of the system planning process draws attention as well to the time factors that are inherent in the process, to the importance of designing for reliability and resilience, and to the central importance of knowledge relevant to man himself, relevant to his social institutions, and relevant to natural and man-made environments. It recognizes also the complex interdependency between policy and decision-making, on the one hand, and human values and aspirations, on the other. And, not least, it recognizes the fact that the planning process must be adapted to the needs of the people (i.e., of Ontario), of future generations, and, of course, of the environment in which they will live.

The Major Issues

A major purpose of the Commission's Preliminary Public Meetings (1975-6) was to provide opportunities for the people of the province to identify what they perceived to be the key issues embodied in our Terms of Reference.[5] Subsequently, during 1976-7, in order to clarify the issues and to obtain more detailed information from Ontario Hydro, provincial government departments, industry, public interest groups, and the general public, the Commission held a series of public information hearings. In all, several hundred specific issues were identified and examined (by expert examiners) during these hearings, and nine issue papers were published, essentially to provide an information base for the debate phase of the inquiry. These issue papers provide, in effect, a detailed interpretation of our mandate. Below, we summarize and explain briefly some of the key issues – they are dealt with in more depth in Volumes 2 to 8. The degree of interdependence between the major areas is exemplified by the fact that topics such as energy conservation, environmental impact, and the end uses of electric power are justifiably included in several major categories of issues (e.g., demand, supply, land use, health and environmental impact, economics and finance, decision-making, and public participation). Furthermore, because of the large number of major issues, and the complexity of their interdependence, we do not pretend to have investigated all of them in the depth they deserve. Nevertheless, we believe we have fulfilled our mandate in so far as study of the fundamental concepts that relate to electric power planning is concerned.

The Demand for Electric Power

Electric power system planning begins with the electricity load forecast, i.e., an estimate of the future demand for electricity. During a period of economic and political uncertainty, and of increasing recognition that an energy crisis, especially in so far as oil production is concerned, is in fact extant and intensifying, load forecasting is a highly complex process. It must take into account the growth of population, the growth in the number of households, the growth of industry to provide an increasing number of jobs, and the growth in the economy as a whole. It is not surprising that the judgemental factor in load forecasting is of increasing importance. Until comparatively recently, for example, energy policies and forecasts were based essentially on the supply of energy. Implicit in the Commission's Terms of Reference is the need to consider the extent to which the supply of electric power can be planned on the basis of demand patterns (i.e., of the end uses of electricity). This constitutes a reorientation of energy planning of considerable significance. It is a truism that reductions in the demand for electric power will lengthen the lifetime of non-renewable primary energy resources and reduce environmental impacts. But to what extent can the demand for electricity be reduced by energy conservation measures,[6] by the efficient utilization of energy, by pricing policies, or by government legislation? There are practical limits to the ability of government to engineer cut-backs in energy consumption without causing reductions in employment and economic activity. What are the major issues related to growth in demand? We have concluded that they are as follows:

- Of all the factors that will influence the future demand for electric power, the growth of population is probably the most fundamental. The population of Ontario, during the period 1958-73, grew at an average annual compounded rate of 2.1 per cent. Recent projections suggest that the growth rate for the period 1980 to 2000 may be as low as 1.2 per cent per annum. On what demographic basis should future electric power needs be based?
- Canada's and Ontario's per capita use of energy is among the highest in the world. To what extent is this due to climatic and geographic factors? To what extent are these levels of consumption essential, and to what extent are they the result of inefficient use of energy?
- During the period 1958-73, the per capita consumption of electric energy increased at a rate of 4.5 per cent per annum, while the primary peak power demand increased at an average annual rate of 6.7 per cent. Since 1973, these growth rates have dropped dramatically. What are the prognostications for the period 1980 to 2000?
- By developing comprehensive end-use patterns[7] in the major sectors of society, will it be possible to obtain more realistic assessments of the future needs for electric power? Can these end-use patterns facilitate the establishment of "demand targets"?
- At present, Ontario Hydro is responsible for both forecasting the electric power load, on the one hand, and supplying the load, on the other. How can the load-forecasting process be put on a much broader base, for example, by obtaining appreciably more public input? What should the government's role be in load forecasting?
- Much has been said about the importance of society embarking upon rigorous and comprehensive programmes of energy conservation. To what extent will these programmes affect the future demand for electricity? Of the many approaches to conservation and conservation technology, which appear to be the most desirable for Ontario? Will a voluntary or a legislative approach to energy conservation be required or a combination of the two? To what extent will it be possible, both technically and politically, to upgrade insulation standards for homes and buildings? How can the cost-effectiveness (in the sense of cost/benefits) of conservation technologies be enhanced?
- What new electricity pricing policies would be effective in ensuring more efficient use of electric power and energy, and hence in facilitating energy conservation? If the prices of oil and natural gas, on an energy-content basis, outstrip those of electricity, will this materially influence the demand for electricity?
- With increasing interest in solar energy for space heating and water heating, to what extent will the demand for electricity be affected? Similarly, how will increasing utilization of energy-efficient appliances, especially electrical motors, affect the future demand for electricity?
- It is well known that the recycling of many materials, especially metals, is energy-conserving. How can these practices be encouraged? Is it fully recognized that a watt saved is equivalent to a watt generated and that a kilowatt hour of electricity saved may be equivalent to 3 kW·h of primary energy saved?
- Food is a vital form of energy, and its production necessitates large-scale consumption of various forms of energy, including electricity. Millions of tons of food are wasted each year. How can "food conservation" be encouraged?
- Some climatologists are suggesting that we may soon begin to experience a more extreme climate. To what extent should Ontarians take such prognostications into account in deciding on an appropriate electric power growth rate?
- The future demand for electric power will depend on the life-styles of the people of the province. It

will depend also on the degree to which the people will accept a stringent environmental ethic. For instance, environmental pollution is related to the inefficient use of energy. Will Ontarians accept less energy-intensive pursuits? Will coercion be required? Or will the change in life-styles be accomplished through education?

The Supply of Electric Power

The availability of an apparently inexhaustible supply of low-cost energy has been a central factor responsible for the increasingly high living standards to which people of the industrialized nations aspire. A major issue, bearing in mind the diminishing supplies of fossil fuels, especially oil and natural gas, is the extent to which growth in supply can be maintained. Energy policies in the past have been based mainly on expanding the supply. How, in the future, can the orientation be effectively turned around? For example, can the supply of electric power be planned on the basis of future demand patterns (i.e., end uses) and demand management?

Although market forces dictate that a scarcity of supply will be dealt with by increased prices, and that a plentiful supply will give rise to lower prices, this is a much too simplistic model to apply to an electricity utility. As mentioned previously, long lead times are involved in putting major bulk power facilities in place and extremely high reliability of supply is essential; and, unlike most products, electricity cannot be put into inventory — it cannot be stored, except to a limited extent. A major threat to society might develop if the supply of electric power failed to keep pace with the demand for electric power (assuming that the demand is "real" and not a result of large-scale inefficient utilization).

The Total Electric Power System

Ontario's electric power system is a highly complex integrated assembly of men, machines, transmission lines, switching and transformer systems, and computers, the purpose of which is to supply electricity on demand, with high reliability, and at the lowest feasible cost, to a very large number of consumers. Ontario Hydro has fulfilled, and is fulfilling, this purpose with a high level of competence. Why, therefore, establish a Royal Commission to inquire into this utility's long-range planning programme and the associated issues outlined in the Terms of Reference? The answer is comparatively simple. Although based on high technology, Hydro's long-range planning programme and the concepts upon which it is based involve many non-technological issues that relate to demand patterns, environmental insults, and economic and financial questions. And, of special importance, there is the question of the decision-making process itself.

The major generation technologies being used by Ontario Hydro are hydraulic, coal-fired thermal, and nuclear thermal. The many complex issues relating to nuclear power were identified and discussed in the *Interim Report*; the latter has been updated in Chapter 5 of this volume. Because of the importance of nuclear power in the inquiry, however, and notwithstanding our previous publications on the subject, we make summary references to it in this discussion of the mandate.

The major issues that relate specifically to the technology of electric power generation and transmission, and to the operation of the system as a whole, are summarized next:
- The principal objective of Ontario Hydro is "to supply all electric demand safely, reliably, within environmental standards and at the lowest feasible cost consistent with financial stability". To what degree can the utility be expected to meet these goals in the future?
- Bearing in mind potential constraints in the supply of primary fuels, to what extent should Ontario Hydro depend on hydraulic power, fossil-fuelled thermal generation, and nuclear power, respectively, to supply, on the one hand, the base-load requirement, and, on the other hand, the peaking requirement?
- The flexibility of an electric power system, in the sense of responsiveness and resilience, depends to a large extent on the nature of the "mix" of generation technologies and their operational characteristics (i.e., on their ability to operate in base-load, intermediate-load, and peaking-load capacities). Furthermore, the lead times involved in the approvals, design, construction, commissioning, etc., stages are particularly relevant in considering the responsiveness of the system. Does Ontario Hydro's long-range plan adequately reflect the need for a responsive system? To what extent would the undoubted economies of scale be sacrificed in order to provide for more flexibility?
- Are the projected reliability of the electric power system, and, concomitantly, the excess margin of generation capacity, adequate? In order to optimize the utilization of costly fossil fuels, are the

load-management practices at present being planned likely to be adequate? Would additional hydraulic storage, to ensure optimum utilization of nuclear power, be cost-effective?

• Is sufficient attention being paid to the alternative generating technologies, especially solar energy, as well as to more efficient methods of converting coal into thermal and electric energy? Is energy-conservation technology being adequately encouraged?

• Is the bulk power transmission system, as planned for the 1980s and 1990s, adequate to supply the potential load on a regional basis?

• In view of the possibility of hydroelectric power being available in Manitoba and Quebec for export to Ontario, are the planned interprovincial interconnections adequate? Ontario Hydro is interconnected with several United States utilities. Are the existing and planned interconnections satisfactory, especially in view of Ontario Hydro's commitments to the Northeast Power Coordination Council (NPCC)? Is direct-current transmission a viable option?

Land Use

Although the losses of farmland attributable to electric power generation, transmission, and other large-scale facilities are relatively small, especially in comparison with the land required for housing and urbanization in general, they are nevertheless significant. To what extent can electric power planning be aimed at facilitating industrial growth in some regions and inhibiting it in other regions? The identification of future load centres is clearly a key requirement, and it is in this respect that provincial planning and electric power planning are closely complementary. Moreover, there is recognition of the desirability of treating comprehensive provincial land-use policies, provincial energy policies, and provincial industrial development policies as an integrated whole.

We have identified the following major issues that relate specifically to the land-use implications of electric power planning:

• How can the Government of Ontario and Ontario Hydro encourage land conservation?

• There is a firmly held conviction in the agricultural community that the siting of major generating facilities and the routing of high-voltage transmission corridors constitute a major threat to Ontario's farmlands. How can this conflict of interest between the province's agricultural needs and its electric power needs be resolved? Can transmission corridors be routed so as to minimize their impact on farmland?

• From the point of view of the conservation of high-quality farmland, would several small decentralized power stations be preferable to a single large power station? Note that smaller stations, e.g., co-generation plants, could be located close to the load centres, thereby reducing the need for high-voltage transmission lines.

• Land banking has been proposed as a means whereby the long lead times that are associated with putting new large-capacity thermal generating stations into service might be reduced by several years. Is this a desirable approach?

• The land-use implications of the disposal of municipal, industrial, and utility wastes are becoming increasingly significant. For example, the long-term disposal of radioactive wastes (uranium mill tailings and spent nuclear fuel) constitutes a unique land-use problem. On the one hand, there is concern about the health hazards of such facilities, and, on the other, there is a need to locate suitable depositories. How can this problem be resolved? With respect to municipal and industrial wastes, what can the "watts from waste" concept contribute to a solution of the disposal problem?

• As the costs of fossil fuels escalate, the attractiveness of biomass energy increases, and, concomitantly, the energy potential of many thousands of acres of abandoned farmland in the province becomes more valuable. To what extent are hybrid (fast-growing) poplar plantations, for example, a viable source of energy and especially of electric power?

Health and Environmental Concerns

There is inevitably a conflict, often a very visible conflict, between the development of energy resources and the protection of the environment. This is not surprising, since there are obvious limits to the tolerance of the natural environment under the impact of assaults associated with the activities of man. Although the primary function of Ontario Hydro is to provide the electricity needed by the province, this purpose must be fulfilled with the least possible threat to the health of the people and to the environment. Also, and of equal significance, all consumers of electric energy have a primary role, albeit an indirect one, in minimizing environmental damage by reducing their consumption of electricity, and consequently in minimizing avoidable losses or wastage of primary energy resources. Wasted energy is synonymous with environmental pollution.[8]

The mandate of the Commission recognizes the need for exploration of the relationship between energy production and utilization and potential threats to health and the environment. In particular, it recognizes the need to establish governmental and social institutions that facilitate an integrated approach to electric power planning and environmental protection. Vital decisions will have to be taken – decisions affecting the natural environment, and ultimately the well-being of man. We summarize below the major health and environmental issues, explicit or implicit, in the Commission's mandate.

- Recognizing that the absolute safety of the CANDU reactor, or, indeed, of any industrial process or human activity, cannot be guaranteed, is the reactor and its associated systems safe within reasonable limits? How can the safety of CANDU reactors be improved, with consequent reduction in the health hazards (due to radiation) associated with nuclear power stations, especially in the event of a malfunction?

- The management of uranium mill tailings and high-level radioactive wastes from nuclear power stations involves serious social and environmental problems. Further, the question of the ultimate disposal of spent fuel is of special significance. Is sufficient attention being paid to these problems? Is the public being adequately informed?

- The siting of large thermal generating stations (i.e., nuclear and coal-fuelled) has important health and environmental implications. Furthermore, because of the need for vast supplies of cooling water, the shores of large lakes and the banks of large rivers are desirable locations from a power generation standpoint, but possibly not from an environmental standpoint. What criteria should be adopted, taking into account such obvious factors as population densities, the availability of adequate water supplies, the need for docking facilities, the probability of earthquakes, and the prevailing wind direction?

- Bearing in mind the degree of dependence of Ontario Hydro on coal-fired thermal generation to the year 1993 and beyond, and the fact that the effluent of these stations includes toxic gases such as the sulphur oxides (SO_x) and the nitrogen oxides (NO_x), as well as carbon dioxide (CO_2), which may have deleterious consequences on future climate, is the monitoring and control of these toxic emissions adequate?

- Secondary and tertiary effects of the emissions from coal-fired stations include the production of ozone, which causes significant damage to a variety of crops, and "acid rain", which damages crops, forests, and freshwater lakes and their fish populations. To what extent can these environmental insults be minimized?

- Associated with all thermal generating stations is the great volume of low-temperature thermal waste that is discharged into large bodies of water (e.g., the Great Lakes). What are the environmental implications of these discharges? To what extent can the thermal wastes be put to good use?

- It has been asserted that high-voltage transmission lines may give rise to certain health problems as a result of the associated high-energy electrical and magnetic fields. Furthermore, hazards of a different kind may arise in the vicinity of transmission lines, when, for example, agricultural machinery is inadequately grounded. Is scientific evidence available to support, or to reject, these contentions?

- During certain weather conditions (e.g., thunderstorms and high-humidity conditions) high-voltage transmission lines in the vicinity of insulators may give rise to so-called "corona discharge" conditions. This leads to the generation of ozone – a toxic gas. Is the level of ozone generated in this way a health hazard and a threat to certain crops?

- How serious are the aesthetic concerns relating to power station sites and high-voltage transmission corridors? Are any new power transmission technologies in the offing that would resolve this problem without introducing additional environmental threats?

- In developing criteria for the routing of high-voltage transmission lines, is the protection of areas of critical environmental value, e.g., marshlands, woodlands, foodlands, lakes, and rivers, adequately taken into account in the making of choices between alternative routes?[9]

- The importance of education in stimulating awareness of environmental concerns (and health-related problems) and the urgent need for conservation measures are fundamental. It is particularly important to implant an understanding of the ecological principles that underpin both the environmental and the conservation ethic. Sadly, most influential teachers, not least in the universities, appear to be totally ignorant of environmental issues. Many, apparently, just do not want to know How can viable educational programmes dealing with environmental and energy conservation be established at all levels, and for all ages? How can Ontario Hydro, with its vast resources, participate in such programmes as part of the electric power planning process?

Economic and Financial Issues

Ontario Hydro's long-range planning, in so far as economic and financial factors are concerned, is based, first, on internal economic efficiency, second, on the impact on the external economy, and, third, on the corporation's ability to finance system expansion. The corporation, in terms of assets, is the largest in Canada apart from the major banks; it has a major impact on the provincial economy, both directly by employing many thousands of people, and indirectly by supplying electricity to millions of homes and to thousands of industries and commercial establishments. Noteworthy is the fact that, of Ontario Hydro's large capital expenditures (averaging during the last four years about $1.5 billion a year), roughly 60 per cent are made in Ontario.

Until 1976 there was a strong link between the growth in demand for electric power and the rate of the province's economic growth. Recently, however, there has been a gradual de-coupling, and this seems likely to continue.

Pricing decisions are crucial. They relate directly to borrowing requirements and hence to the magnitude of long-term debt, and indirectly, through the practice of conservation and the more efficient use of power, to the demand for electricity. Such factors as the relationship between demand and price, and the allocation of costs between various categories of consumers, are particularly significant – they are the special concern of the Ontario Energy Board. The Commission's mandate calls for an assessment of the underlying concepts, and especially of the interrelationships that are inherent in the socio-economics of electric power. For example, it is difficult to quantify the cost of electric energy, because there are increasing external costs that relate particularly to health, education, and environmental protection needs.

The major issues relating to the economics and financing of the electric power system are:
- Can Ontario Hydro's system expansion programme be used deliberately and effectively as a tool of regional development?
- Electricity utilities are probably the most capital-intensive of all economic enterprises. Because of the high cost of bulk power facilities and the need to raise large sums in the market-place, the utility is very sensitive to the cost of money. Is this fact reflected in long-range system planning and, on a short-term basis, in the rate structure? Also, to what extent should competing demand for capital (e.g., for reforestation, the development of biomass energy, conservation technology) put limits on Ontario Hydro's capital expansion programme?
- Task Force Hydro, in considering the basic principles of rate-setting, made the following recommendation:

  Ontario Hydro [should] adopt a pricing policy that will more accurately reflect the supply cost of electricity and that will give effect to government policies for the allocation of capital within the energy sector.

  To what extent should Ontario Hydro's current customers bear the cost of future expansion (not least to ensure that the debt-equity ratio does not continue to increase)? A related issue is – should bulk energy and retail rates vary with the time of day and the season of the year? The latter issue is related closely to the role of load management and indirectly to the conservation of electric power.
- To what extent, if at all, does low-priced electric power and energy encourage industry to locate in the province?
- In view of the large hydraulic power potential available in Quebec and Manitoba, should Ontario Hydro, for the 1990s, negotiate the purchase of large quantities of firm power from those provinces' utilities?
- Should Ontario Hydro pursue vigorously the export of power to the United States?
- With respect to alternative energy resources, is the Ontario government (and Ontario Hydro) doing enough to encourage energy conservation and the development of renewable energy such as solar and biomass? The co-generation of thermal and electric power appreciably enhances the efficiency of utilization of the non-renewable energy resources. Are adequate steps being taken to encourage industry to pursue such developments?

Decision-Making and Public Participation

Perhaps the most pervasive requirement in the Commission's Terms of Reference is that it "examine the long-range electric power planning concepts of Ontario Hydro... so that an approved framework can be decided upon for Ontario Hydro in planning and implementing the electric power system in the best interests of the people of Ontario". Because major thermal generating stations have lead times in the order of 10-15 years, and a subsequent useful lifetime of 30-40 years, the scope of the decisions involved in system planning is substantial, both in time and space.

The major decision-making regulatory and advisory bodies in the field of energy are Ontario Hydro's Board of Directors, the Ontario Cabinet, the Atomic Energy Control Board, the National Energy Board, the Ontario Energy Board, and the Environmental Assessment Board. Each of these has been involved to a greater or lesser extent with decisions relating to the planning of Ontario's electric power system. However, the role of Ontario Hydro has, not surprisingly, been central. In particular, the load forecast (the basis for long-term planning) and the technical and financial decisions relating to additional bulk power facilities are the responsibility of the utility.[10] This is in accord with the recommendation (1973) of Task Force Hydro, set up by the Committee on Government Productivity in September 1971, to the effect that while policy formulation is the prerogative of the government, Ontario Hydro has the responsibility for "programme delivery". But, because of the social, economic, and environmental implications of programme delivery, the decision-making framework is increasingly diffuse and complex. A cental issue is — how can a more integrated approach to electric power planning, within the framework of provincial planning, be achieved?

How can the public participate effectively in such a complex decision-making environment? In addressing this question, the Commission's mandate draws attention, peripherally, to the complementary issues of energy education and access to information. Without adequate knowledge, participation in decision-making, especially in an area as complex as energy policy and planning, would be abortive. Nor is it clear how an energy-education programme should be structured. Nevertheless, there are compelling reasons why they should be undertaken in the future. Furthermore, such a programme should emphasize both technical and qualitative aspects. Prescription rather than description should be the norm, and the programme should seek to involve people, directly and personally, through participation in decision-making, in helping to shape the future. Three basic issues are: What is humanly desirable? What is technologically possible? What is politically possible?

Additional issues that we have identified and addressed are:
- To what extent can Ontario Hydro's long-range planning programme be integrated with provincial planning, taking into account the differing time-frames for decision and implementation?
- How, most effectively, can the future "need" for electricity be determined? Who should be involved in the decision-making processes?
- How can Ontario Hydro and the public utility commissions collaborate effectively in the development of long-range planning concepts and, in particular, in the development of provincial energy conservation strategies?
- Should the initiative in electric power planning, within the framework of provincial planning, be Ontario Hydro's or the government's? To what extent should Ontario Hydro participate in government-initiated planning?
- Should there be more emphasis on decentralized power system planning on the part of municipalities and public utility commissions, as opposed to centralized planning by the government and Ontario Hydro? These processes would not, of course, be mutually exclusive.
- How can an integrated approach to land-use planning be developed? For example, how can local/ regional/provincial government concerns be integrated in such a way that the planning of additional bulk power facilities, and the subsequent implementation of the planning, proceed without unnecessary, and often very costly, delays?
- Should there be a periodic public evaluation of Ontario Hydro's long-range plans? In particular, should a permanent agency, or commission, be established to consider such generic issues as need, alternative demand and supply scenarios, and specific issues relating, for example, to rate structures and the siting of major power facilities?
- Is there a need for Ontario Hydro's mandate to be reviewed periodically to ensure that it continuously reflects the changing needs of the province? Would the permanent agency suggested above be the appropriate review body?

- What steps should be taken to ensure that local and regional energy study groups are encouraged and, in some measure, financed?
- How can public hearing processes be improved, perhaps through diversification, so as to facilitate public participation and the education of the public in energy matters?

There appears to be a consensus to the effect that the decision-making process, in the sense of the nature and structure of planning organizations, is more important than the specific plans that emerge, and that managerial innovation and co-ordination are essential in the assembling, structuring, and disseminating of information in an understandable form.

Power Technology Assessment

Inevitably, as Ontario's electric power system grows, its social and environmental impacts also grow. The Commission's mandate may be thought of as requiring an assessment of this growth and, of critical importance, of how its various impacts can be monitored and controlled. This process has been referred to as "technology assessment"[11] — attempting to fit technology to society rather than vice versa.

The purpose of technological assessment is to improve the use of technology (the challenge is "to technology" not "of technology"), and in particular to anticipate potentially harmful side-effects. The concept has become universally accepted (note, for example, the Office of Technology Assessment, Washington, D.C.), because there have been growing misgivings in all sectors of society concerning the increasing threat of technology to the quality of life. Indeed, some of the most concerned people are scientists and engineers.

If man's survival for many generations to come is to be assured, man and his technology must be complementary, and this will increasingly necessitate the assessment and control of technology for the benefit of mankind as a whole. It is in this sense that we have interpreted our mandate. This is why our inquiry has been oriented so markedly towards people and how people can participate in power technology assessment.

Figure 1.1 Terms of Reference of the Royal Commission on Electric Power Planning

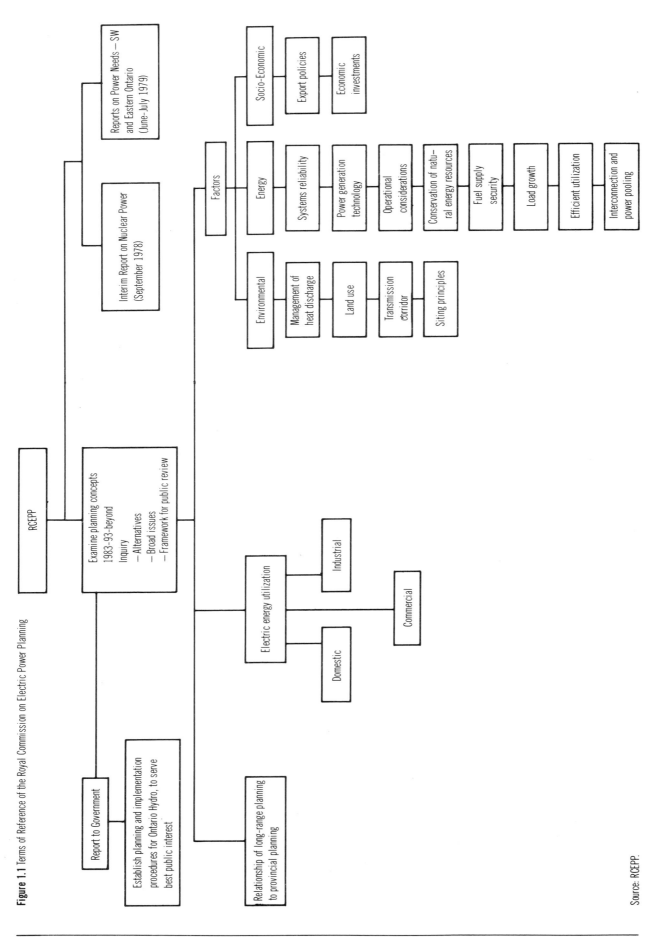

Source: RCEPP.

Figure 1.2 Evolution of a System

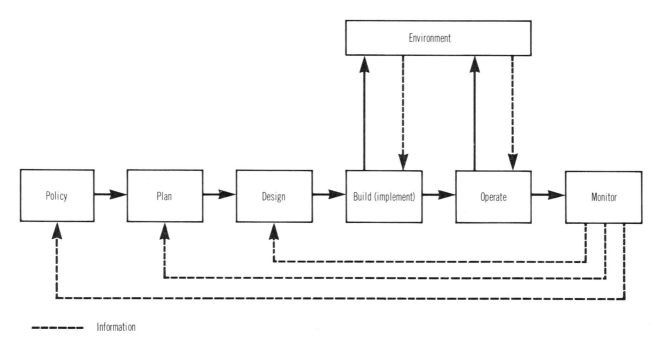

Information

Source: RCEPP.

CHAPTER TWO

Ontario Hydro: Its Impact on the Province

> The co-operative partnership between Ontario Hydro, the municipalities, and the Government of Ontario has been a dramatic success story. One of the most rapid rates of industrialization in the world has been served and facilitated, and Ontario residents have been provided with electricity at very low rates compared with other provinces and the United states, without the inconvenience and economic loss experienced through brown-outs. At the same time, Ontario Hydro has achieved a reputation among its peers as a world leader. It has been of immeasurable service to the province of Ontario. – Task Force Hydro, Report No. 1

In Ontario, "hydro" is virtually synonymous with "electricity". Very few, we suspect, of Ontario Hydro's almost 3 million customers refer to their "hydro bill" as their "electricity bill", or their electricity supply as anything other than "hydro". Not surprisingly, therefore, the history of Ontario Hydro is closely identifiable with the history of 20th century Ontario. Today, the utility is the second largest electricity utility in North America[1] with assets of about $13 billion and with approximately 27,000 employees. The growth of Ontario Hydro during the period 1910-79, in terms of the growth in December-January peak demand and in dependable capacity, is shown in Figure 2.1.

Fig. 2.1: p. 23

Ontario Hydro, a Crown corporation, was created in its present form essentially in response to the recommendations of the Advisory Committee on Energy and of Task Force Hydro.[2] The findings of Task Force Hydro were reported in five reports published between August 1972 and June 1973. The recommendations were accepted and endorsed by the Committee on Government Productivity and submitted to the Executive Council of the Government of Ontario in August 1972.

The impact of these advisory bodies on Ontario Hydro, as well as on all aspects of its operations, has been appreciable. Indeed, Task Force Hydro, in particular, anticipated most of the administrative, economic, financial, social, environmental, and technical problems with which the corporation was faced during the 1970s. Its imprint remains firmly embedded in the psyche of Ontario Hydro. This chapter is, in effect, a restatement and, in so far as operational factors are concerned, an updating of the Task Force Hydro Report No. 1 – "Hydro in Ontario – A Future Role and Place". Ontario Hydro's mandate can be stated simply as:

> Ontario Hydro's prime responsibility is to supply the demands of the people of Ontario with electric energy at the lowest feasible cost consistent with safety to employees and the public, a high quality of service to its customers, and subject to the social, economic and environmental concerns of the people of Ontario.

The extent to which Ontario Hydro has fulfilled the part of its mandate that requires it to provide the people of the province with electricity at the lowest feasible cost, but nevertheless consistent with a high level of reliability, is illustrated in Figure 2.2. Although, over the last 16 years, the real price of electricity in the province kept pace with the consumer price index, it has consistently lagged behind the average real wage level in Canada. It is important to note, on the other hand, that the debt/equity ratio of the corporation consistently increased during the 1970s, as shown in Table 2.1. This raises two questions: Is the price of electricity truly equitable from the point of view of future generations? To what extent are we mortgaging the future?

Fig. 2.2: p. 23

Table 2.1 Ontario Hydro's Debt to Equity Ratios and Capital Borrowings (1969-78)

| Year | Debt/Equity ratio | Capital borrowings ($10^6$$ Cdn.) | | |
		Canadian	Foreign	Total
1969	0.731/0.269	185	215.4	400.4
1970	0.742/0.258	310	175	485
1971	0.752/0.248	335	165.7	500.7
1972	0.762/0.238	300	256.7	556.7
1973	0.77/0.223	375	160.5	535.5
1974	0.782/0.218	400	300	700
1975	0.825/0.175	450	1,115.5	1,565.5
1976	0.844/0.156	600	897.3	1,497.3
1977	0.844/0.156	750	631.2	1,381.2
1978	0.853/0.147	800	839.9	1,639.9

Source: Ontario Hydro

To ensure minimum cost, coupled with a high standard of reliability, Ontario Hydro introduced, in the period ending in 1974, a variety of ideas and measures aimed at maximizing the consumption of electricity. This policy was implicit in various direct and indirect subsidies and in sales promotion programmes that encouraged consumption and thereby lowered costs. During the 1960s and early 1970s this policy obviously served the province well and resulted in a steady reduction in the real unit cost of electric energy. But there has been a dramatic volte-face during the last few years; instead of maximizing consumption, the policy has been increasingly to stress the importance of conserving electricity.

A decade ago, Ontario Hydro's policy of encouraging the use of electricity was justifiable, because at that time a high proportion of the generation capacity was hydroelectric, and the thermal generation that did exist was based on low fuel (coal) costs. Furthermore, it is a truism that electricity is taken for granted by virtually all consumers, since, unlike other forms of energy, it is on instant demand – lights are switched on and an instantaneous response is expected. Electricity cannot be stored except in small quantities, and it is used productively, or wastefully, at the instant when it is generated. The situation pertaining to oil and coal is very different.

The uses of electricity are legion. It has changed the life-styles, and indeed, the quality of life, of literally hundreds of millions of people around the globe. Noteworthy, too, is the fact that electricity, *per se*, has created many new industries, and consequently many millions of jobs; the most striking and significant have been in the field of communications and computers – probably no industry has had a more profound impact on civilization and culture than the electricity-based communications and computer industries. The associated products are, in a real sense, manifestations of the extension of man's perceptions, indeed extensions to man's central nervous system. Also, many human activities and traditional industries have been completely changed by the advent of cheap electricity. (Not only leisure activities involving, for example, television and radio, immediately come to mind, but also sporting activities involving man-made environments such as sheets of ice for hockey and curling, and evening baseball and tennis, requiring artificial daylight.) The metallurgical industries, notably aluminum and iron and steel, have been, and are being, transformed as a result of the availability of electricity on a large scale; those industries require thousands of megawatts and billions of megawatt hours.

As well, during the last two decades, we have witnessed amazing changes in the average home, all due to the "electrical revolution" – the refrigerator, the electric oven, the electric washer and dryer, the dishwasher, the vacuum cleaner, the floor polisher, the electric kettle, the electric frying pan, and so on. These electrical household appliances have minimized drudgery in the average home and increased the time available, especially to the housewife, for more creative pursuits. They have dramatically improved the home environment as well as the social environment.

Reflect also on the contributions of electricity to the performing arts, especially the theatre, to the fine arts, notably the art galleries, to the operation of libraries, to the supermarkets and shopping plazas, to the operation of airlines and airports to maximize the safety and convenience of passengers, to the whole of transport technology (e.g., traffic signals and environmentally desirable electrical transportation in the form of trains, trolley buses, and, hopefully in the future, electric automobiles). During the 1960s there was undoubtedly justification for Ontario Hydro's slogan "Live Better Electrically". But since then the situation has changed and society is becoming increasingly conscious of the compelling need to conserve energy, especially energy from the earth's non-renewable resources.

Historical Background

The supply of electricity to Ontario consumers, albeit originally on a greatly reduced scale compared with today, and to only a few people, dates from the early 1880s. In 1883, for example, the Municipal Act of 1858 was revised to empower municipalities to operate electricity utilities.

But it was not until the 1890s that the transmission of electric power over reasonable distances became practicable. This technological advance encouraged several municipalities located some distance away from hydraulic power stations to switch from local coal-fired generation of power. Of particular significance was the first hydroelectric project based on Niagara Falls – in 1903 the Government of Ontario gave municipalities the legal right to transmit power from Niagara Falls, and it was at about this time that serious consideration was given to the possibility of a centralized public power system. Subsequently, in 1906, largely as a result of the enthusiastic support of Adam Beck, M.P.P. (later Sir Adam

Beck), the Hydro Electric Power Commission (HEPC) of Ontario, forerunner of Ontario Hydro as we know it today, was created. The Power Commission Act, under which Ontario Hydro still operates, was passed in 1907. Clearly, at the time, these were bold and imaginative pieces of legislation. During the ensuing years they were largely instrumental in facilitating the industrial and economic growth of the province.

(Although Ontario's electric power system is usually identified with Ontario Hydro, there are small but important contributions to the system from the Great Lakes Power Corporation, the Canadian Niagara Power Company, and other privately-owned companies — especially pulp and paper companies. Further, as will be discussed in Chapters 6 and 7, Ontario's electric power system is interconnected with several neighbouring utilities.)

Fig. 2.3: p. 24

During the first 10 years of Ontario Hydro's life, many municipalities, notably Guelph, London, Waterloo, Windsor, and Sarnia, as well as the concentrations of populations centred in Toronto and Hamilton, were coupled into the system. This was made possible by the construction of 115 kV transmission lines. With the expansion of hydroelectric facilities across the province, there was continued expansion of the high-voltage transmission grid. In the meantime, there were important developments in transmission technology, and 230 kV transmission lines were introduced during the 1920s — see Figure 2.3. The limits of the capability of the 115 kV bulk power network were reached during the 1950s when the 230 kV transmission network was put into place.

Until 1958, when the last major hydroelectric development in Ontario, the Robert H. Saunders Generating Station on the St. Lawrence River, was opened, Ontario's electric power system had been essentially hydraulic. But during the 1950s, with the continuing rapid growth of the load, it was clear that the hydraulic power potential of the province was inadequate, and a major thermal (coal-fired) power plant construction programme was initiated. The first thermal plants were located in the Toronto and Windsor areas.

For planning and administrative purposes, Ontario's power system is divided into two major parts: the "East System", which is by far the larger and which covers the geographic regions of southern, eastern, and most of northeastern Ontario, consists of a bulk power transmission system in the form of a network or grid, and the "West System" which serves the western and northwestern areas of the province. The latter comprises an essentially linear transmission system in contradistinction to a network, and is connected to the East System by a single transmission line. The demarcation line between the East System and the West System is, roughly, a north-south line drawn through the community of Wawa.[3]

Because of major advances in the design of large steam turbine electricity generating units, the capacity of thermal generating stations has increased markedly since Ontario Hydro's first major thermal generating station — the first 100 MW unit of Hearn Generating Station — was placed in service in 1951. Today, the work of designing an 850 MW unit is well in hand. The economy of scale made possible by the introduction of large thermal generating stations has helped to keep the cost of power generation at comparatively low levels. However, during the last few years, increasing attention has been paid to certain social and environmental costs; it is clear that these must now be weighed in an increasingly complex cost/benefit balance.

The frequency standardization programme, involving conversion of the power system from 25 Hz to 60 Hz, constituted a landmark in the history of the utility. It was undertaken during the period 1948-59; the only 25 Hz supply remaining serves a few industrial consumers in the Niagara area. Frequency standardization was a necessary step because, at the time, most of North America with the exception of Ontario was supplied with 60 Hz power, and consequently all electrical appliances and machinery for use in the home and in factories were designed for this higher frequency. An important consequence of frequency standarization was that Ontario's electric power system could be operated in synchronism with neighbouring systems. Note, however, that interconnections with Hydro-Québec constitute a special case.

Another important system development, whose purpose was to strengthen the bulk power transmission network, was the introduction of 500 kV transmission circuits.[4] These anticipated a time when the bulk power transmission requirements in the province would exceed the capability of the 230 kV circuits. Ontario Hydro introduced 500 kV lines into the system in the mid 1960s. In 1979, 1,177 km of 500 kV circuits were in operation and 930 km were committed or under construction, while about

13,000 km of 230 kV circuits were in operation. The existing and committed bulk power transmission system and the major generating stations are shown in Figure 2.4. Fig. 2.4: p. 25

The Ontario Hydro Corporation

In 1973, legislation was passed amending the Power Commission Act.[5] The Hydro-Electric Power Commission of Ontario was renamed The Hydro-Electric Power Corporation of Ontario (commonly referred to as Ontario Hydro), and was designated as a Crown corporation whose structure and operation would closely parallel that of a large private institution, with special emphasis on flexibility.

The main thrust of the legislation was to clarify and strengthen the relationship between Ontario Hydro and the Government of Ontario, on the one hand, and to strengthen the ties between the corporation and the people of Ontario, on the other. Accordingly, the Board of Directors of the corporation was to be responsible to the government for the operation of the corporation and would receive policy direction from the government. The membership of the board was to be broadly based to ensure reasonable representation of various walks of life in the Ontario community, as well as adequate expertise relating to Hydro's operations. Such a board would enhance the mutual understanding between Hydro and the government and vice versa, especially with respect to the social goals of the government and the goal of supplying the people of Ontario with reasonably priced electricity and acceptable environmental, health, and safety standards. Because of the high technology component, especially in connection with nuclear power, of Ontario's electric power system, the corporation would be responsible for ensuring that these technologies were not in conflict with the government's social objectives.

The chairman of the board of Ontario Hydro would have very special responsibilities. He would be appointed on a full-time basis and his orientation would be, in the words of the Task Force Hydro report, "outward to the Ontario community and to the Government". In particular, he would act as the main communication channel between the government and Hydro, and in so doing would ensure that the board was provided with all relevant information necessary to keep Ontario Hydro policy consistent with that of the broad framework of government policy.

The president of Ontario Hydro would be responsible for managing the affairs of the corporation in order that corporate objectives and policies would be translated into appropriate action.

Ontario Hydro and the Public Utility Commissions (PUCs)

The historic role of Ontario's municipal utilities — the public utility commissions — was introduced in Chapter 1. Today, about 2 million customers in Ontario are served by 334 PUCs, which undertake the distribution and sale of electricity generated and transmitted (via the bulk power system) by Ontario Hydro. The larger municipal utilities, such as Toronto Hydro with 196,000 customers, which sold almost 7,500 billion kW·h of electricity in 1978, are comparatively autonomous, while the smaller PUCs rely heavily on Hydro for assistance in the operation of their systems. All PUCs are subject to regulation by Hydro in respect of matters such as retail rates and financial and accounting practices. Hydro also has its own direct customers. These are mainly rural customers, but they include major industrial users in the province.

For the majority of electricity consumers, the main interface, and hence communication channel, is between the individual consumer and the PUC. Ontario Hydro, *per se*, being chiefly the supplier of bulk power, only comes into the picture with its direct customers (residential, commercial, and industrial) and the lines of communication between people and the utility are necessarily through regional offices, which are remote from many of these customers. Hence, the PUC has not only the important practical role of distributing electricity in the province but just as important, a social role, especially in encouraging energy conservation as adviser and consultant to the public at large.

The role of the PUCs in Ontario Hydro's load-forecasting procedures has been considered in some detail in the Commission's reports on the need for additional bulk power facilities in southwestern and eastern Ontario.[6] We pointed out there that the contributions of the PUCs to the load-forecasting process is a one-year forecast and a more speculative five-year forecast. While this information is obviously of importance to Hydro, we believe that, because of the increasing emphasis on "end-use forecasting", the role of the PUCs in developing load-growth patterns, with special emphasis on end use, could provide an opportunity for much more input from the public than hitherto.

There is another area in which, we believe, close co-operation between certain PUCs and Ontario Hydro will become increasingly important. Although the recent, and anticipated, evolution of Ontario's electric power system is based essentially on large-capacity generating stations (both nuclear and coal), and on extensions of the 230 kV and 500 kV transmission networks, there is increasing interest in the potential role of dual-purpose plants (thermal and electric energy), that is, in co-generation and district heating plants.[7] If the development of dual-purpose generating plants in the vicinity of major load centres proves to be socially, environmentally, and economically acceptable, and indeed desirable, it is obvious that close co-operation between Hydro and the appropriate PUCs would be essential.

This brief reference to the role of the PUCs would be incomplete without some mention of the work of the Ontario Municipal Electric Association (OMEA). The OMEA, whose members are the commissioners of the "public hydro utilities", i.e., the PUCs, was formed in 1912; it has played a central role in the evolution of Ontario's electric power system. In addition to co-ordinating the objectives and representing the views of individual utilities with respect to the establishment of uniform standards and procedures, it has provided, very effectively, the checks and balances that relate to the operation of Hydro as a provincial "delivery institution".

Social and Economic Implications

Apart from the social, cultural, and technological benefits that the people of Ontario have derived from a plentiful supply of comparatively cheap electricity, many other benefits have accrued from the actual generation and distribution of electric power. We refer, in particular, to the creation of jobs. Clearly, Ontario Hydro's expenditures on labour, materials, and goods and services contribute both directly and indirectly, through the well-known multiplier effects, to the economy of the province. (For details, see Volume 5 and Chapter 11 of this volume.) Noteworthy too is the fact that a large proportion of these expenditures is made for "high technology", and high-technology industries employ comparatively highly educated people and produce goods and services characterized by a large "added value".[8] It has been estimated, in the case of Hydro, that an increase of $1 million in expenditures, through the multiplier effect, gives rise to a $1.5 million increase in provincial income and to 48,000 man-years of employment within the province. For example, in 1978, capital and operating expenditures by Hydro each amounted to about $1.7 billion. Assuming that 70 per cent of these expenditures were made within the province, and taking into account direct, indirect, and induced multipliers, the total income effect in the province was about $5.1 billion, which is approximately 6 per cent of the gross provincial product (GPP).

In connection with the employment opportunities provided by Ontario Hydro, it is important to note that Canada's nuclear industry is dependent, in large measure, upon the utility's nuclear power programme. And, because of the highly sophisticated nature of the industry, this means that several thousand engineering and science graduates, as well as graduates of the community colleges, have been provided with jobs commensurate with their scientific and technical qualifications.

Until comparatively recently, in all western industrialized nations, there has been a close correlation between the rate of economic growth and the rate of growth in the utilization of electricity; this certainly applies to Ontario. For example, between 1955 and 1974 the primary electric power peak, as well as the primary electric energy generated, increased by a factor of about three, while over the same period the GPP increased by a factor of about 2.5. However, probably because of the increase in the real price of electricity and the impact of energy conservation, the link between economic growth and growth in electric power consumption appears to be weakening. (The same phenomenon is occurring in most western nations.) In this regard, it is important to note these words of Ontario Hydro:

> Clearly, any changes in the historical relationship between economic development and the use of electricity will have far-reaching implications. These may impact upon both the optimum future power system configuration and the growth and structure of the economy. Ontario Hydro, in planning the power system, is cognizant of these possibilities.[9]

As shown in Figure 2.2, and as mentioned previously, increases in the price of electricity in Ontario have only just begun to catch up with trends in two basic economic indicators — the consumer price index and the average wage level in Canada. Although there is inadequate information relating to the relationship between price increase and level of demand (the so-called price/demand elasticity), there is some evidence that, for example, residential demand is, in fact, responsive to price increases. However, the indications are that, since the proportion of family income spent on electricity decreases as the income increases, increases in the price of electricity adversely affect the below-average-income family more than the above-average-income family. This applies also to other key "necessities of life" such as food; it is a highly complex social problem and one that is exacerbated by high levels of inflation. Certain it is, in the words of Task Force Hydro, that

> the fundamental relationship between aggregate rates, economic development and the standard of living of the people clearly makes power pricing a most important policy issue for Government and Hydro.[10]

Ideally, the incremental social costs and the incremental social benefits associated with a growing electric power system should be equal. In the case of Ontario Hydro, because there has been increasing reliance during the last 20 years on large thermal generating plants and an associated need for additional high-voltage transmission corridors, the added social costs of generating and distributing electricity could outstrip the incremental social benefits. Some obvious social costs are those related to the undesirable impacts on:

- human health, due to environmental pollution
- Ontario's, and indeed the earth's, ecosystem, as a consequence of pollution
- agricultural products and fish populations, as a result of air and water pollution
- food production, as a result of the loss of high-quality agricultural land due to urbanization. the siting of generating stations, and the routing of transmission corridors
- buildings, due to air pollution

However, it is important to note that, as well as being major contributors to environmental pollution, and hence to the growth of social costs, the electricity utilities have also been major forces in counteracting the impacts of such pollution. For instance, electrical devices are used widely for the abatement of pollutants, notably electrostatic precipitators for collecting particulates, and electrolytic techniques for isolating noxious and toxic chemicals. But, in spite of breakthroughs in some areas, quantification of the social costs and benefits of electric power remains an extremely difficult problem. We must rely on the experience and judgement of knowledgeable technical groups and, most important, on the opinion of society as a whole in deciding whether to pursue a particular branch of power technology.

Financial Factors

Ontario Hydro's financial policies are continuously under review. They are based on the need to:

- finance new facilities at the lowest feasible cost consistent with a financially sound operation
- allocate the cost of capital facilities equitably among present and future customers
- be financially independent, remaining at arm's length from government in financial matters, except for the provincial guarantee of Ontario Hydro's debt
- maintain a level of liquidity sufficient to achieve the above objectives[11]

Throughout its history, the utility has preferred to finance its capital construction programme by issuing long-term bonds in the Canadian bond market, but because of the large capital requirements of recent years – see Table 2.1 – this has proved increasingly impractical. And during the last decade there has been increasing reliance on foreign sources of capital, especially the United States bond market and the European market. The point is that Ontario Hydro's capital needs have far outstripped the potential of the Canadian market to supply them. There is clearly increased competition for capital from foreign markets and this has a deleterious impact on interest rates as well as on the security of future capital resources.

The Ontario government's guarantee of the corporation's debt has important ramifications for the people of the province. Indeed, it is responsible for the so-called "triple A" rating of Ontario Hydro in the United States bond market (this is the highest possible credit rating), which ensures minimum interest rates. In turn, "as low as possible" interest rates favourably affect electric power costs, and hence prices to the consumer. Similarly, because of the province's prime credit rating, the corporation can negotiate long-term contracts for primary fuels, including coal and uranium, at lower than world rates. Again, this is clearly beneficial to Hydro's customers today, and probably increasingly in the future.[12]

Ontario Hydro and Regional Development

Prominent in the Commission's Terms of Reference is the requirement that we "inquire comprehensively into Ontario Hydro's long-range planning programme in its relation to provincial planning." In many respects, this single clause encompasses the philosophy of the whole inquiry. Several chapters in this volume address various aspects of the relationship, and Volume 7, "The Socio-Economic and Land-Use Impacts of Electric Power in Ontario", and Volume 8, "Decision-Making, Regulation, and Public Participation: A Framework for Electric Power Planning in Ontario for the 1980s", deal with it in depth. That Hydro has had a profound impact on the quality of life in the province is undeniable – it has affected all walks of life. Among many examples, there is probably none more dramatic than the rural electrification programme (anticipated in the government's amendment, passed in 1921, to the Power Commission Act, under which the government agreed to pay 50 per cent of the cost of all rural transmission lines that had been constructed or were to be constructed). The Rural Hydro-Electric Distribution Act of 1927 provided the comprehensive legislation whereby electric power could be supplied to most of rural Ontario. The effect of this legislation on, for example, the agricultural industry of the province has been particularly significant. During the last 50 years, the productivity of Ontario's farms has increased many-fold because of advances in agricultural science and technology coupled with the availability of artificial fertilizers, advances in agricultural machinery, and widespread use of electrically-powered equipment. But it was essentially the rural electrification programme that triggered this "agricultural revolution".

Increasingly during the last 40 years, beginning with The Planning and Development Act of 1937, and more recently with the regional development legislation passed in 1973, the Government of Ontario has demonstrated its awareness of the need to co-ordinate local, regional, and provincial planning. And much of this planning must be closely correlated with the province's electric power system. Noteworthy is the fact that Ontario Hydro is subject to the provisions of the Planning Act, which was passed in 1946 and, with amendments, remains in force. For example, Ontario Hydro must consult municipalities that would be affected by the construction of new power facilities, to ensure that such expansion plans are in accord with a municipality's official plan and with its "restricted area by-laws". In the recent past, problems have arisen due to conflict of interest between a municipality and Ontario Hydro. The completion of the 500 kV tranmsission line from the Bruce Generating Station to the Milton Transformer Station is the most notable example.[13] The authority to resolve such problems is vested in the Ontario Municipal Board.

With the adoption of the "Regional Development Policy for the Province of Ontario" in April 1976, a further major step was taken to ensure more effective co-ordination of Hydro's planning with that being undertaken by various government ministries. The responsibility of the Government of Ontario was stated by the Honourable J.P. Robarts, then Premier of Ontario:

> It is the responsibility of the Ontario government to assess the present and future requirements of the province relating to social, economic, and governmental development. The provincial government also has the responsibility to carry out and give direction to regional land use and economic development planning. It has the duty to ensure that, when development occurs in any part of the province, it shall take place as a result of good regional planning.[14]

In June 1972, to facilitate regional planning, and especially to ensure co-ordination between government ministries and agencies concerned with planning, the government established the "Advisory Committee on Urban and Regional Planning". The purpose of this committee, composed essentially of deputy ministers, is to review and make recommendations relating to regional developments, the siting of new towns, and development strategies for the province as a whole. To ensure that the plans of Ontario Hydro interface adequately with provincial plans, the president of Ontario Hydro is a member of this committee. The arrangement has the dual advantage of keeping Hydro informed of medium- and long-term government planning, and keeping senior civil servants in touch with Hydro's long-range planning.

Three major pieces of provincial legislation were passed in 1973:
- The Ontario Planning and Development Act
- The Niagara Escarpment Planning and Development Act
- The Parkway Belt Planning and Development Act

These three acts supplement the original Planning Act. Each "strengthens the government's potential for achieving orderly growth". Each ensures that municipal plans are respected and that growth, especially in the metropolitan Toronto and Hamilton regions, is controlled through the provision of leisure regions, e.g., the Parkway Belt. Each act has special implications for Ontario Hydro. The corporation must, for example, co-ordinate its land requirements with the requirements and criteria of the Niagara Escarpment Commission.

To complement this legislation, a broad range of studies relating to land use in the province have been

undertaken during the last decade. These studies have important implications for Ontario Hydro, in four specific areas:
- the food lands of the province
- the conservation areas and provincial parks
- the planning of new towns
- transportation and communication planning

During many hours of public hearings concerned with the land-use implications of Hydro's intermediate and long-range plans, the Commission heard compelling arguments to the effect that, at all costs, Ontario's prime food lands should be conserved. We will refer to the central problem of the stewardship of the land in another chapter. The siting of large generating stations and the routing of major bulk power transmission corridors clearly have important land-use implications. The concern of many agricultural public interest groups has been manifest not only during the hearings of the Commission but in several research studies.

The need to preserve land areas dedicated to recreational activities, e.g., the provincial parks, will never be in doubt. Procedures are in place to ensure that there is minimum interference with these areas when power transmission corridors are being planned. For example, Ontario Hydro has a right of way in the Algonquin Provincial Park, but the Park Plan prohibits any future power lines from traversing the area. There is clearly a conflict situation developing. For example, if additional bulk power facilities are required and if transmission lines are prohibited from traversing either prime farmland or park lands, how and where can lines be located? Perhaps there is a partial answer in the design of power line towers so that, as far as possible, they are aesthetically acceptable even though their cost may be appreciably greater than that of conventional lattice towers.

The planning of new towns is inextricably linked with the availability of electricity, water, sewage plants, roads, etc. During the preliminary planning phase, Hydro is consulted concerning the availability of an adequate electricity supply to the new town site. A comparatively recent example is the North Pickering town site – Ontario Hydro has been involved in the planning process for several years.

It is clearly desirable to co-ordinate planning of the routing of bulk power transmission and new highway routes. Similarly, other linear users of land, such as oil and natural gas pipelines, television and radio microwave links, and railways, co-ordinate their planning programmes with those of Ontario Hydro. Obviously, the multiple use of rights of way is a desirable goal, and studies involving provincial government ministries, Hydro, the railway companies, the oil and natural gas pipeline companies, and the communications companies are continuously in progress to ensure optimum co-ordination of land use.

The role of Ontario Hydro in facilitating the growth of industry in Ontario dates back to the beginning of the century. The availability of cheap hydroelectric power here attracted many industries. The degree to which Hydro could or should be used as an instrument for the management of industrial growth is debatable.

It has been argued that the availability of competitively priced, reliable electricity attracts industries to southern Ontario and that Hydro facilities should therefore be used to encourage new industrial growth. However, the utility has indicated that decisions to locate industrial plants are usually independent of the siting of generating stations, for the following reasons:
- The cost of electric power to industrial users does not depend on location.
- In all but a very small number of industries, electricity is a minor component of production.
- Factors such as proximity to markets in southern Ontario and the United States, availability of transportation infrastructures, and tax regimes are the key considerations.

This matter is discussed in detail in Volume 7.

The Impact of Major Ontario Hydro Projects on Local Communities

During the construction of a nuclear power station, such as the Bruce Nuclear Generating Station, there is a profound social, economic, and environmental impact on neighbouring towns and villages. For instance, there is an increased demand for housing, schools, health services, police and fire protection, shopping centres, and recreational facilities. Also, there is a need for additional municipal administration facilities, communications, roads, and the associated infrastructures.

Some of these activities will be beneficial to the communities in so far as local labour will be required and there will be an influx of capital into the area, stimulating the local economy. However, there are several undesirable aspects. For example, people employed in the local agricultural industry, in other local industry, and in commerce may be attracted by the higher wages associated with the construction of, for example, a power station. Furthermore, because the construction phase of the station may have a duration of only a few years (in the case of the Bruce nuclear complex, involving two large nuclear power stations and three heavy-water plants, this phase may stretch to 15 years), there is likely to be a large efflux of workers and their families as the construction is gradually phased out. Note, in particular, that the number of personnel required to operate the plants is appreciably smaller than the number required to construct the plants; furthermore, the spectrum of skills and educational requirements involved is quite different. Accordingly, unless new industries requiring the skills of the construction workers are established in the neighbourhood, there will be a surplus of homes, schools, health and social services, etc. The economic impact of this transition from a comparatively large population of workers and their families to a comparatively small population has undesirable economic consequences. As far as we know, Ontario Hydro has not consciously raised the economic expectations of the communities in the vicinity of large-scale construction projects. Nevertheless, the problem remains, those most affected being the non-transient workers who may have difficulty finding new employment, and who are burdened with high taxes.

Ontario Hydro has commissioned, during the last few years, several studies to ascertain the nature and consequences of the building of a large power facility in a community. These studies have considered the problem in four phases:

• During site selection, the socio-economic factors mentioned previously, as well as public input, are taken into account. Indeed, bearing in mind the broad-ranging implications of the Environmental Assessment Act of 1975, with the concomitant broad interpretation of "environment", such detailed studies will be an essential component of Ontario Hydro's submission to the Environmental Assessment Board when approval of a site is sought. As project development studies proceed and detailed information relating to the work force and service requirements becomes available, this information is incorporated in an environmental assessment of the site that is considered to be most appropriate.

• Following approval of the project, the next phase involves close collaboration between the communities concerned, the appropriate levels of government, and Ontario Hydro. For example, in the early phase of development of the Darlington Generating Station, discussions involving elected representatives of the Town of Newcastle and the regional municipality of Durham and representatives of Hydro took place over a period of three years, during which financial and economic agreements were worked out.

• During the construction phase at the Bruce Generating Station, the community impact of the project is being monitored continuously, and the initial community impact assessment is being reviewed and where necessary updated.

• The final phase is the period when construction has been virtually completed, and the transition from construction to operation is taking place. Such studies are being undertaken at the Pickering Generating Station to ensure smooth transitions and minimum community impact.

Interconnections with Neighbouring Utilities

With the growing realization (largely as a result of escalating oil prices and diminishing security of supply) that the industrialized nations of the world, with few exceptions, are facing a real energy crisis, it is probable that increasing importance will be attached to policies that encourage energy exchanges between friendly neighbours. The concept of regional energy developments may become increasingly attractive.

In subsequent chapters of this volume, and in Volume 2, we consider the importance, from economic, social, and environmental viewpoints, to Ontario Hydro and the province of electric power interconnections with neighbouring utilities. Such interconnections can improve system reliability, provide means of assisting a neighbouring utility in the event of an emergency, capitalize on diversity in both reserves and loads, and enhance frequency stability of the network. Also, energy interconnections, manifest in natural gas pipelines and high-voltage transmission lines that traverse international and provincial boundaries, facilitate regional energy exchanges and, in turn, especially during emergencies, emphasize the human value of "the good neighbour."[15]

Public Involvement in the Affairs of Ontario Hydro

Because of Ontario Hydro's central importance and high level of visibility in the province, and in spite of its excellent public service record, it has become a focal point of public concern. This situation is exacerbated because, in the eyes of many people, the corporation epitomizes the highly centralized technology-based institution. Furthermore, amongst Canadian electricity utilities, Ontario Hydro is the main proponent and most active user of nuclear power. In a period when the "small is beautiful" approach to energy questions is being advocated on a broad front, the concern is not surprising.

Since the early 1970s, largely as a result of initiatives taken by the farm community, Ontario Hydro has held public meetings to acquaint property owners with its property acquisition policies, and with the procedures associated with the provisions of the Expropriations Act. The utility has also organized public seminars on specific projects, to inform people of the planning processes and to seek public input. These seminars have resulted in formally established "needs" and "site and route selection" committees – membership of which has included, in addition to Ontario Hydro personnel,[16] representatives of municipal councils, public interest groups, chambers of commerce, local industry, and private citizens. These committees have been only partially successful, and they have been subjected in some areas to considerable criticism. One basic difficulty is the obvious confusion in the minds of many people concerning the respective roles of committees sponsored by Ontario Hydro, on the one hand, and the several tribunals including this Commission that appear to have closely related purposes, on the other hand.

Although the concept of direct citizen participation in decision-making is not universally accepted, it has nevertheless gained considerable favour during the last decade. Few would deny the efficacy of such participation in policy formulation concerning such matters as regional planning and environmental impact. That this has been recognized by the Government of Ontario is demonstrated by the establishment of the Ontario Municipal Board, the Ontario Energy Board, the Solandt Commission, the Environmental Assessment Board, and the Royal Commission on Electric Power Planning. Each of these tribunals, *ipso facto*, is based on the concept of public participation.

This Commission, virtually since its inception, has recognized that more public involvement in the affairs of Ontario Hydro is essential. In the search for equitable solutions to many electric power planning problems, only the people of the province, as a whole, are in a position to advise the utility of the nature of their "best interests".

Figure 2.1 Growth of Ontario Hydro's System

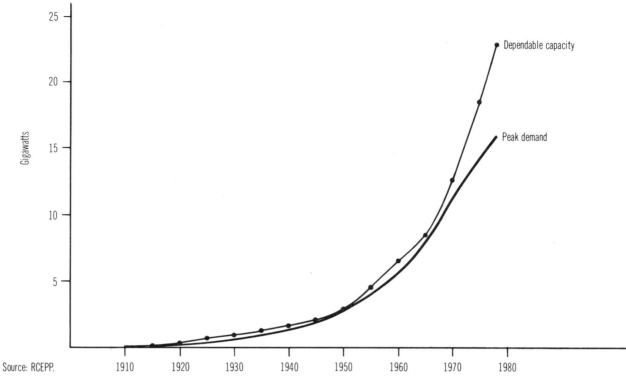

Source: RCEPP.

Figure 2.2 Growth in Labour Compensation per Person Employed (Wage Level), Consumer Price Index, and Ontario Hydro's Revenues per Kilowatt Hour (1961 = 100)

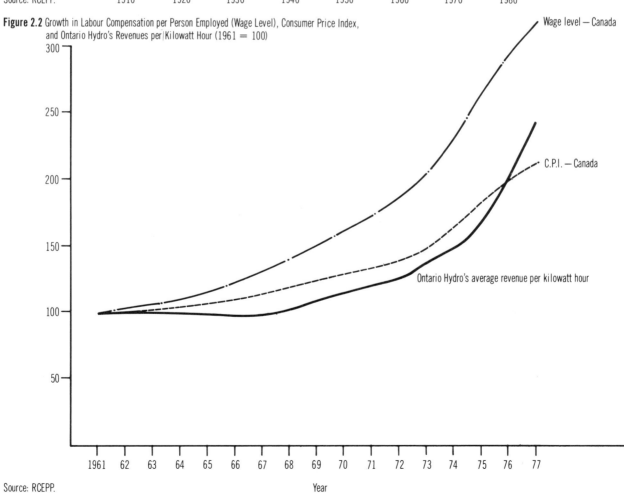

Source: RCEPP.

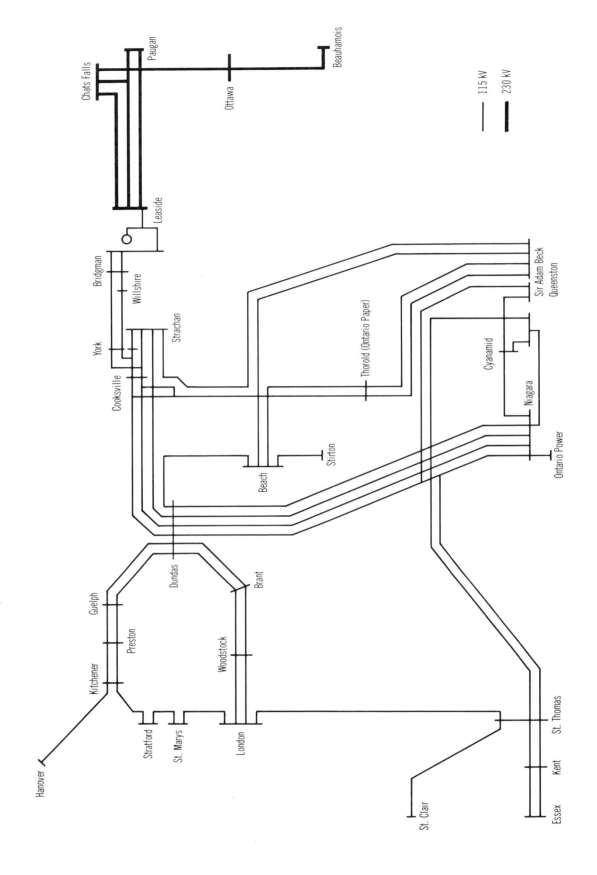

Figure 2.3 Hydro Electric Power Commission of Ontario 230 kV and 115 kV 25 Hz System — 1932

Source: Ontario Hydro.

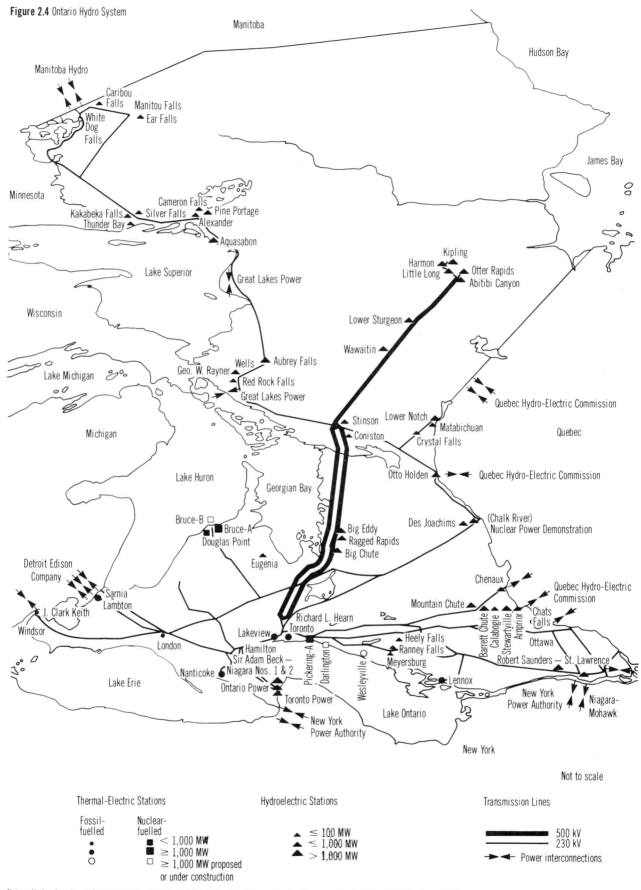

Figure 2.4 Ontario Hydro System

Manitoba

Hudson Bay

James Bay

Manitoba Hydro

Caribou Falls

Manitou Falls
Ear Falls

White Dog Falls

Minnesota

Cameron Falls
Kakabeka Falls   Silver Falls   Pine Portage
Thunder Bay   Alexander

Aquasabon

Wisconsin

Lake Superior

Kipling
Harmon
Little Long   Otter Rapids
Abitibi Canyon

Great Lakes Power

Lower Sturgeon

Wawaitin

Lake Michigan

Wells   Aubrey Falls
Geo. W. Rayner
Red Rock Falls
Great Lakes Power

Michigan

Lower Notch
Stinson
Coniston   Matabichuan
Crystal Falls

Quebec Hydro-Electric Commission

Quebec

Lake Huron

Georgian Bay

Otto Holden   Quebec Hydro-Electric Commission

Bruce-B
Bruce-A
Douglas Point

Big Eddy
Ragged Rapids
Big Chute

Des Joachims   (Chalk River)
Nuclear Power Demonstration

Detroit Edison Company

Eugenia

Chenaux   Quebec Hydro-Electric Commission

Mountain Chute   Chats Falls

Sarnia
Lambton

J. Clark Keith

Windsor

London

Lakeview
Richard L. Hearn
Toronto

Hamilton
Sir Adam Beck —
Niagara Nos. 1 & 2
Nanticoke
Ontario Power
Toronto Power
New York Power Authority

Pickering-A
Darlington
Wesleyville

Heely Falls
Ranney Falls
Meyersburg

Lennox

Barrett Chute
Calabogie
Stewartville
Arnprior
Ottawa

Robert Saunders — St. Lawrence

New York Power Authority   Niagara-Mohawk

Lake Erie

Lake Ontario

New York

Not to scale

| Thermal-Electric Stations | | Hydroelectric Stations | Transmission Lines |

Thermal-Electric Stations

Fossil-fuelled
•
•
○

Nuclear-fuelled
■ < 1,000 MW
■ ≥ 1,000 MW
□ ≥ 1,000 MW proposed or under construction

Hydroelectric Stations
▲ ≤ 100 MW
▲ ≤ 1,000 MW
▲ > 1,000 MW

Transmission Lines
━━━ 500 kV
─── 230 kV
►—◄ Power interconnections

Notes: Hydroelectric stations generating less than 4MW are excluded. Transmission lines carrying less than 230 kV are excluded.

Source: "Ontario Statistics", Central Statistical Services, Ministry of Treasury and Economics, 1979.

CHAPTER THREE
Ontario's Electric Power Requirements

When we analyse the reasons why the demand for electric power has increased so rapidly all over the world since the begining of this century, it is clear that the key factor has been electricity's flexibility. Electric power adapts very readily to the demands of modern technological society. Since its advent, electricity technology has given rise to a massive "speeding up" of many industrial, commercial, and domestic processes. In Chapter 2 of the *Interim Report on Nuclear Power in Ontario* we introduced the key factors that have influenced the growth in the demand for electricity in Ontario; because this material is not repeated here, reference should be made to the earlier report.

The demand for energy by all living organisms including man, and by society, is predicated on two, and only two, factors – the maintenance of the organism or system, and its growth. The future demand for electricity, for example, has two basic components; first, the level of consumption required to maintain all existing services supplied by electricity, and second, the requirement associated with growth in capital equipment and stock that consumes electricity. We believe that the demand for electricity arising out of the first component is likely to decrease in the future, essentially because of the impact of higher electricity prices, and the increased efficiency with which electricity will be utilized. Furthermore, existing end uses of electricity, notably electrically heated homes, electric motors, and lighting systems will inevitably be replaced by more energy-efficient systems and appliances, and the quantity of electricity required to meet these end uses will consequently decrease in the future.

However, the replacement process is likely to be comparatively slow. We anticipate, moreover, that a broad range of retrofit options will probably become cost-effective; these will improve significantly the efficiency with which existing equipment and processes perform their tasks. A notable example is insulation, and there are other conservation measures that could significantly reduce the quantity of electricity required to heat homes and buildings. Indeed, conservation, in the simplistic sense of less wasteful consumption habits (i.e., lowering thermostat settings) or in the sense of price-induced efficiency improvements, is likely to be more substantial in the maintenance component than in the growth component of demand.

The growth component of the demand for energy (i.e., fuel and food) is determined essentially by demographic factors, and by the increase in per capita income made possible by productivity gains. Additionally, in the case of electricity, its share of the new energy market will depend on, first, the new uses for electricity that are created by technological advances, second, the displacement by electricity of other fuels from their traditional end uses, and third, the improvements in efficiency that will characterize new electricity-consuming capital goods and services.

During the last few years, especially since the Commission was established in July 1975, the growth in Ontario Hydro's annual peak demand for electricity has departed significantly from the historical trend. For instance, the long-term average annual growth rate of about 7 per cent, which had characterized peak load growth in Ontario (and elsewhere in North America) for over half a century, dropped to an average growth rate of 3.2 per cent over the period 1975-9. This marked deviation from the norm might, with the aid of hindsight, be attributed in part to the repercussions of the oil crisis in 1973. For example, higher oil prices acted to depress economic activity and, in turn, to lower electricity demand in the short term. On the other hand, the price relationships between oil and other energy sources changed markedly, and stimulated the demand for electricity in Ontario in spite of the increased electricity costs. Furthermore, higher fossil-fuel prices, in general, create new conservation opportunities. How long the adjustment to higher fossil-fuel prices will take, what form it will take, and whether or not there are other factors tending to dampen the load growth are questions that have added to the uncertainty surrounding the load-forecasting process.

The substantial "over-forecast" of the demand for electricity, since 1974, has prompted Ontario Hydro, the Ministry of Energy, the Ministry of Treasury and Economics, and the Commission to reassess the factors that determine the electricity load growth in the long term. The major conclusion of these independent studies is that future growth in electricity demand will continue to remain appreciably below the historical trend.[1] In particular, the Commission has concluded that the factors largely responsible for the recent drop in load growth are not transient. We believe that the province is entering an era of slower population growth, and perhaps slower productivity gains as well, and that higher energy costs will probably accelerate the structural changes that reduce the energy intensity of the production of goods and services. Furthermore, saturation of major energy-consuming goods and services may reduce energy needs from the consumer's point of view, and the recognition that oil supplies are being rapidly depleted, and the effect this will have on future prices and security of supply, promises a long-term transition away from this non-renewable fuel.

The higher cost of oil enhances the economic desirability of other energy options indigenous to Ontario. For example, just as the relative cost-effectiveness of nuclear generation has been improved relative to coal-fired generation by higher fossil fuel prices, similar assumptions apply to such options as co-generation, district heating, home insulation levels, etc.; all show increasing promise. The quantitative impact of these "alternate technologies" on the demand for "centralized electric power" is difficult to predict, except that the direction will be negative, i.e., there will be a reduction in demand for electricity. As well, the situation is confused further by the apparent surplus of natural gas, a fuel which is a substitute for both oil and electricity in certain key end uses such as space heating, although its availability beyond the 1980s is in question.

We have concluded that the lack of information, especially relating to energy end uses, makes it uniquely difficult to quantify the impact of the above factors on long-term load growth; moreover, the problem is exacerbated by present supply uncertainties. We believe, however, for reasons to be developed subsequently in this volume, that the deficiency of the data base should be adequately dealt with during the 1980s. In consequence of this deficiency, however, the load-forecasting process has been transformed during recent years from a fairly routine exercise into one characterized by great uncertainty. This can only be overcome, in the short term, by the exercise of judgement, based on energy-policy choices and explicitly open to scrutiny.

It should be noted as well that, because higher fossil-fuel prices have enhanced the economics of nuclear power, there has been an associated change in the generation mix. This has had important consequences for load forecasting. The lead times associated with nuclear stations, including the public hearing phase of the approval process, is at present about 14 years; this is at least four years longer than that of an equivalent coal-fired station. Further, the longer the lead time, the greater the complexity of load forecasting.[2]

Another significant factor is the high capital cost of nuclear power relative to fossil-fuelled generation. In effect, this increases the cost penalty associated with "over-forecasts" of electricity demand. The interest and depreciation charges of a nuclear station accrue regardless of the quantity of energy sold. On the other hand, the capital intensity and low fuelling costs of a nuclear station result in its unit energy costs being relatively inflation-proof once the station is in service.

In the preparation of the long-term peak-load forecast, the load-duration curve (see Volume 2, Figure 2.6) is assumed to change little. This was a justifiable assumption during the 1950s and 1960s when fossil-fuelled and peaking hydraulic stations constituted the bulk of the generation capacity because of the excellent load-following capabilities of these stations. However, as the nuclear component in the generation mix increases, it becomes necessary to re-examine this assumption, since the current design of CANDU reactors has only limited load-following capability. The forecast of load shape will become more critical some time in the 1990s when the installed nuclear capacity is increased to the point where nuclear and hydraulic capacity fulfil all base-load requirements. If the load forecast could be predicated on the major end uses of electricity, it should be possible to determine the future load-duration curve as a weighted average of the load shape for each end use.[3] This would assist system planners in specifying the incremental nuclear:fossil mix. And, in particular, it would have an important bearing on the timing of the nuclear power station that will follow the Darlington Generating Station.

It has been recognized for many years that demand is sensitive to rate levels and structures and could be influenced by advertising. Because electricity supply was a "declining cost industry" from the early 1950s until the early 1970s, promotion of the use of electricity led to a lower unit energy cost system. Furthermore, Ontario Hydro rationalized its involvement in the market-place on the grounds that increases in load growth, and concomitant system expansion, allowed the system to take advantage of increasing economies of scale and the decreasing real cost of coal.[4]

Because of the extreme uncertainty surrounding load forecasts for the 1990s we have concluded that a "demand-scenario" approach should also be pursued. Moreover, the cost of major errors in matching supply with demand (which arise because of the inflexibility of advancing the in-service dates of major generating stations, and because of the capital charges for surplus generation capacity) makes it desirable, we believe, for Ontario Hydro and the provincial government to pursue policies of demand management that will guide load growth towards efficient utilization of the capacity committed to meet a selected scenario.

We recommend that:

3.1   Through the development of demand scenarios based on end-use data, future planning philosophy should be reoriented to emphasize demand management increasingly rather than maintain the focus on supply expansion, as is traditional.

Hydro's role in demand management could take the form of changes in rate structure or financial policies that change the relative price of electricity and its competitiveness in some or all end-use sectors or, alternatively, demand management could be achieved through responsive pricing or load control.

Clearly the provincial government can influence electricity demand by providing tax incentives and/ or low-cost financing, both to assist conservation measures and to encourage inter-fuel substitution. The topic is considered in some detail in Chapters 10 and 11. Additionally, the government could introduce dis-incentive schemes to discourage energy-consumption patterns that are inconsistent with its overall strategy, e.g., the demand scenario.

The role to be played by demand management would depend also on the extent to which the selected scenario corresponded in practice with the "natural" demand for electricity, i.e., the normal growth patterns that would be expected in the absence of intervention by Ontario Hydro or the government. Conversely, the demand scenario itself could be selected on the basis of a "business-as-usual" forecast, adjusted by estimates of the net effect on peak load and load shape of measures to manage demand that lie within Hydro's mandate and are consistent with the province's energy strategy. The Commission has concluded that the people of Ontario, through the government, should begin to shape their energy future, e.g., by participating in demand-scenario selection (see Chapter 12). Indeed, the first report of the Commission, based on the identification of energy issues by the people, was entitled *Shaping the Future.* And shaping the energy future, especially in the case of electricity, is much more than a mere forecasting exercise. We believe that by making policy choices regarding the desired mix of energy forms and conservation, some of the future uncertainty relating to electricity demand growth can be eliminated.

It is particularly important to stress that the supply scenario chosen to complement a demand scenario should compensate for uncertainty by building in as much flexibility as possible. This might be achieved by reducing the planning lead times for generating stations of all types, principally by streamlining the environmental assessment, public-participation, and land-acquisition phases of project approval. This should not imply superficial examination of such critical issues but rather minimization of repetitive arguments that usually characterize public hearings during these phases. We discuss this topic further in Chapter 12.

Ontario Hydro and the government should attempt to minimize the uncertainty surrounding future electricity demand growth by guiding demand in directions that are consistent with the province's overall energy objectives and that efficiently utilize existing and committed generating capacity. Demand or load management, rate policies, tax incentives to assist in financing conservation measures or desirable inter-fuel substitution – these are legitimate means to this end, as are dis-incentives to discourage energy-consumption patterns that are inconsistent with the overall strategy.

Another approach (see Chapter 7) is to build a bank of smaller-scale, shorter-lead-time technologies that effectively smooth out supply/demand gaps by providing more flexible and smaller increments to the system in order to respond more effectively to the uncertainty of the load forecast. Co-generation and small hydraulic projects fall into this category. Furthermore, strengthened interconnections with Manitoba and Quebec to facilitate power imports, and with neighbouring United States utilities to permit increased power exports, would result in greater system resilience, which is a primary goal of our planning philosophy. Kenneth Boulding summarized this philosophy in these words:

> The way to prepare for the future should rather enable us to recognize and take advantage of good luck and to anticipate bad luck so that we are not destroyed by it. Any strategy which assumes a known future involves delusions of certainty which can lead to catastrophic decisions.[5]

Load Forecasting

The generic approach to load forecasting used by Ontario Hydro (and many other North American electricity utilities) has been outlined in the Commission's southwestern Ontario regional report. Only the basic concepts of this and other methodologies that have been brought to our attention will be introduced in this section. Moreover, to avoid the complexities of regional load-forecasting methods, we restrict the discussion to the forecasting of the utility's primary peak demand for electricity.[6] The primary peak demand forecast does not take into account the effects on peak capacity requirements of load management, interruptible contracts, and the demand that might be met by additional private co-generation facilities. Further, the temperature-sensitive component of the load is projected on the assumption of "the typical maximum degree-day of the year".

We review three approaches to load forecasting. First, the methodology employed by Ontario Hydro in its 1979 load forecast; second, the approach adopted by the Ministry of Treasury and Economics, formerly the Ministry of Treasury, Economics and Inter-Governmental Affairs (TEIGA), in its 1976 submission to the Commission; third, the scenario approach developed by the Commission and presented more fully in Volume 3. We emphasize that all three approaches suffer from a lack of relevant data for the period of critical concern, i,e., the post-1973 period. Consequently, all of the approaches depend on a high level of judgement, in contradistinction to hard data. However, the introduction of a major judgemental component into a model means, *ipso facto*, that the model loses its air of scientific authenticity.

The role played by judgement, and the degree to which this is explicitly recognized, differ markedly in the three forecasting approaches. However, on account of the paucity of data, the Commission cannot recommend an objective forecasting methodology upon which to base decision-making. On the other hand, we suggest that the data base, especially as it relates to electricity end uses, should be built up as quickly as possible, and that subsequently a blend of forecasting techniques should be used. We have concluded that any load forecast developed at the present time must inevitably incorporate a large element of value judgement, and depend, at least in part, upon "second-guessing" the energy strategy choices that will be made by politicians, either overtly or by default. It is in this context that planning decisions will continue to be made — based, however, on more comprehensive information.

Ontario Hydro's 1979 Load Forecast

The role of econometric models in the preparation of Ontario Hydro's long-term load forecasts has evolved rapidly during recent years.[7] For instance, as recently as 1976, the so-called "forecasting equation" expressed the future electric load as a function of gross national product (GNP) and the independent variable, time; it was used simply as a check of the "reasonableness" of the forecast. On the other hand, in explaining the 1978 load forecast, Hydro stressed that econometric models, although not used specifically in the production of the forecast, were important in the validation of the forecast and "as a guide to the use of judgement". The 1979 load forecast is much more sophisticated. Essentially, the forecast equation models the East System peak demand as a function of employment levels, productivity, electricity rates, the price of oil, and the price of natural gas. Electricity demand is represented by an estimated statistical relationship between the above variables which, in effect, provides the analytical tool for developing the load forecast. The 1979 load forecast itself was one of a set of scenarios generated by varying the projections of the independent variables (e.g., the prices of oil and of natural gas).

The social and political repercussions of the continuing energy crisis, and not least the rapid increase in fossil-fuel prices since 1973, have called for economic adjustments in all industrialized countries. This is the main reason why load forecasting is extremely difficult. Indeed, it has been suggested that the econometric model used in load forecasting should be regarded as a device that, in effect, provides a quantification of the judgement of the load-forecaster. For example, if data appropriate to the years prior to 1973 had been used in the most recent forecasting equation, it would have been found that the nature of the relationship between electricity demand and the prices of fossil fuels would apply essentially to a "business-as-usual" view of the energy future from a 1960s perspective. Consequently, we have concluded that the extrapolation of trends established during a period of real decline in energy prices, coupled with a comparatively stable relationship between electricity prices and fossil-fuel prices, would not be helpful in assessing the major implications of a transition period to substantially higher-priced energy.

On the other hand, because only five years of "post-OPEC crisis" information is available (i.e., 1973-8), the additional "five data points" do not provide the forecaster with an adequate information base. The "time-series" is just not sufficiently long. This means that it is virtually impossible to evaluate statistically the long-term implications of substantially higher oil and gas prices, and the narrowing gap (based on equivalent end uses) between electricity and fossil-fuel prices. While the incorporation of recent information into the model moderates the historical 7 per cent growth rate in electricity demand, the information base is still not sufficiently well established to provide a realistic forecast of the demand 10 to 15 years hence. It is for this reason that the forecaster's judgement is required.

Without going into the detail that is provided in Volume 3 of this Report, it is important to comment on the validity of the extant forecasting equation for East System peak demand. Notwithstanding the fact that the coefficients in the equation appear to be statistically significant, we do not believe that the equation adequately models the long-term "environment" for electricity demand. In particular, the negative co-efficient estimated for the oil price variable[8] suggests that, in the long run, the equation would predict that higher oil prices will lead to lower electricity consumption. This does not appear to be plausible. In the short term, although it can be argued that higher oil prices depress economic growth and indirectly dampen the demand for electricity, this cross-dependency of two significant coefficients in the forecasting equation constitutes, we have concluded, a critical flaw in the model.

However, the information needed to correct this deficiency is not available, and consequently the long-term cross-price elasticity of demand between oil and electricity cannot be estimated with any degree of confidence.[9] Furthermore, because estimates of future oil prices and of the GNP are also involved, it is unlikely that simple solutions will be found. A basic requirement, clearly, is the development of a new generation of whatever macro-models are required to estimate the long-run impact of oil-price increases on the economic growth potential of Ontario.

Because it may take at least five years to accumulate an adequate data base to correct the above deficiencies and to build models involving the energy variables, it is clear that a high level of judgement will be required during the transitional period in selecting the appropriate load forecast upon which to base system planning. This judgement, we argue, must be gleaned (see Chapter 12) from various sources in addition to Ontario Hydro.

The Ministry of Treasury and Economics Submission

The submission in 1976 by TEIGA, entitled "Growth in Ontario's Demand for Electric Energy", provided the Commission with its first insights into the load-forecasting concepts introduced in the previous section. In particular, the submission introduced the Commission to the concept of disaggregation in terms of electricity end uses. Because of the unavailability of relevant information, the study could not take into account the impact of higher oil and gas prices on electricity demand. However, the broad implications of conservation and inter-fuel substitution (especially in the residential and industrial sectors, in which the incremental demand for electricity could be estimated by making specific assumptions about the increased market share of electrical space heating and about changes in appliance stocks) were taken into account. An attractive feature of this approach to forecasting is the fact that departures from historical end-use consumption trends (the study was based essentially on statistics for the period 1960-72) may be assessed quantitatively and, of special significance, the judgements involved in the forecasting process can be documented and made available for scrutiny.

The TEIGA 20-year load forecast for the period 1975-95, based on demographic and per capita income assumptions that closely parallel those of the Commission, averaged a 3.5 per cent load growth per year. Bearing in mind the limitations of the data base, and the unpredictable developments in the energy sector, especially concerning energy prices, which have occurred since the forecast was developed in 1976, this appears to be a remarkably accurate forecast of trends in load growth. Indeed, it anticipated, by about two years, the marked drop in Hydro's load forecast.

The RCEPP Model

Recognizing the significance of load forecasting in electric power planning, the Commission has, during the last three years, developed its own approach. We acknowledge, in this connection, the collaboration of B. C. McInnis of Statistics Canada. In particular, we have developed hypothetical but nevertheless plausible relationships between the demand for electricity and the potential consumption of other forms of energy. To provide an historical perspective, we show in Figure 3.1 the trends in the consumption of the major forms of energy in Ontario during the period 1959-75, together with total

Fig. 3.1: p. 38

energy end-use projections based on three average annual total energy growth rates (1.2 per cent, 1.9 per cent, and 2.6 per cent), and a 4 per cent average annual growth rate in the demand for electricity, to the year 2000. The dominance of the demand for the fossil fuels (especially oil and natural gas) is noteworthy, but not surprising. Indeed, even on the basis of a 4 per cent per annum load growth rate (which we consider to be high) the market share of electricity to the year 2000 will probably not exceed 25 per cent.

Our model is essentially a framework in which the *physical* potential to increase electricity's share of the Ontario energy market can be consistently worked out. It analyses energy assumption patterns by end use, to categorize their demand into ones in which substitution is feasible at expected price levels, and ones in which a fuel is captive to specific uses. In effect, we postulate various levels of average annual growth in demand for secondary energy in Ontario to the year 2000. The model (Volume 3, Figure 5.8) then allows for determinations of oil, gas, coal, and electricity supply growth rates that will meet the secondary energy requirements. A desirable feature of the approach is that it incorporates a visual presentation of these combinations. For example, Figure 3.2, which is based on a 1.75 per cent per Fig. 3.2: p. 39 year energy growth and a 2 per cent renewables penetration, illustrates the problem of energy substitution. Although the model may appear complicated, we cannot overemphasize the fact that the problem of assessing energy requirements in the future is highly complex. This is essentially because so many variables are involved and because of the interrelationships between them. Indeed, Figure 3.2 is the simplest form of statement of the energy problem as a whole that we have been able to devise. We summarize below how the model can be used.

Given a particular primary and secondary energy growth rate, and once oil and gas consumption growth has been "fixed", the growth rates for coal and electricity can be identified with particular levels of penetration into markets that are now predominantly served by oil and gas, i.e., areas in which significant substitution is feasible. We believe space heating and industrial process heat are particularly significant in this regard. Using the diagram, the rate of penetration required to meet the total energy demand can be readily determined for both coal and electricity. Note that the scales in Figure 3.2 have been calibrated in terms of the Ministry of Energy end-use model. We have developed the model in such a way that "captive" demands for coal and electricity (in the sense that these end-use demands are unlikely to be replaced by other fuels) are given first priority, and second priority is given to the most likely markets for substitution (e.g., the use of coal for process heat and of electricity for space heat).

The use of the RCEPP model to assess the implications of changing the mix of secondary energy forms in Ontario over the next 20 years necessitates several sets of assumptions. These are required, essentially, to correlate the province's economic growth potential (as inferred from the expected labour force and productivity growth rates) with secondary energy demand. Furthermore, rather than relying on the extrapolation of historical trends in developing the economic-growth/energy-demand relationship, we have attempted to develop an appreciation of the forces that modify this relationship. Suffice it to add that fundamental changes in the nature of society (e.g., life-styles) are gradual and extremely difficult to model statistically.

In particular, a reduction in energy intensity (i.e., the energy required to produce a product or service) of economic activity will result in lower energy consumption per unit of output. We note also that energy intensity may be reduced through a structural change in the economy, through saturation in the demand for energy-consuming goods and services, and through conservation. These possibilities are considered below.

- Structural change (e.g., expansion of the government sector as opposed to the private sector) can produce a reduction in overall energy intensity by altering the mix of economic activities in favour of those with lower energy requirements. The persisting shift from manufacturing to service industries normally has this effect. Similarly, more rapid growth of high-technology industries (e.g., communications) relative to resource-processing industries (e.g., metal fabrication) will give rise to reduced energy intensity. This was exemplified during the period 1959-75, when secondary energy consumption in Ontario grew at an average annual rate of 4 per cent, while during the same period the gross provincial product (GPP) grew at a rate of 5.1 per cent. This was a period of declining real energy costs and increasing purchases of consumer goods. During the same period, moreover, there was a rapid expansion of the government sector.
- Saturation in the demand for energy-consuming goods (e.g., refrigerators, television sets) may not have a major effect on energy requirements for at least a decade. The impact of saturation on

energy requirements could occur either because reduced spending on energy-consuming goods and services slows economic growth or, should consumer demand and growth in production continue, because time constraints and finite human needs limit the practical ability of society to utilize the goods and services available.

• Energy conservation (see Chapter 10) is synonymous with increased efficiency in the utilization of energy. We anticipate a slow process of societal adjustment, in some cases motivated by higher real energy costs and in other cases by the introduction of conservation technology, and this will increasingly affect future energy demand. In Volume 4, we provide a set of estimates of energy savings through conservation, considered feasible by the year 2000, in the major end-use categories. These estimates, which are sensitive to increasing energy costs, presuppose the removal of certain institutional barriers that could delay implementation.

Our insights into the impact of structural change and saturation on energy demand have been obtained from scenarios prepared especially for the Commission by Dr. McInnis. These were based on the Statistics Canada "Long-Term Simulation Model". Two scenarios were studied in depth. In the first, the so-called "business-as-usual" scenario, characterized by a 3.6 per cent per annum economic growth rate, the secondary energy growth rate was computed at 2 per cent. The second, perhaps even more plausible, scenario assumed that consumer saturation would reduce economic growth to an average of 2.2 per cent per annum over the same period; in this scenario the secondary energy growth rate was computed to be 1 per cent. Neither of these scenarios reflects the impact of conservation.[10]

A secondary energy demand growth rate may be computed once a projection of the GPP growth rate, together with assumptions regarding structural change, saturation, and conservation potential, has been established. Furthermore, assumptions relating to the penetration of renewable energy resources must be made. In effect, we have reduced the energy supply problem to a determination of the mix of conventional fuels that could physically satisfy the demand. To reduce the range of feasible substitutions to a manageable level, the approach we have taken (see Volume 3 of this Report) is to select a "preferred" projection for the combined oil and gas availability of 1 per cent per annum to the year 2000. The inevitable energy gap is assumed to be filled by electricity and coal. We have noted, in particular, that adjustments to the secondary energy demand growth rate markedly affect the growth rates of electricity and coal. Furthermore, if a policy is established whereby the use of coal is minimized for environmental reasons, electricity becomes the critical factor.

End-Use Forecasting

An important application of the Commission's model is for analysing energy consumption patterns according to end use. However, such analyses are at present severely curtailed because of data limitations.

We have concluded that such a data base would result in a powerful blend of engineering and economic analysis being brought to bear on the load-forecasting problem. The engineering evaluation of the cost-effectiveness of electricity in comparison with other options, for a range of future energy prices, could be combined with econometric modelling of the growth of each end-use sector, together with its price sensitivity, in order to generate a detailed perspective of the future market for electricity.[11]

Furthermore, econometric models of each end-use sector could be applied to the estimation of the effects of demographic trends, structural change, and saturation on specific components of future electricity demand, and aggregation of the end-use projections should provide a more viable load forecast than the single-equation forecasting model at present in use by Ontario Hydro. As well, this total electricity demand equation for the province would provide an important cross-check on the results of the "micro-econometric" approach.

In so far as the potential of end-use forecasting is concerned, we have noted that several utilities in the New York Power Pool, as well as the New York Public Service Commission, employ a dual-forecasting approach similar to the one we are recommending. For example, both a total-system forecast based on a variety of complementary techniques, and an end-use forecast oriented specifically towards the residential sector, are prepared and rationalized in the determination of the load forecast that is used for planning purposes.[12]

Assuming that the province wishes to set goals for energy consumption, as part of an overall energy strategy, an expanded end-use energy data base such as the one we have advocated will be essential in order to establish realistic goals. Indeed, the realism and cost of a particular strategy can only be assessed adequately when its implications for the modification of systems and processes utilizing energy on a large scale can be taken into account.

Recognizing the problems of macro-economic models that attempt to extrapolate the trends of the past through the discontinuity of 1973-4 and the increased uncertainties arising from the impact of structural changes in the economy, saturation in the demand for energy-consuming goods and services, increased energy efficiency arising from higher energy prices, and the impact of specific energy conservation programmes,

We recommend that:

3.2 A comprehensive energy end-use data base for the province should be developed as soon as possible and Ontario Hydro, in addition to macro-economic or "top down" forecasting models, should develop complementary models based on the detailed building up of electrical demand on an end-use basis. Ontario and federal government ministries and agencies should support Ontario Hydro's efforts to fill the remaining data gaps.

This approach corresponds essentially to a disaggregation of the load-forecasting problem. We believe that it will reduce in some degree the uncertainty created by the impact of changing relative prices in the energy sector and associated impacts on the rest of the economy.

3.3 Ontario Hydro should employ, as a useful analytical device for load-forecasting purposes, the distinction between "captive" and "competitive" end uses of electricity.

3.4 Because of increasing emphasis on end-use forecasting, the role of the public utility commissions in developing load growth patterns should be enhanced to provide opportunities for more input than hitherto by the public.

Ontario's Load Growth in Perspective

The historic trend in the demand for electricity in Ontario is typical of many countries in the industrialized western world. According to a recent international review, 7 per cent per annum electricity growth was widespread in these countries from the 1950s through the early 1970s. Like Ontario, utility capacity expansion plans were based on the expectation that these growth rates would continue at least to the end of the century.[13]

The study cites the familiar factors for these lower growth rates: gloomy prospects for the world economy in the 1980s, improvements in the energy efficiency of electricity-consuming technologies, recognition that it is technically and economically efficient to double the energy efficiency of household appliances and substantially improve that of electric motors operating at less than full load, use of heat pumps for space heating, insulation, industrial co-generation, and district heating, etc. The authors found considerable uncertainty in the electricity forecasts to the end of the century, with the range extending from 2 to 6 per cent per annum. At the upper end of the range were countries still undergoing rapid industrial development and those most worried about vulnerability to imported oil (e.g., France and Japan); at the lower end were the mature industrial economies (the United States) and those with good alternatives to the use of electricity in non-captive markets (e.g., the United Kingdom, with its North Sea gas).

Considering Ontario as a mature industrial economy, with some good alternatives to the increased use of electricity in the non-premium space-heating, water-heating, and industrial process-heating markets, we think this international review lends additional support to our selection of a target Ontario electricity growth rate for the balance of this century, to which we now turn.

The Commission Scenarios

We hope that the analytical model, developed by the Commission to study the feasible range of electricity demand growth rates to the year 2000, will be helpful to energy planners in formulating an energy strategy for the province. Utilized to its full extent, we believe, it could demonstrate the implications of a wide range of the possible energy futures open to Ontario. It is essential, as we have emphasized, to examine the role to be played by electricity in the broad context of an integrated energy policy. Consequently, the ultimately selected electricity supply scenario should be consistent with the expected availability of alternative energy sources and with the level of implementation of energy-conservation measures. It should also reflect the priorities the province wishes to assign to the minimization of the impact of potential fossil-fuel-supply interruptions or price increases, or to the value of helping to keep the nuclear option open beyond the year 2000.

Our object, in the following discussion, is to show how the feasible range of electricity growth rates can be narrowed down in order to focus on viable system-planning alternatives that recognize the uncertainty inherent in the current forecasting environment. In the series of assumptions that are required for the task of specifying a secondary energy demand growth rate and an appropriate mix of energies, it is clear that each assumption is subject to controversy.

In developing the demand scenarios, we have assumed a population growth forecast for Ontario somewhat lower than the one adopted in the *Interim Report*, which estimated Ontario's population in the year 2000 at 10.6 million — the result of a 1 per cent per annum average rate of increase. In contrast, during the period 1957-76 the rate of population growth in the province averaged 2.4 per cent per year. On the other hand, we have estimated that the growth rate of the labour force, essentially because of the changing age structure of the population and the increased participation of females, will be 1.6 per cent per year. This estimate does not anticipate a return to the high levels of immigration into Canada of a decade ago. We have assumed that immigration into Ontario will be offset by out-migration from Ontario to the western provinces.

Between the early 1950s and 1973, productivity (i.e., output per employee) in Ontario grew at an average rate of 2.2 per cent per year. But since 1973, largely because of the rapid growth in the labour force and the weak economic performance, productivity has been reduced to 0.5 per cent. We do not expect the experience of the last five years to represent a long-term trend. Nevertheless, we believe that productivity growth will probably be lower in the next 10 to 20 years than it was prior to 1973. This will be due in part to increased expenditures in the industrial sector resulting from higher safety and environmental standards, and in part to the diversion of major investments into the energy sector. These investments, which constitute one component of the structural change mentioned previously, do not contribute to a net increase in the output of the economy although they are likely to induce changes in the mix of products.

Our estimate of the average productivity growth to the year 2000 of 1.7 per cent per annum is an average rate between the years 1961 and 1978. This corresponds to a 0.5 per cent per year reduction in productivity growth relative to the long-term historical trend. Combining labour force and productivity growth rates yields an estimate of potential GPP growth for the province of 3.3 per cent per year (see Volume 3, Chapter 2).

It is extremely difficult to quantify the impacts of structural changes (i.e., improved urban design, new investment patterns, government legislation), and the effect of saturation in the major energy-using appliances and processes, on the energy intensity of the economy. However, a consensus[14] suggests that the combined effect of structural and saturation effects is likely to continue to be a reduction in the 0.5 to 1.0 per cent range, as for the last two decades. To some extent, this reduction in the energy intensity of the economy has already been accounted for in the projection of lower growth in productivity. We have estimated, assuming a 3.3 per cent average economic growth rate and a net reduction in energy intensity of 0.7 per cent (due to structural factors), that the projected secondary energy average growth rate will probably lie in the range of 2.6 to 3.3 per cent. This estimated range of energy growth rates does not take into account the effect of conservation measures or specific efficiency-improvement measures in each end-use sector.

To assess the potential for conservation in various end-use categories, we have analysed the findings of numerous studies (see Volume 3, Chapter 6). The consensus appears to be that the total energy saving (weighted in accordance with the existing mix of energy end uses) resulting from measures that could be implemented by the year 2000 is in the order of 30 per cent of the secondary energy demand. This corresponds to a 1.55 per cent reduction in the average rate of growth of energy, and suggests that average secondary energy growth rates of between 1.0 and 1.75 per cent are plausible for the next 20 years.[15] Note that the lower end of the range corresponds to zero per capita end-use energy growth. However, we have chosen a range of secondary energy growth rates of 1 to 2 per cent (which corresponds to slower adoption of conservation measures) as the most plausible range, and our "preferred" scenario assumes a total energy growth rate of 1.75 per cent. Figure 3.2 provides a total end-use energy model of technical substitution possibilities consistent with a 1.75 per cent growth rate and a 2 per cent renewable energy penetration by the year 2000. The critical factor is the expected consumption of oil and natural gas, which could, in the event of extremely high prices or energy policies aimed at discouraging their use, be less than the supply currently available to Ontario. Our basic assumptions are summarized below.

- Even assuming a marked escalation of the oil-sands programme and success in the "frontier areas", Canadian oil production after declining in the intermediate years will reach its present level again by the year 2000. To augment domestic supplies, oil will continue to be imported from the OPEC countries – but we note that the cost per barrel could quadruple by the end of the century. Taking into account price increases of this magnitude, we believe that the average growth rate of oil consumption over the next 20 years will approach zero.

• We are optimistic about the availability of natural gas. We assume an average growth rate to the year 2000 of 3 per cent per annum. This growth will depend markedly on the oil/natural gas price relationship.

• Combining the growth rates of oil and natural gas, we assume that the secondary energy supplied by these sources could grow at a rate in the 1.0 to 1.5 per cent range.

The above assumptions are considered in detail in Volume 3. They suggest that the average annual growth in demand for electricity to the year 2000 could range between 1.5 per cent and 3.7 per cent, depending on the future role of coal in the production of heat and process-steam (see shaded sector of Figure 3.2). If coal-combustion technologies that are comparatively benign environmentally become commercially viable (e.g., fluidized-bed combustion), and if methods for the reduction of toxic gaseous effluents (especially the sulphur and nitrogen oxides, which give rise to acid rain) become available at reasonable cost, the use of coal would probably burgeon. However, as we show in Chapter 9, there are other environmental hazards associated with coal combustion that might lead to minimization of its use in the province.

The Commission's model implicitly assumes that the future growth in the use of coal will be determined largely by the degree to which it is substituted for oil or natural gas in the direct production of industrial process-steam. We assume, moreover, that there will be no increase in its use for space heating and water heating. However, through co-generation plants coal combustion could provide direct industrial heat and also contribute, though on a small scale, to the generation of base-load electricity. This component is not shown in Figure 3.2. Most important, the trade-off between electricity and coal will determine the amount of oil and natural gas that could be released from the industrial sector for space heating or water heating in the residential sector. Therefore, any policy aimed at minimizing the direct use of coal in industry would tend to increase the level of electrical space heating and water heating required.

Referring again to Figure 3.2, it may be seen that a 3 per cent average annual increase in electricity demand corresponds to approximately 50 per cent saturation of the space-heating and water-heating market by electricity or an equivalent amount of electric space-, water-, and industrial process-heating use in the year 2000, while a 3.7 per cent average growth rate would correspond to electricity achieving 75 per cent of the space- and water-heating market, or the equivalent, by the end of the century. However, the present penetration rate (30 per cent of the new housing stock and a small number of conversions) would lead to only 20 or 25 per cent of the market share over the next 20 years. Furthermore, there is a strong likelihood that natural gas will continue to capture a large proportion of new housing heating requirements during the 1980s, when two-thirds of the new housing construction required to the year 2000 is expected to take place. This suggests that electricity would have to make substantial inroads into the existing oil-heating market in order to achieve even a 50 per cent saturation level.

The above figures are based upon the scenario portrayed in Figure 3.2, which is an approximation based on several assumptions. In arriving at a demand scenario on which to base system expansion, we have concluded that it is neither prudent nor necessary to postulate a precise growth rate figure extending growth far into an unpredictable future. For planning purposes, there is an obvious need to project system expansion requirements beyond the anticipated lead times needed to construct major facilities – say 15 to 20 years. Since the relevant parameters cannot be established with any degree of precision, it is most appropriate to establish a load growth range.

The high end of the range will be influenced by contingencies and uncertainties in the basic assumptions used in the model; the low end might be set on the basis of current end-use patterns continuing into the future. The swing between the high and low end of the range will be determined largely by electricity's share of the space-heating load, which will depend to a large extent on the availability and price of heating oil and natural gas.

For the high end, the scenario in Figure 3.2 suggests a maximum electricity demand growth of 3.7 per cent, which would be achieved only if coal does not penetrate further into the existing oil and gas market – which we believe is unlikely. However, it is possible that electricity will be used in the future to supplement oil or gas in space heating, which would optimize the use of Hydro's base-load capacity. For example, a hybrid electric-fossil space-heating system in which 2 or 3 kW of electrical heating operates continuously throughout the heating season, and is supplemented when necessary by a storable back-up fuel, would not require Hydro to draw upon peaking capacity associated with conventional electrical resistance heating.[16]

Accordingly, we recommend that:

3.5 As part of a larger objective of planning for an improved annual load shape and higher load factors and as a means of increasing the resiliency of the electric power system and reducing Ontario's dependence on crude oil, Ontario Hydro should give high priority to demonstrating the technical and economic feasibility of new and retrofit hybrid electric/fossil space-heating systems.

While there is little evidence to support the availability of substantial export markets for electrical energy to the United States over the next 20 years, the general energy crunch that is anticipated could ultimately place unexpected demands on American utilities. This might create a need for reserve electrical capacity in a strategic North American sense, to which Ontario Hydro might be required to contribute. On the basis of such factors, in our judgement, the upper bound of electricity demand growth over the next 20 years should be set at 4 per cent per year.

The lower limit is based largely on current end-use patterns and is related essentially to population growth. Therefore, assuming some additional penetration by electricity of competitive end-use markets, the lower limit should, in our judgement, be in the range of 2.5 per cent per year. Thus we conclude that the growth in demand for electrical energy in Ontario[17] will fall within the range 2.5 to 4.0 per cent up to the year 2000. For system-planning purposes, this energy forecast must be converted into peak capacity (i.e., megawatts), which depends upon trends in system annual load factor and weather conditions over the winter months. While weather conditions are random and unpredictable, they do follow statistical patterns that exhibit some consistency over long periods. Corrected for weather, Ontario Hydro's peak demand growth has closely paralleled that of total energy demand growth over the last several decades. However, while annual load factors have shown a tendency to rise slowly in the last few years, this trend could be arrested in the future with an increase in conventional electrical resistance space-heating penetration. Avoidance of this potential drop in load factor will require an intelligent combination of initiatives, including hybrid heating systems and load management.

We recommend that:

3.6 For system planning purposes, Ontario Hydro should base its system expansion plan on a growth range for peak capacity to the year 2000 of 2.5 to 4.0 per cent per annum.

It must be emphasized that the load forecast over the next 20 years must be reviewed and updated continuously to reflect world and local events. In Chapter 12, recommendations will be made as to how such updating could be accomplished.

We expect that, by the early 1990s, demand management (e.g., load control) techniques and shorter lead-time technologies (e.g., co-generation) will make Ontario Hydro's system more responsive to uncertainties in the load forecast. However, during the 1980s, we believe, Hydro should plan its system on the basis of a load growth towards the upper end of the 2.5 to 4 per cent range.

Given the large number of parameters affecting both the demand and the supply sides of the energy problem facing Ontario over the next decade, we cannot specify the precise role nuclear energy might play in any system-expansion scenario based on a load growth range of 2.5 to 4 per cent to the year 2000. As we discuss in more detail in Chapter 5, the future health of the nuclear industry remains a cause for concern. However, we emphasize the importance of keeping all feasible energy options open, particularly during the next decade of uncertainty. Therefore, we consider it to be in Ontario's interest that the CANDU option remain viable.

Figure 3.1 Ontario Total End-Use Energy Requirements, 1975-2000

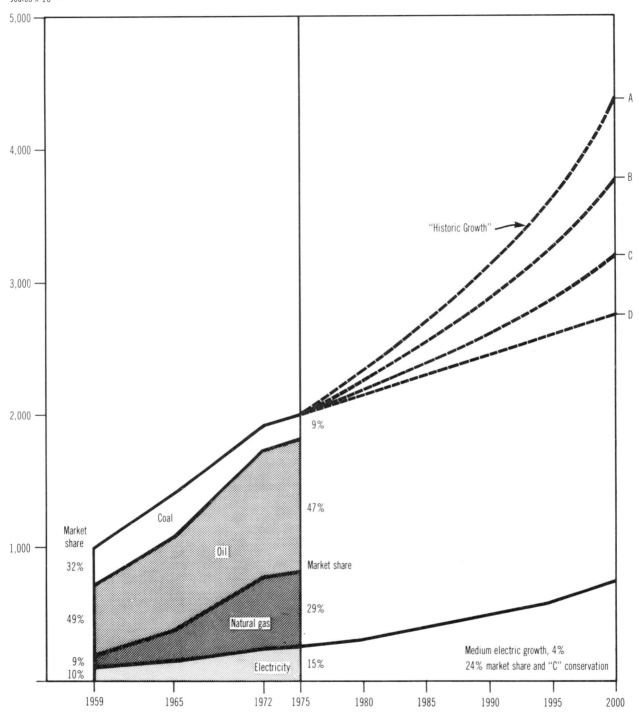

End-use energy consumption
Joules x 10^{15}

A No conservation: growth rate 3.2 per cent

B Low conservation: growth rate 2.6 per cent

C Medium conservation: growth rate 1.9 per cent

D High conservation: growth rate 1.2 per cent

Note: Assuming medium economic growth, 4%.

Source: Projections by RCEPP.

Figure 3.2 Preferred Projection. Secondary Energy Growth 1.75% per Year to 2000 and 2% Renewables Penetration in 2000

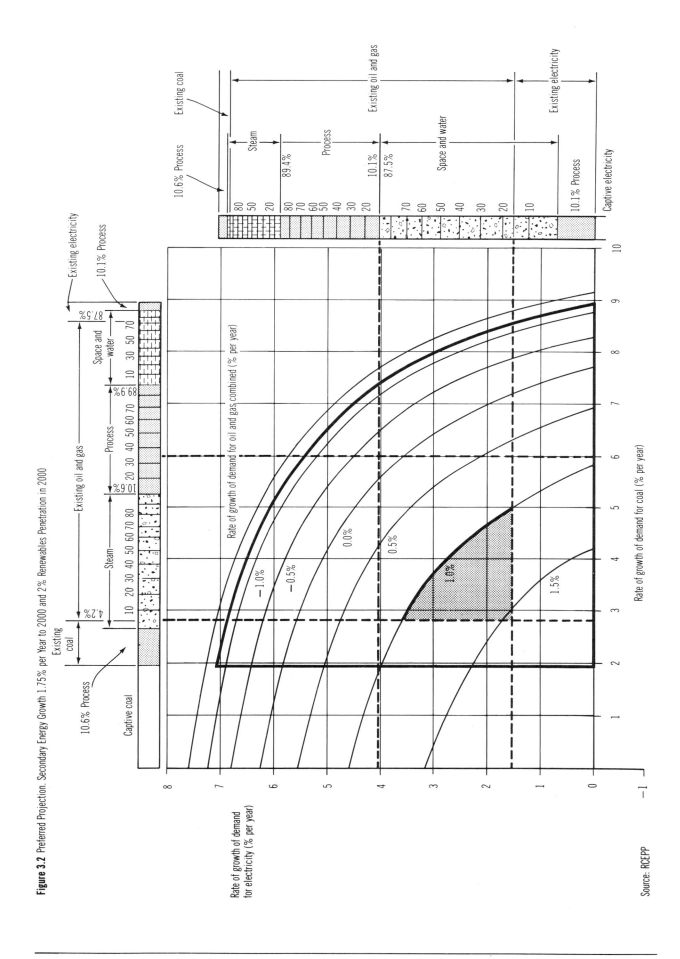

Source: RCEPP

CHAPTER FOUR
The Technology of Power Generation and Alternative Energy Sources

Ontario Hydro, in common with all electricity utilities, is required to provide a continuous and reliable supply of electric power. This is a stringent requirement, not least because to all intents and purposes electricity cannot be stored and yet must be instantaneously available on demand to all consumers. While virtually every other industry establishes inventories and can control its production or service within reasonable limits, this is quite impracticable in the case of electric power supply. Indeed, the instantaneous electric power needs of Ontario are determined by the switching on or off of literally millions of switches. This extremely important and fundamental characteristic of an electric power system should be borne in mind, not only in the planning of the system, but, just as important, in the public discussion of it. The generating technologies in use, and potentially available in the future, must as a consequence have unique characteristics, the most important of which is high reliability.

The supply of power is a highly technical process. Ontario's electric power system comprises many generating stations that convert such primary sources of energy as falling water, fossil fuels, and uranium into electricity, and each of which may incorporate a number of generating units. The output of these units is stepped up to high voltage (in some cases to 500,000 V – i.e., the voltage equivalent of 42,000 automobile batteries connected in series) for bulk power transmission. The high-voltage bulk power transmission lines that interconnect generating stations with transformer stations and switching stations provide the power for 334 municipal utilities, for large industrial users, and for several hundred thousand rural customers. Each basic facility and component of the system is itself complex. For example, a single generating unit may have a capacity of 850,000 kW (Bruce A incorporates four such units) and be capable of supplying a city of about 400,000 inhabitants (e.g., the Borough of Scarborough).

Most laymen refer to electric power and electric energy synonymously. However, there is an important distinction. We generate electric power (measured in kilowatts (kW) and megawatts (MW) by producing at a generating station an instantaneous flow of electricity – a rate of flow of electric energy. On the other hand, electric energy (measured in kilowatt hours (kW·h) or megawatt hours (MW·h)) is the amount of energy that has flowed or has been consumed over a period of time. For example, when a power of 1 kW is applied to an electric heater, for one hour, the electric energy consumed is 1 kW·h – the familiar unit of electricity. The maximum level of power that can be generated at any instant by Ontario Hydro depends solely on the design capacity of the generating units in service at that particular instant. On the other hand, the energy generated by the utility during a year is related to the amount of water passed through hydraulic turbines and the quantities of fuel (uranium, coal, oil, and natural gas) burned in thermal generating stations during the year. If a major source of power is hydraulic or nuclear, then in spite of the high capital cost of the facility, the electric energy is likely to be comparatively cheap because of the comparatively low "fuel" cost for these sources. As the capital invested in the system escalates, it is clear, effective utilization of the facilities becomes more and more important, to reduce the unit energy cost. The higher the "load factor" – the ratio of the average power supplied over a given period to the maximum power demand – the more effective the use of the facilities. (Ontario Hydro's annual system load factor in 1978 was approximately 0.67.)

Generating stations capable of operating continuously with high reliability, and for which the cost of fuel is virtually zero (hydraulic stations) or comparatively little (nuclear stations), are used for "base-load" generation. These stations are sometimes supplemented in this role by coal-fired stations.

When required to produce power, generating stations may or may not be available, depending on their state of readiness. Stations under repair or maintenance are, of course, unavailable. Forced or planned outages are very infrequent for hydraulic stations, but more frequent for thermal stations (nuclear- and fossil-fuelled). Typically, hydraulic stations have availabilities exceeding 95 per cent and thermal 77 per cent. Ontario Hydro classifies stations with annual capacity factors – the ratio of the average power generated over the period of a year to the maximum capacity of the station – exceeding 55 per cent as base-load generation. This corresponds to a station with a 77 per cent availability operating for five days a week which is 71 per cent of the time. The base-load stations are not expected to "follow" the hourly variations in system load. On the other hand, stations that have low capital but high fuel costs (such as oil- and gas-fuelled stations) and rapid response characteristics are used for supplying rapidly changing loads during peak periods; these are termed "peaking stations". Some hydroelectric stations, because of their excellent load-following properties and capability for storage, are also used for peaking. Ontario Hydro classifies peaking stations as those with annual capacity factors of less than 10 per cent. The utility refers to stations with annual capacity factors in the 10 per cent to 55 per cent range as "intermediate-load stations"; all of these are either hydroelectric stations or coal-fired thermal stations.

The Primary Sources of Power

As implied in the previous section, electricity is not a primary source of energy (unless means were to be devised to capture the electric energy released during thunderstorms), and we depend for its generation on the use of primary energy sources such as coal, oil, natural gas, uranium, and the kinetic energy of falling water. As a secondary source of energy, electricity currently accounts for 32.5 per cent of the province's overall primary energy consumption. Because the generation of electric power depends upon an assured primary fuel supply, we summarize in Table 4.1 the major sources of primary energy and their contributions.

Table 4.1 Mix of Primary Energy Sources in Generating Electric Energy 1975-90 (percentages)

	1975	1980	1985	1990
Hydroelectric	48	34	28	24
Coal	28	29	22	12
Residual oil	1	2	2	2
Natural gas	7	1	1	0
Uranium	16	34	47	62
Total	100	100	100	100

Note: These estimates are based on a 3.5 per cent average annual load growth and on Ontario Hydro's currently committed programme. Although purchases accounted for about 18 per cent of the toal energy made available by Ontario Hydro in 1975, future forecasts of purchases are difficult to make because of its inadvertent nature. Only firm purchases under contract can be forecast, but this is expected to be less than 1 per cent in 1980 and zero in 1985 and 1990.
Source: RCEPP.

Primary energy sources are said to be either renewable (e.g., hydraulic, direct solar, wind, biomass) or non-renewable (e.g., fossil fuels, uranium, thorium, deuterium, and lithium).[1] The current global energy debate is concerned essentially with the rapid rate of depletion of the non-renewable fuels (especially oil and natural gas) and with the extent to which these fuels should be replaced by renewable sources of energy or, more controversially, by nuclear fission and eventually by nuclear fusion.

Hydraulic Energy

Hydraulic energy is a renewable source of energy — it is a manifestation of solar energy. Based on the hydrologic cycle it is only useful for energy production when the climate and the topography are right. For example, rivers, which depend for their flow on rainfall and snowfall, can be dammed to provide a head for hydroelectric generation. Or, alternatively and preferably, if a natural waterfall exists such as Niagara Falls, the kinetic energy of the falling water can be tapped.

Until 1951, when the R.L. Hearn coal-burning station was put into service, Ontario's electric power system was exclusively based on hydraulic energy. At present, the proportion of hydroelectric generation in the total electric power system is about 34 per cent. Of the total available hydroelectric capacity, i.e., 7,000 MW, about 91 per cent is owned and operated by Ontario Hydro and the remainder by municipal utilities and private companies of which the Great Lakes Power Corporation is the largest. By the end of the century the percentage contribution of hydraulic energy is expected to decline, assuming an average annual load growth of 3.5 per cent, to about 18 per cent of the provincial electricity demand.

Fossil Fuels

Ontario's only coal resource of reasonable magnitude is the Onakawana lignite deposit in northern Ontario, with an estimated reserve of about 190 million tonnes. This would fuel a 1,000 MW base-load generating station for about 25-30 years. However, there are significant social and environmental concerns that relate to such a development — see Volume 7, Chapter 8, and Chapter 8 of this volume.

The extent of Ontario Hydro's reliance on coal as a primary fuel is shown in Table 4.1. Until comparatively recently virtually all the coal was imported from Pennsylvania and West Virginia (coals with a comparatively high sulphur content). However, the utility will burn an increasing amount of western Canadian coal during the next 20 years.

Natural gas is an excellent boiler fuel, especially from an environmental standpoint. However, it is anticipated that its use by Ontario Hydro to provide peaking power will decrease. The basic reason for this is that natural gas is much too valuable a primary fuel to burn with thermodynamic efficiencies of only 35-40 per cent, even for peaking power. However, as we show in the following section, if gas-fired co-generation plants with efficiencies of 75 to 80 per cent were to be introduced in the industrial sector, a very small portion of Hydro's load could be supplied by gas in the future.

Ontario's electric power system already incorporates a residual oil-burning power station (Lennox); construction of a second station (Wesleyville) has been suspended indefinitely. Because the supply of residual oil for Ontario Hydro's requirements is regarded as reasonably secure to the end of the century, it is anticipated that oil-fired stations will continue to play an important part in providing peaking power. As in the case of natural gas, oil is a comparatively clean primary fuel compared with coal, though a costly one.

Uranium

As considered in some detail in the *Interim Report*, uranium is a primary energy source that is indigenous to Ontario. Uranium-235 can be burned in a nuclear reactor, just as coal is burned in a conventional furnace, to generate steam and hence electricity. An update of the security of uranium supplies is provided in the next chapter.

Renewable Sources

During a period of escalating prices of conventional non-renewable fuels, it is not surprising that increasing attention is being focused on solar energy and its many manifestations, one of which, hydraulic energy, has already been mentioned. Other manifestations such as biomass in the form of wood, vegetation, and organic wastes, direct solar energy for space heating and water heating, and wind energy all have potential limited roles in satisfying Ontario's future primary energy needs to the year 2000. Unfortunately, the solar energy flux is diffuse, and comparatively large collectors are needed; moreover, the source is intermittent and some form of storage is required. While it is very adequately stored in the form of vegetation, trees, and indeed all forms of life, the use of solar energy for heating purposes is costly and from the standpoint of cost effectiveness (as will be shown) it is not likely to be competitive with other sources in the near future in Ontario. Most important, home and building design (including insulation) must be considered, in part, as a means of capturing solar energy. Particularly in Chapter 10, we will discuss the potential of this ubiquitous primary energy source.

Energy Conversion

During the inquiry it became increasingly clear that energy policy formulation, and particularly the analytical work related to it, must be based on some understanding of the related science and technology. This should not be interpreted, as we will demonstrate in Chapter 12, in the sense of energy policy being exclusively in the hands of the experts − far from it. But we emphasize that it is not possible to debate meaningfully the merits and demerits of renewable and non-renewable sources of energy without some knowledge of the concept of energy conversion, and this, in turn, necessitates some acquaintance with laws of thermodynamics.

These laws explain why the efficiency of thermal generating stations (i.e., fossil-fuelled and nuclear), in spite of excellent design, is always comparatively low (30 to 40 per cent). Conventionally, thermal stations are used solely for the generation of electricity, but they can be designed for the co-generation of both electricity and heat (which can be used for district heating or provided to industry in the form of process-steam). The overall efficiency of utilization of the primary fuels can, in this way, be increased twofold (to 70 or 80 per cent).

At first sight, it would appear that the concept of co-generation contravenes the laws of thermodynamics, in that the levels of efficiency achievable far exceed those characterizing conventional thermal generating stations. However, as shown in Volume 4, Chapter 5, through the use of special steam-turbine configurations (back-pressure turbines) it is possible in effect to split the flow of steam energy so that a proportion of the steam (perhaps 50 per cent) can be used directly for industrial processes (or for district heating), and in this way a high level of efficiency in the utilization of the primary energy source is realizable. This extremely important concept will be expanded upon later.

The first law of thermodynamics states quite simply that, in a conservative (i.e., isolated) system, energy can neither be created nor destroyed, although it can be converted from one form to another. For example, the primary energy in free-falling water (i.e., gravitational energy) can be converted to electric energy with a conversion efficiency of about 90 per cent (the rest of the primary energy is converted into heat through friction effects and is thereby lost).[2] In simple form, the second law of thermodynamics states that in all natural and man-made processes the conversion of energy, or the transfer of energy (e.g., the flow of electricity through a transmission line), is always accompanied by a degrading of energy to the point where it is no longer usable. In other words, it is impossible to convert a given quantity of energy (e.g., the thermal energy in a kilogram of coal) completely into an equivalent amount of another form of energy (e.g., electricity). Suffice it to note, here, that this inherent loss of energy during all energy conversion processes is synonymous with the concept of irreversibility.

Another interpretation of the second law of thermodynamics associates "quality" with a specific source of energy. What do we mean by "quality of energy"? Formally, we define it as follows: The quality of an energy source is determined by the efficiency with which the energy can be converted into useful work. For example, some forms of energy, such as electricity and gravitational energy, are high-quality forms because they can be converted into other energy forms of comparatively high quality (e.g., high-temperature thermal energy) and, concomitantly, into useful work with a high level of efficiency. On the other hand, the massive amount of thermal energy contained, for instance, in Lake Ontario, or in the Arctic Ocean, is low-quality energy because it cannot readily be converted into useful work. Indeed, it is paradoxical that most of the earth's massive sources of energy (e.g., the thermal energy stored in oceans and lakes) are low-quality sources and therefore not readily usable by man.[3]

Now, consider the conversion of the chemical energy in coal, oil, and natural gas into electric energy. Four basic stages are involved:
   • by the combustion of the fossil fuel in a furnace, the generation of high-temperature thermal energy
   • by means of a heat exchanger in the form of a boiler, the conversion of the heat of combustion into high-temperature and high-pressure steam
   • using a steam turbine, the conversion of the energy available in the steam into mechanical energy[4]
   • using an electric generator (e.g., a three-phase alternator), the conversion of the mechanical energy of the turbine rotor into electric energy

The overall efficiency of the conversion of the energy available in the fossil fuel into electric energy is the sum of the conversion efficiencies at each of the four stages mentioned above; it is approximately 35 to 40 per cent. In the case of nuclear power, the uranium is, in effect, burned in a reactor, which is equivalent in principle to a fossil-fuelled furnace; the remaining energy conversion stages are the same as for fossil-fuelled stations. However, essentially for safety reasons, the temperature of the nuclear-generated steam is somewhat lower than that of the fossil-fuel-generated steam, and in consequence the overall efficiency of the nuclear generating station is only about 30 per cent.

Furthermore, the same general thermodynamic principles are involved when the chemical energy of gasoline is converted into mechanical energy by means of an internal combustion engine; the efficiency of this conversion process rarely exceeds 25 per cent. The fact that all energy conversion processes inherently involve an irretrievable loss of valuable available energy underlines the importance of choosing the best process from a set of alternatives. The lesson to be learned, and it is an important one, is that to ensure optimum efficiency of utilization of energy sources, the conversion processes should wherever possible be fitted to "end uses". In effect, we are introducing the concept of "thermodynamic thrift".

No discussion of elementary thermodynamics would be complete without some mention of "entropy". As well as the available energy, say in a piece of coal, there is also energy that is bound up with the products of combustion but so dissipated that it is unavailable for use. Energy conversion processes are always such that the quantity of available energy in the total system always decreases while the quantity of unavailable energy (i.e., the entropy) in the system always increases. Indeed, the definition of time itself can be based on the fact that entropy increases continuously. Entropy has been referred to as "time's arrow".[5] Detailed descriptions and the *modus operandi* of a broad range of both conventional and non-conventional energy conversion processes and how they are utilized in the generation of electric power are given in Volume 4.[6] Because our main concern in this volume is to introduce concepts and to state conclusions, the technological aspects of power generation are not included.

Hard and Soft Energy Technologies

The debate on the world's energy future, and on the multiplicity of alternative strategies that have been proposed, is in reality a debate on the future life-styles and values of society. The subject is characterized by a high degree of uncertainty; its ramifications are incredibly complex. In large measure, the debate has been polarized into advocacy of "hard path" technology, on the one hand, or of "soft path" technology, on the other hand. This was particularly evident during many of the Commission's public hearings. The designations "hard" and "soft", and the distinction between these strategies, were first propounded, and associated scenarios developed, by Amory Lovins.[7] The scenarios are purported to demonstrate the desirability on social, political, economic, and environmental grounds of the soft-energy path. Because the planning concepts related to the hard- and soft-path strategies have special significance for this inquiry, we outline below their respective characteristics. However, we have restricted the discussion in the chapter, for the most part, to the technological aspects of the two paths. In subsequent chapters dealing with the total electric power system, environmental implications, energy conservation, and economic and financial factors, we consider the broader implications.

The Hard Path

The hard-technology path is defined in the sense of a continued expansion of centralized (i.e., large-scale, capital-intensive, and complex) energy technologies, especially electric energy. The future is viewed as a mirror image of the past. In the short term, there would be continuing, and indeed increasing, reliance on the non-renewable energy resources — oil, natural gas, coal, and uranium. In the longer term, there would be reliance on synthetic fuels derived from coal, on second-generation nuclear fission reactors (in which uranium is used much more efficiently), and, perhaps within 50 years, on large-scale solar energy systems based on satellite technology and nuclear fusion.

It is argued by proponents of the hard path that centralization of electric power generation and the concept of the integrated power system, necessitating the transmission of bulk power at high voltages, are necessary for economic reasons. Ontario Hydro has argued, and we agree, that probably to the end of the century there is no technology on the horizon that could produce electricity as cheaply as the existing system. One extremely important reason is that a massive and complex infrastructure is in place and has, from both social and technological standpoints, withstood the test of more than 50 years of operation. This, we believe, is a particularly significant factor.

Nevertheless, according to the proponents of soft energy, energy strategies based exclusively on the hard-path option are fraught with major difficulties. Many of these are considered in this Report, especially in Volumes 5, 6, and 7, and in several chapters of this volume. They relate to such issues as reactor safety, the safe disposal of nuclear wastes, the high capital costs of nuclear power, the environmental impacts of coal-fired and nuclear stations, the low efficiency of conversion of primary energy sources to electricity (and consequently the high level of waste of available energy), and the land-use and environmental implications of bulk power transmission. Lovins argues convincingly, as well, that the hard path encourages the growth of bureaucracy, reduces the potential for public participation in key energy decisions, concentrates economic and political power, and puts considerable power in the hands of an élitist technocracy.

The Soft Path

As defined by Lovins, soft-path technology relies heavily on renewable energy resources — i.e., direct solar energy and the essential by-products of solar energy (hydraulic, wind, and biomass) — and includes technologies based on non-renewable resources that are efficient and environmentally benign, such as small-scale decentralized[8] co-generation plants and fluidized-bed combustion.

Solar collectors and associated storage systems for space heating and water heating, and wind-power generators, have been demonstrated in Ontario on a very small scale. However, hydraulic power, on both a large scale and a small scale, has been proving its value for more than 80 years, while efficient co-generation plants (especially in Europe) have been in comparatively widespread use for about 50 years. Fluidized-bed combustion is showing promise from an environmental standpoint; it is expected that the technology will be commercially viable in Ontario for small-scale applications during the 1980s.

While the reserves of non-renewable natural gas and oil are being seriously depleted, and are unlikely to provide more than a comparatively small proportion of the world's energy needs by, say, the year 2030, the potential of the renewable energy resources is virtually limitless. Further, there is an inherent quality in the soft path that emphasizes social, cultural, and environmental values associated with the environmental and conservation ethics. The emphasis is on meeting society's energy needs through an end-use approach, and, above all, on efficient utilization.

Soft Technologies

Although the case for the soft-technology path is theoretically indisputable, there remain major problems of implementation. Most of the "hard evidence" available concerning the use of renewable energy resources in Ontario, especially active solar space heating, suggests that these systems are not at present cost-effective, nor are they likely to be for many years.[9]

As well as depending on the area and efficiency of solar collectors, the viability of active solar space-heating systems depends to a considerable extent on the volume and nature of the thermal storage required to deal with the problem of the intermittency of solar energy. If only short-term storage is provided and if the system depends upon electricity to back it up, it may tend to reduce the utility's load factor, which may not be desirable. On the other hand, the appreciably more expensive long-term storage may not require as much electricity back-up on peak days[10] and this may help marginally in reducing the seasonal variation in the electrical load of the utility. Any such reduction in annual load factor will lower the need for peaking fossil fuels, and may reduce or defer the need for costly new capital facilities. However, at present, the pay-back period for long-term-storage solar energy systems is about 20 to 30 years, and the extent to which these systems will be adopted in Ontario within the next decade or two is highly problematical. Even though the economics of active solar space heating do not demonstrate cost-effectiveness, there is nevertheless considerable scope for innovation. Moreover, there appears to be little doubt that solar hot-water heating and passive solar energy systems (e.g., in homes designed to capture and store solar energy during the winter months without utilizing solar collector panels and ancillary equipment) are already cost-effective.

Active solar systems may be more appropriate for geographical regions other than southern Ontario. It is obvious, for instance, that the states of Arizona and New Mexico have a more appropriate climate for the widespread introduction of solar-energy systems than the province of Ontario. Furthermore, the concept of "worst credible conditions" (that would be experienced by solar- and wind-energy systems located in specific areas) is obviously important. Indeed, at the request of the Chairman of the Commission, the Canadian Atmospheric Environment Service (AES) undertook the task of assembling the basic meteorological information required for designing an active solar heating system or a wind generator system at five locations in Ontario.[11] While it is not possible, or appropriate, to discuss the detailed findings, that excellent report has an important bearing on the potential for solar space heating in Ontario. For example, the report shows that there is a finite probability of eight sequential virtually sunless days occurring in southern Ontario during the months of December and January. If electric power were used as back-up for solar energy, then, unless the majority of solar space-heating installations included essentially year-round solar energy storage, there would be minimal saving in peak-power requirements. Accordingly, the annual peak-power demand would be unaffected, and the power system load factor would deteriorate.

We urge that the AES report be used as a basis for the design of active solar energy homes in specific urban areas in the province. Assuming that the design specifications of a solar heating system are known, the report can be used to determine the number of days on which supplementary energy will be required for any particular location. Similarly, assuming the worst credible conditions for solar heating, based on the "100-year return periods", estimates of the extreme requirements for conventional energy back-up can be made.

In the case of wind-energy systems, the number of days with available wind energy below specified thresholds has been analysed for the same five locations. Because the kinetic energy available from the horizontal motion of the air is proportional to the cross-sectional area of the column of air (i.e., the area swept out by the wind generator), and proportional to the cube of the speed of the air, the performance of wind generators is particularly sensitive to wind velocities. With few exceptions (Sudbury is a case in point) the report provides evidence that the average wind speed in southern Ontario is marginal in so far as the cost-effectiveness of wind energy is concerned.

Another promising alternative is co-generation. Co-generation[12] may be regarded as meaning both generating a watt and saving a watt. It should play an increasingly important part in Ontario in the generation of electric and thermal energy. Co-generation based on coal and wood gasification technologies holds considerable potential for application as base-load generation.[13] The concept is based on the functional integration of coal and wood gasifier and power generator systems. Such systems may be used in a co-generation mode with a correspondingly high thermodynamic efficiency of about 75 to 80 per cent. As "fluidized-bed gasifiers", they have the additional advantage of low environmental impact, as mentioned in Chapter 9; they are expected to be commercially available within five to 10 years.

The development of other so-called soft technologies, especially biomass energy (i.e., wood-fired generation, ethanol, and methane from organic wastes), which are of high potential importance in the long term, and especially if fossil-fuel prices continue to escalate, should, we believe, be encouraged. Appropriate research, development, and demonstration projects are considered in a later section of this chapter.

The widespread practice of energy conservation is advocated by proponents of both strategies. However, it is clear that the level of conservation demanded by the soft-energy path will necessitate major changes in life-styles. The extent to which the public will accept such changes, and in particular embrace enthusiastically the conservation ethic, remains in doubt. However, there is an increasing recognition that conservation measures must be stepped up regardless of the energy strategy that is ultimately chosen.

The level of dependence of the province, to the year 2000, on soft-energy technologies will depend upon how quickly the associated industrial, commercial, and institutional infrastructures can be put in place. Unfortunately, past experience (and indeed history) suggests that transitions from one energy technology to another usually take several decades.

Are the Paths Mutually Exclusive?

There is an implicit recognition in the writings of Amory Lovins[14] that the hard- and soft-energy paths are mutually exclusive. However, during his appearance before the Commission on October 19, 1977, in response to a question by the Chairman relating to the mutual exclusivity issue, Lovins stated:

> You mentioned, Chairman, a statement that hard and soft technologies are exclusive, and I haven't said that and, in particular, I haven't said that they are technically incompatible, because they are not.

> During the transition that a soft path represents, we would have, for about 50 years, hard and soft technologies co-existing, big power stations alongside solar collectors, and their mix would change.

> It would start with hard technologies and end up with soft ones.

> What I did say was exclusive was the two patterns for smooth, internally consistent evolution of the energy system within its social and political context; that is not technologies but paths. I argued that these were institutionally, culturally and logistically exclusive.[15]

To put the question into the perspective of Ontario, we summarize below some of our conclusions that relate to the planning of the electric power system, with special reference to the choice between hard and soft:

• During the last 60 years (and more) Ontario Hydro has served the province well by supplying electricity very reliably and at the lowest feasible cost. The province's electric power system is an integrated system that incorporates hydraulic, nuclear, and fossil-fuelled generation.

• The infrastructure of the province's electric power system (i.e., generating stations and bulk transmission system) has evolved over several decades. We have concluded that the existence of the system should in no way jeopardize the development of the soft path. Indeed, the two may be complementary.

• If major research, development, and demonstration projects are completed successfully during the 1980s, we believe, there is a reasonable probability that soft technologies might contribute at least 2,000 MW to 3,000 MW of the utility's additional power needs by the year 2000.

• The planning horizons of Ontario Hydro are in the order of 10 to 20 years. It would be unrealistic, and indeed irresponsible, for the utility to rely on the availability on a large commercial scale of appropriate soft (decentralized) technologies that would replace a major proportion of existing and projected conventional generation capacity by the early 1990s.

• We believe it would be impracticable to establish an infrastructure, of sufficient scope and scale to support a major soft-technology industry in Ontario, that could replace even 25 per cent of Ontario Hydro's generation capacity by the end of the century. This would necessitate the infrastructure being in place within the next 10 years.

• Largely as a result of the introduction of new metering technologies, energy-pricing policies and innovative methods of storing electricity during off-peak periods (e.g., by the year 2000, it is possible that thousands of electric automobiles, based on storage batteries, will be on the roads of North America), we believe that load-management techniques will enhance appreciably the overall efficiency of the province's electric power system.

With regard to the issue of centralization versus decentralization, therefore, it is clear that neither the hard path, essentially because of the uncertainties relating to the future of nuclear and fossil-fuelled power, nor the soft path, because of uncertainties relating to reliability, economics and overall social acceptance, offers an exclusive approach to energy policy.

The first step is for the province to decide how much energy will be needed during the next 20 years and to plan accordingly. Soft-path technologies such as solar energy, including hydraulic, biomass, and direct solar radiation, will obviously have an increasingly important role to play in the energy mix and this development should be stimulated and encouraged. However, extravagant claims, especially over the short term, concerning the capabilities of these technologies and their future role could do irreparable harm to the soft-energy technologies, essentially because of public disenchantment with hastily conceived designs and poor performance and economic return.

Consequently, our overall conclusion is that the hard- and soft-energy paths are not mutually exclusive. The basic energy-planning concepts we recommend are predicated necessarily on keeping all credible options open. Even conceptually it is a mistake to talk about a hard-energy path and a soft-energy path because these paths have already coalesced and in our opinion will continue to do so for many years to come.

In the previous chapter we stressed the urgent need for a comprehensive data bank relating to the end uses of energy, and, particularly, electric energy. This is required not only for load forecasting but also to make effective decisions relating to how energy can be utilized more efficiently. For example, is electric space heating an efficient end use for electric power? When energy end-use patterns are available it will be possible to answer such crucial questions and to assess the extent to which alternative technologies, especially those based on solar energy, can be utilized.

Research, Development, and Demonstration

During a period of energy uncertainty, research, development, and demonstration programmes in energy generation technologies, as well as their social, economic, and environmental implications, are virtually mandatory. R&D programmes that relate specifically to nuclear power, bulk-power transmission and environmental impact are treated in Chapters 5, 6, and 9, respectively.

Demonstration programmes based on the findings of research and development undertaken in the past will be aimed at the short-term future, while complementary research and development programmes will be directed towards the long-term future needs of society. In this section we consider both short-term demonstration projects and research and development areas of special relevance to Ontario's long-range electric power planning. A Science Council of Canada report[16] provides an excellent assessment of the short-term energy research and demonstration requirements. Of the non-nuclear energy technologies considered in that report that have relevance to the Commission, in some cases peripheral, we strongly endorse the recommendations with respect to the demonstration projects described below. They relate to the combustion of coal, to renewable energy sources (e.g., biomass, solar energy, and energy generation from solid wastes), and to the co-generation of electricity and heat. Although all of these technologies are considered in some depth in Volume 4, we believe it is important to provide some brief explanatory notes in this volume.

In our consideration of soft-technology research, development, and demonstration projects which we have concluded are important, bearing in mind our strong conviction that all options (both hard and soft) should be kept open as long as they remain viable, in the planning of Ontario's energy system, especially for the late 1980s and the 1990s, we have been impressed by some of the proposals of the Science Council of Canada.

The soft-technology demonstration programmes recommended by the Science Council, and their tentative completion dates, are: fluidized-bed-technology pilot plant and full-scale demonstration (1991), feasibility of generating gaseous and liquid fuels from forest and agricultural residues (1991), solar water- and space-heating systems (1992), energy generation from solid wastes (1991), and co-generation of electricity and heat (1989).

Demonstration of Fluidized-Bed Combustion Technology

Because of its versatility, its minimal environmental impact, and most important its flexibility from the point of view of the range of fuels that can be utilized, fluidized-bed combustion appears to have great promise. The basic principle, first demonstrated in 1921, incorporates the burning of fuel particles, such as crushed coal, in a "fluidized bed" consisting almost entirely of granular inert particles of sand, limestone, or ash supported by a rising stream of air. Fluidized-bed gasifiers, with coal and wood as fuel and using gas turbines, are also under development. An advantage of this technology is that it makes it possible to control sulphur and nitrogen oxide emissions[17] within reasonable limits. The formation of nitrogen oxide is inhibited by the considerably lower temperature of fluidized-bed combustion through the use of ballast materials mixed with the fuel, while the sulphur oxide emissions are minimized by the use of limestone as the ballast material. The sulphur is deposited in the bed in the form of non-volatile calcium sulphate. Consequently, even when comparatively high-sulphur coals are used as fuel, the combustion process is comparatively benign environmentally. Furthermore, as stressed in the Science Council report, fluidized-bed combustion is achievable with a range of fuels, including low-quality coal; indeed, a major objective of the demonstration project is the development of boilers with adequate versatility.

Also, it is important to note that fluidized-bed systems could achieve a 4 per cent higher efficiency and could probably be used with boilers having a broad range of capacities. The United States Department of Energy plans to build a 250 MW demonstration plant, and we understand that it may be practicable and economic to develop boilers ranging from a few megawatts to 500 MW.

We believe that this technology holds considerable promise for Ontario.

Therefore, we recommend that:

4.1 During the next decade the Ontario government and Ontario Hydro should actively support the demonstration of fluidized-bed combustion with special reference to its future role in the generation of electric power.

Generation of Gaseous and Liquid Fuels from Forest and Agricultural Residues

The utilization of agricultural and forest industry wastes, coupled with the possibility of growing plants and trees (e.g., the hybrid poplar) for fuel, constitutes a potentially significant source of energy indigenous to the province. The technology is already highly developed in Brazil, where liquid fuels (i.e., methanol and ethanol) produced from biomass are being used as additives to gasoline for the transportation sector. It has been estimated that biomass already represents about 10 per cent of the world's primary energy consumption.

The biomass energy demonstration project suggested by the Science Council lists the following desiderata:
- the gasification of forest and agricultural residues
- the testing of methanol technologies
- the evaluation of ethanol potential
- the generation of biogas

Although not directly related to electric power generation, except perhaps through fluidized-bed combustion, these technologies could make a significant contribution to the province's total energy requirements by the year 2000 — assuming, of course, the successful completion of the demonstrations and their cost-effectiveness within reasonable limits.

Ontario's forest industries are already taking steps to utilize the combustible wastes of pulp and paper operations; this practice is widespread in Sweden. The utilization of forest industry wastes directly as fuel for small-scale power generation is under study. An example is the Hearst Study[18], the object of which was to study both the feasibility and the cost-effectiveness of burning wood wastes from lumber mills to generate process-steam and electricity. We understand that there is no intention of proceeding with the demonstration project, on the grounds that the economics are not sufficiently attractive. We believe this decision is unfortunate, especially in view of the Science Council recommendations. A Hearst pilot plant would complement (or even replace) the aforementioned biomass demonstration project and would provide the detailed information so urgently required before biomass energy applications can receive the attention they deserve.

Therefore, we recommend that:

4.2 The Ontario government should support the demonstration of biomass energy projects, including gasification of forest and agricultural residues, testing methanol technologies, evaluating ethanol potential, and generation of biogas.

Solar Water Heating and Space Heating

We have made repeated reference, in the previous section, to the potential of solar space heating and water heating, especially passive space-heating and active water-heating systems. The conservation implications of these soft technologies will be considered further in Chapter 10. However, although an increasing number of solar-energy projects have been completed and more are being initiated, and although the National Research Council is undertaking an assessment of the technology, many uncertainties remain. The primary purpose of the demonstration project recommended by the Science Council is to resolve many of these and, in particular, to evaluate the industrial, economic, legal, and social feasibility of large-scale active solar-energy technologies. The emphasis is on large-scale technologies because, although active solar space heating may be uneconomical for a 50-100 household unit, it may be quite economical for a 500 household unit, especially bearing in mind the probable escalation of energy prices in the 1980s. Of special interest, also, is the proposed side-by-side evaluation on a sector basis, in an urban area, of electrical resistance heating combined with heat pumps, natural gas heating, and solar heating. Only in this way will it be possible to evaluate the economics of solar heating in the "real world".

In addition to evaluating technical performance and cost-effectiveness, the demonstration will also assess environmental impact on a total-fuel-cycle basis.[19] The urgent need for an in-depth scientifically based evaluation of active solar-energy heating systems is perhaps best summed up in the words of Professor John Holdren of the University of California at Berkeley:

> With respect to solar energy, too little is known of the economic and (for some forms of solar) ecological impacts of deployment on a truly large scale for a prudent society to abandon everything else.[20]

We recommend that:

4.3 During the next decade the Ontario government should continue its programme to demonstrate the suitability of solar space heating and water heating in the Ontario context with special reference to its potential role in energy conservation.

Solid Municipal Waste as Fuel

A surprising amount of combustible solid municipal waste is being disposed of daily, chiefly at large landfill sites, and probably thousands of tons of ferrous and non-ferrous metals that could be recycled are lost each year in the form of municipal "garbage". The Science Council demonstration project recognizes that, "as energy costs and garbage disposal costs rise, the trade-off might swing gradually in favour of energy recovery from solid wastes". Although unlikely to be a major source of energy, the combustion of appropriate solid wastes, coupled with the recapture of thousands of tons of newsprint and other recyclable materials, is probably desirable on environmental grounds alone. Existing disposal methods are increasingly energy-intensive. Disposal sites are more and more remote from urban centres, the energy cost of transportation is significant, and large tracts of land that might be used more productively are used as landfill sites. Furthermore, there are associated health and environmental hazards through the leaching of heavy metals into ground water and the presence of vermin.

Notwithstanding the obvious advantages of using garbage as fuel for generating electricity, interest in Ontario appears to be waning rather than waxing — for example, the "Watts from Waste" project, which was to burn "refuse-derived fuel"(RDF) mixed with coal at Hydro's Lakeview Generating Station, and which was boosted during our public hearings, does not appear to be making much headway. (We understand that the cause is unanticipated cost escalation at the refuse-separation and processing stage.) On the other hand, in Milwaukee, Wisconsin, such a facility was brought "on stream" in less than five years and, except for some problems with glass slag, appears to be operating satisfactorily. Other cities in the United States are moving in the same direction.

The Science Council demonstration project proposes an in-depth feasibility study that takes into account, in addition to the technology, the economic, social, and environmental implications. The programme envisages the development of three demonstration plants backed up by a research and development programme in such areas as materials, combustion, emission control, and waste management.

In Toronto, the province and Ontario Hydro are considering the modification of one or more of the natural gas boilers at Hydro's Hearn Generating Station, to burn refuse-derived fuels (RDF) with gas as an auxiliary fuel or, more likely, replacing the existing boilers with refuse-burning units. (The Hearn units are currently mothballed.) Installation of new units designed to burn refuse with minimum pre-processing would overcome the high costs of refuse separation and processing. Because of its downtown Toronto location, the plant could also be designed to co-generate electricity as well as steam or hot water for the existing or an expanded district heating system in that city.

Further, it has been estimated that conversion of Metro Toronto's 2 million tons per year of garbage to useful electricity and heat would save the equivalent of over 2 million barrels of oil per year.

Accordingly, we recommend that:

4.4   The Ontario government and Ontario Hydro should make every effort to convert the "mothballed" gas-fired boilers at the R.L. Hearn Generating Station to burn refuse or refuse-derived fuels.

Co-Generation of Electricity and Heat

The prospects, in Ontario and indeed in all industrial nations, for the co-generation of useful heat as well as electricity are improving as the price of fossil fuels increases. This is inevitable, considering that the overall efficiency of utilization of available energy by large-scale conventional power plants is in the 30 to 40 per cent range, while the potential efficiency in co-generation plants is in the range of 70 to 80 per cent. The technology (see Volume 4, Chapter 5) is already well advanced, and has stood the test of time in several European countries. Furthermore, a major Canadian chemical company in Sarnia (the Dow Chemical Company of Canada) has been using co-generation technology successfully on a large scale (i.e., 170 MW of electrical output and 550,000 kg/hour of process-steam output) for several years. The proposed Science Council demonstration programme relating to co-generation envisages potential applications of the technology in the three basic sectors – industrial, commercial, and residential. Subsequently, as well as large fixed installations, the development of transportable units with capacities in the range of 1 MW to 10 MW would be undertaken. The objectives of the co-generation demonstration programme may be summarized as follows:

- development of a Canadian modular approach to the evolution of district heating systems
- use of small transportable units
- exploitation of the flexibility of steam turbines
- examination of financing, ownership, administrative, and operational matters relating to co-generation plants
- study of the critical question of lead times and their economic implications for industry
- study of the institutional barriers to co-generation that exist at present through the physical demonstration programme
- assessment of the diversity of suitable fuels and their availability at specific sites
- an overall assessment of the cost-effectiveness of co-generation and its social and environmental acceptablity

This is one of the most comprehensive, and costly, demonstration programmes recommended by the Science Council. District heating, in which hot water or steam is supplied to residential and commercial premises, can be a form of co-generation if the heat source is otherwise "waste" heat from an electricity generating station. At a relatively small sacrifice in electrical output, overall generating plant efficiency can be increased from the 33 per cent range to 80 per cent and overall fuel requirements can be cut by up to one-third (see Volume 4, Chapter 5). Other advantages include a reduction in air and water pollution since less fuel is burned, and an increase in user convenience and safety, as home furnaces are no longer required.

CANDU nuclear stations are particularly efficient steam sources. Ontario Hydro studies demonstrate that steam can be extracted from the Pickering B nuclear reactors and transported as hot water 10 km to the North Pickering townsite at lower cost than the cost of generating it from a natural-gas-fired boiler at the townsite.

A better plan may be to recover some of the 6 per cent of a CANDU reactor's thermal energy that is rejected by the heavy water moderator's cooling water system. If fully utilized, the production efficiency of a CANDU generating station would increase from about 30 per cent to 36 per cent. In an analysis of alternative designs for supplying hot water to a hypothetical 9,000 home town located 16 km from the Darlington nuclear power station, Hydro found that extracting heat from the moderator heat exchangers at between 50°C and 65°C was cost-effective compared with the more conventional use of turbine-extraction steam and a hot water temperature of 100°C.

Much is made in the several studies to date of the institutional, regulatory, and financial obstacles to the implementation of district heating in Ontario, even in cases where the basic economies seem sound. This is understandable for a new technology, and the Commission can only urge governments at all levels to be imaginative in offering appropriate tax and regulatory incentives, and making a field trial possible. We would not wish to see Ontario Hydro become directly involved in the ownership and management of a district-heating utility unless it proved impossible to interest private- sector corporations, such as the natural gas distributors or oil companies or even the municipal electric utilities, in taking a leading role. In our opinion, the co-generation and district heating concepts are of extreme importance, especially to Ontario. Indeed, as we stress in Chapter 10 in connection with the urgent need for energy conservation, the time is ripe for major demonstration projects to be launched by the federal government, the provincial government, the utilities, and industry, working in collaboration.

Accordingly, we recommend that:

4.5   The Ontario government and Ontario Hydro should assign high priority to the demonstration of industrial co-generation.

4.6   The Ontario government should expand its efforts to put in place a low-temperature hot-water district heating system, to demonstrate its energy efficiency under Ontario conditions, and to test the use of conventional as well as renewable or non-conventional fuels for the combined generation of heat and electricity.

Long-Term Research Projects

We have emphasized the need, in connection with short-term system planning, for a range of demonstration projects that could lead to implementation on a commercial scale in the 1990s. As well, and in parallel, long-term planning of the province's electric power system must take into account research and development activities that may not be commercially viable for at least 25 years. Failure of the nation and the province to demonstrate far-sightedness in this respect, through active participation, could have catastrophic consequences. Indeed, the survival of future generations, when stocks of non-renewable fuels (especially oil and natural gas) have virtually disappeared, could be compromised unless major research programmes are undertaken over the next two decades. R&D areas that show promise are:

- Fusion research – we were attracted by the arguments of the Institute for Aerospace Studies at the University of Toronto (Exhibit 152, June 1977) that Canada should take advantage of the invitation of the United States and other advanced western nations to join in their fusion research efforts. The institute proposed this as an insurance option – an opportunity to gain access to the fusion area at relatively small cost. By concentrating on the materials, engineering, and related technologies, Canada and Ontario might eventually be able to subcontract on fusion undertakings elsewhere. The risks might be modest because of the potential industrial "spin-offs" of these activities.
- Electric energy storage systems – for example, batteries for electric automobiles and for the storage of electric energy on a comparatively large scale in order to enhance the efficiency of utilization of generating facilities. Note that if "time-of-day pricing" were to be introduced it might be profitable for large industrial and commercial consumers to store off-peak energy.
- Studies of a "hydrogen economy" – this concept calls for generation, transportation, and distribution of energy in the form of hydrogen (a non-polluting, multi-purpose, and readily storable fuel), which could be produced by using nuclear-generated electricity to electrolyse water.
- Magnetohydrodynamics (MHD) – an emerging technology whereby the energy of combustion of fossil fuels is converted directly into electric energy. Metallurgical research aimed at the development of special ceramics that will withstand very high temperatures and a high-energy flow of

corrosive particles is a most pressing requirement. Note that both the Soviet Union and the United States have virtually reached a stage where commercialization of MHD generators with capacities of 50-100 MW appears to be practicable.

- More reliable heat pumps (and associated novel thermal energy sources) with appreciably higher capacity than those now available should be developed.

The above list of R&D projects, some of which are already under way, gives some indication of the scope of the research that will complement Ontario Hydro's power generating facilities. Each project is related to energy conservation, to the extent that it can improve the efficiency of energy conversion.

Clearly, there is a need for diversity in energy research, development, and demonstration programmes, not least in the generation of electric power. But there is also a need to be selective. In the words of John Holdren:

> In the pursuit of long term energy options, then, resort to some substantial amount of diversity as a hedge against uncertainties, as large as today's, is justifiable, indeed essential. Part of the value of paying for diversity for a while, of course, is having the privilege, as information improves, of finally rejecting options that turn out to be unsuitable.[21]

CHAPTER FIVE

Nuclear Power

In September 1978, in response to Order-in-Council No. 3489/77, the Commission published its *Interim Report on Nuclear Power in Ontario*. Because the Commission's public hearings were incomplete at the time of publication, no specific recommendations were presented. During the last year, moreover, there have been significant developments in the field of nuclear power, notably the Three Mile Island (TMI) nuclear power station accident. The purpose of this chapter is twofold. First, to update the *Interim Report* in the light of the TMI accident and other important developments in the field of nuclear power, and, second, to present our general conclusions and recommendations that relate specifically to the CANDU nuclear fuel cycle. With few exceptions, which will be identified subsequently, the major findings of our earlier report (see Compendium of Major Findings) may be treated as final conclusions and will not be reiterated in this Report. However, for the sake of continuity, there are unavoidable overlaps between the two reports.

The Risks of Nuclear Power

All energy technologies involve risks, but the most serious of those associated with the nuclear fuel cycle are unique and occur in all phases of that cycle, in particular, in the mining and milling of uranium, in the operation of the reactor, and in the management and disposal of spent nuclear fuel. Each of these topics has been treated in depth in the *Interim Report* and in Volume 6.

Society is faced with three basic questions. First, how can risk be assessed? Second, how much risk is acceptable? And, third, how can risk be regulated? These questions have physical, psychological, and social dimensions. They relate to the basic issue – how can society cope with uncertainty? Especially when, as in the case of nuclear power, the interpretation of data frequently involves value judgements. Decision-making has special significance in connection with nuclear power, not least because the view appears to be widely held that nuclear power is a high-risk technology – indeed, it is a symbol for technological dissent. To counter this view, the comparative risks of various activities, including "living at the boundary of a nuclear power station", have been described in the reports of several commissions, and the matter was raised during our public hearings. We have concluded that it is not meaningful to compare activities involving voluntary risks (e.g., cigarette-smoking, car-driving, scuba-diving) with activities involving involuntary risks (e.g., exposure to pollution caused by thermal power generation, or exposure of the non-smoker to tobacco smoke). The basic reason is that such comparisons do not take psychological factors into account. We agree with Otway, who recognizes ... the futility of seeking to reassure people that, for example, nuclear power is safe by comparing its low levels of risk with the number of hours of skiing that would provide an equal risk – in a psychological sense they are not comparable.[1]

The risks of nuclear power, as manifest in the TMI accident, are related exclusively to the risk of exposing people (directly and/or indirectly through the food chain) to the ionizing radiation that results from the products of fission contained in the reactor core. People are also exposed to above-average (i.e., background) radiation in uranium mines, in the vicinity of mill-tailings ponds, in managing nuclear wastes, in living at high altitudes, and in flying at high altitudes, and there are many other radiation risk situations, such as medical X-rays.[2]

Radiation Risks

Although the association between man's exposure to ionizing radiation and the incidence of cancer and genetic damage has been known and studied for more than 50 years, it was the advent of nuclear weapons and, subsequently, nuclear power that accelerated the research work. To date it has been estimated that "literally billions of dollars" have been expended on these studies.[3]

The mechanism whereby cancer and genetic harm are induced is related to the damage caused to the DNA molecule by radiation (similar damage may be caused by other toxic agents). The DNA is modified in a way that can give rise to spontaneous and uncontrolled cell reproduction. During recent years it has been assumed that the level of biological damage is proportional to the energy released within the tissue. This is referred to as the linear hypothesis. However, by no means fully understood by the public is the scientifically accepted hypothesis that mankind and his forerunners have been exposed to ionizing radiation (from outer space, from terrestrial sources, and even from internal sources in the body) for millions of years. This natural background radiation, in the order of 100 millirem (mrem) per person per year, causes DNA damage in susceptible sub-groups of people, but the overall effects are barely perceptible because of the many other potential carcinogenic agents in the environment. Consequently, human exposures of several times the natural background radiation have been assumed to be acceptable. For the general population, the acceptable level is taken to be 500 mrem per person per year. A breakdown of radiation sources, including medical diagnostics, is given in Annex D of the *Interim Report*.

Under normal operating conditions the radiation resulting from the routine release of radioactive isotopes from a CANDU power station is in the order of 1 to 2 per cent of the combined average level of natural and medical radiation doses; this is generally assumed to be negligible.

Workers in nuclear power stations, as well as in other occupations that are "nuclear-related" (e.g., medical radiography), are exposed to higher levels of radiation than, for example, the public living in the vicinity of a nuclear power station. For these workers radiation dose limits are at present set at 5 rem per person per year — all workers are continuously monitored and the exposure of each is aggregated. The dose limit is 10 times that set for the general public. Although few workers receive aggregated doses at these limits, there remains cause for concern.[4] The problem of exposure of nuclear reactor workers to comparatively high levels of radiation may become critical if a major programme of "re-tubing" (i.e., replacing the pressure tubes that are subject to intense radiation) of CANDU reactors is undertaken. However, note that re-tubing is currently expected to be required only for the Pickering A and Bruce A generating stations, because tube expansion was taken into account in the design of later stations. This process will involve cadres of skilled workers each of whom will be subjected to comparatively high doses; but these must not exceed the 5 rem annual dose limit per worker.

We recommend that:

5.1 Ontario Hydro should publish a report as soon as possible on the expected exposure levels resulting from any reactor re-tubing operations, addressing, in particular, the following questions:

How many workers (Ontario Hydro employees and others) will be subjected to the 5 rem annual dose limit in connection with the re-tubing of a single reactor?

Will workers be subject to high dose levels on a continuing basis when the re-tubing of the Pickering A and Bruce A reactors begins on a sequential basis?

A worker could receive an aggregated dose of 50 rems over, say, a 15-year period. Is this medically acceptable? Should these exposures be age-dependent?

What is the total number of workers required, on a continuing basis, to undertake re-tubing operations? Are that many adequately skilled workers at present available?

To what extent can the re-tubing operation be undertaken by "remote control", thereby minimizing the aggregated exposures of workers?

Will workers who may be subjected to higher-than-normal radiation doses, and their unions, be fully informed of the nature of the risk?

In respect of the above, we note especially that the problem of assessing radiation risks at comparatively low dosage levels is compounded by the long latency periods involved in the induction of cancer (and the even longer periods required to assess genetic damage, if any). There is also the extremely important question of the age of workers. It seems reasonable to suggest that annual radiation dose levels should be related to specific age groups — the younger the worker, the lower the exposure dose.[5]

Central to this whole question is the so-called ALARA principle which states that the design and operation of all facets of the nuclear fuel cycle should be predicated on radiation doses "as low as reasonably achievable". In other words, the dose limits set by regulatory agencies should be regarded as absolute maxima.

The most serious postulated nuclear accident is one that gives rise to the complete melting of the reactor core, and the subsequent breaching of multiple containment systems. We stress, however, that the probability of such an accident occurring is very small. In spite of major studies of the potential effects of such an accident,[6] there remains considerable diversity of scientific opinion concerning "radiation-induced human carcinogenesis". This is largely due to the inadequacy of the information base. For example, one U.S. review group concluded in 1978: "The key difficulty with any realistic assessment of the effects of radiological exposure is the paucity of human data."[7] And a related conclusion, in the same report, is:

> The data on radiation-induced human carcinogenesis are still too sparse and too contradictory to provide a basis in themselves for these values. As a result of the indefinite character of the data base, there is a spectrum of opinion among radio-biologists as to the use of dose-reduction factors.

Notwithstanding these major uncertainties, it is important to put into perspective, on the basis of the most recent scientific information available, the relative hazards of low-level radiation.[8] Assuming that one million people are exposed over the whole body on a one-time basis to ionizing radiation of one rem each(this is 10 times the average annual exposure of each one of us to a combination of natural radiation and medical diagnostic and therapeutic radiation), the estimated number of cases of cancer induced in males would be in the range 192 to 756, and in females 344 to 1,306. The probable number of cancer deaths associated with this exposure would be in the range 70 to 353. Furthermore, if a population of parents were exposed to one rem of radiation each, there would be an expected increase in serious genetic disorders in the range of 5 to 75 per million live births (in the first-generation offspring).[9]

But these estimates must be viewed in terms of the incidence of all cancers in humans in North America, which is in the order of 180,000 per million people on a lifetime basis. We note also that naturally occurring genetic disorders may be in the order of 100,000 per million births.[10]

The Safety of Nuclear Reactors

The *Interim Report* considered the categories of reactor malfunction to which the CANDU reactor may be subject, and introduced various estimates of their frequency of occurrence and their possible consequences. However, since that report was published, additional material relating to reactor safety has been brought to the attention of the Commission.[11]

To date, the most comprehensive study of the risks of operating a nuclear reactor was undertaken by a team headed by Norman Rasmussen, Professor of Nuclear Engineering at the Massachusetts Institute of Technology, sponsored by the United States Nuclear Regulatory Commission. This team's report was published in 1975. Not unexpectedly, because of the extreme complexity of the analysis (a nuclear reactor is one of the most sophisticated technological achievements of man) and the novel approach, the report was subjected to considerable criticism. Consequently, the U.S. Nuclear Regulatory Commission established an independent group of scientists and engineers to assess it. Although the Rasmussen study and the subsequent assessment by Professor H.W. Lewis relate exclusively to the operation of light-water reactors, the basic findings of both studies are relevant to the safety of CANDU reactors. The following points from the Lewis report conclusions have particular relevance to CANDU safety studies.

- The methodology, based on fault-tree and event-tree analysis, represents an important advance over earlier methodologies applied to reactor risks. However, one member of the study group as well as a panel of radiological experts set up by the U.S. National Academy of Sciences concluded that it is doubtful that the approach can be implemented "so as to give a high level of confidence that the probability of core melt is well below the limit set by experience".
- The group was "unable to determine whether the absolute probabilities of accident sequences in the Rasmussen study are high or low" — it expressed the belief that the error bounds are underestimated. The main reason given for this conclusion was the inadequate data base, as well as some questionable statistical procedures.
- Although the Rasmussen report concluded that comparatively minor loss-of-coolant accidents and human errors are important aspects of reactor safety, these were not adequately covered in the analysis.
- Mathematical studies of highly complex systems such as nuclear reactors can never be complete — they will always be subject to revision.
- The Rasmussen report was the first to assess the potential consequences of large releases of radioactivity into the atmosphere. However, there is a continuing need for the associated biological models to be updated in the light of new information.
- It is extremely difficult to incorporate human-error information into fault-tree analysis. Furthermore, the Lewis group concluded that the Rasmussen study had significantly "underrated the role of operators and other employees in mitigating or controlling some potential accident sequences".

It is not widely known that the Whiteshell Nuclear Research Establishment, in co-operation with the Sheridan Park Power Projects Division of AECL, has a comprehensive research programme concerned with the safety of CANDU reactors. Although the primary purpose of these studies is to improve analytical procedures, it is important to note that the programme is producing a better understanding of the performance of the process and safety systems of existing reactors. Special attention is being devoted to:

- the performance of process and safety systems in removing heat and maintaining a flow of coolant, and fuel behaviour, during a postulated loss-of-coolant accident
- the fuel-to-moderator heat transfer, the hydrogen production, and the fission product release during a postulated loss-of-coolant accident
- the performance of containment systems during a postulated loss-of-coolant accident

An important feature of the research is the validation of the reactor accident analysis computer codes, at present under development; these are based on experiments involving a range of scales in space and time. This research, we understand, will be completed in 1985.

Because it reviews current safety principles and criteria and recommends changes in existing standards, the controversial 1979 report of the Inter-Organizational Working Group to the AECB has special relevance to CANDU reactor safety. Under an independent chairman, Dr. W. Paskievici of L'Ecole Polytechnique, University of Montreal, this group consisted of representatives of AECB, AECL, Hydro-Québec, Ontario Hydro, and New Brunswick Power. The level of independence of the working group has been challenged by several anti-nuclear public interest groups. In effect, the report recommends that the allowable radiation releases for accidents of very low probability but high consequences should be increased, while the allowable dose levels to the population resulting from less serious accidents with higher probability of occurrence should be decreased. Thus, the allowable radiation dose level for an individual member of the public living in the vicinity of a nuclear power station would, in fact, be increased (i.e., the standards would be relaxed) in the event of an admittedly very improbable major reactor accident. It is important to note that the recommendations of this report had not, at the time of writing (December 1979), been approved by the AECB.[12]

Three Mile Island — Some Lessons

No single incident relating to the commercial use of nuclear power has had a more profound impact on the general public and on nuclear-power-oriented electricity utilities than the TMI accident.[13] In the view of the proponents of nuclear power, the accident demonstrated that, in spite of its serious nature, and essentially because of the reliable operation of key control systems (i.e., shut-down systems) and effective "defence in depth", not a single fatality or injury occurred. Furthermore, they assert, the lessons to be learned will facilitate the safe operation of nuclear power stations in the future and improvements in reactor design. On the other hand, those opposed to nuclear power contend that the accident proved that nuclear power stations are basically unsafe, that the situation at TMI was potentially extremely dangerous, that the regulatory procedures and licensing processes are inadequate, and that the public is ill-informed with respect to the threats of nuclear power. Because of the obvious lessons that have been learned as a result of the accident, and their implications for the safe operation of the CANDU reactor, we summarize in this section the sequence of events that occurred on March 28, 1979, and the conclusions of the Presidential Commission of Inquiry (the Kemeny Commission) that relate specifically to our inquiry, and we identify the key causes of the accident and comment with special reference to CANDU.

The Accident. To put the accident that occurred at Reactor 2 of the TMI station in Pennsylvania into perspective, we outline below the sequence of events that occurred:

- The accident was initiated essentially by an inadvertent shut-down of several feedwater system pumps supplying water to the steam generators. Within one second, this caused the steam turbine to trip (as designed). Due to the loss of feedwater supply, the reactor coolant system pressure began to rise, causing electromagnetic pressure relief valves to open (as designed), and allowing coolant to be released in order to lower pressure. Because of the pressure build-up, the reactor safety systems responded automatically and the reactor was shut down (as designed). This sequence of events happened within 10 seconds of the initial pump failure; reactor cool-down and reduction of reactor coolant system pressure resulted. Up to this point everything worked according to design.
- As the reactor pressure began returning to normal levels, the electromagnetic relief valve should have closed, thereby ending further loss of coolant. However, that valve failed to reclose (the

control room indicators showed that the signal to close the valve had, in fact, been given, but there was no indicator to show the actual status of the valve). This was the critical event in the entire sequence. If the relief valve had closed, the accident could probably have been avoided. For example, if the operators had known the status of the "stuck open" valve, they could have stopped the drain of coolant by closing a "block valve".

• In the meantime, the water level in the steam generators dropped to such a level that the auxiliary feedwater valves opened automatically (as designed). However, feedwater did not flow because the auxiliary feedwater block valves had been mistakenly left closed (following servicing) some time during the previous 48 hours.

• As the reactor coolant system pressure continued to drop because of the open relief valve, the high-pressure emergency core cooling system was automatically activated (as designed). But, and this was another critical factor, some four and one-half minutes into the accident, the operators mistakenly "throttled back" the emergency pumps. Shortly thereafter, the operators restored the feedwater flow and consequently the reactor cooling system temperature and pressure were reduced. However, within minutes, because of the increasing pressure of the coolant under drainage conditions, a protective disc on the drain tank ruptured. This resulted in radioactive coolant spewing into the reactor building sump and thence into the auxiliary building. Notable is the fact that this disastrous sequence of events could have been avoided if the drain tank pressure indicator had been examined. Unfortunately, this indicator is located on a panel behind the two-metre-high primary control room panels on which all critical instruments were placed.

• Within one hour and three-quarters of the initiation of the accident, apparently because of dropping reactor coolant flow and pressure, and because of increasing coolant pump vibration, the operators shut down the coolant pumps. This was another critical step, and it seems to have sealed the reactor's fate. The resulting loss-of-coolant circulation probably caused boiling in the vicinity of the fuel rods, reduced heat transfer from the fuel, and gave rise to substantial fuel sheath failure. This in turn resulted in the release of radioactive fission products contained in the fuel matrix, and in the production of hydrogen gas.

• Because the top of the reactor core was no longer covered with water, temperatures as high as 2,200°C ensued. At such temperatures, the zirconium that encases the fuel rods reacted with water to produce zirconium oxide and free hydrogen. The expanding hydrogen began to accumulate as a "bubble" at the top of the reactor vessel. This was of concern because, as the "bubble" expanded, it might force the level of cooling water down even further, thereby exposing the core to a greater extent. Also, hydrogen and oxygen are an explosive mixture if the oxygen reaches a certain concentration. In retrospect, it is clear that there was never any real danger of a chemical explosion; oxygen is known for its readiness to combine with zirconium, so that the "bubble" probably never contained sufficient quantities of oxygen.

• Within approximately 16 hours of the initiation of the accident, the operators isolated the faulty relief valve (which had stuck open), brought one reactor coolant pump into operation and restored feedwater flow to the steam generators.

• There were very small releases of radioactivity into the atmosphere during the accident, mostly from an auxiliary building that was outside containment and into which radioactive water had been pumped.

The above summary of events is intended to demonstrate that the basic causes of the accident were extremely simple. A relief valve failed to close; an indicator in the reactor control room showed that a signal to close the relief valve had been given but there was nothing to indicate that the valve had not closed. Two isolating valves that should have been open had been left closed for a considerable time before the accident. Paradoxically, this resulted from a routine monitoring programme that was carried out periodically to check reactor safety systems and other operations.

Summary of the Presidential Commission's Conclusions. The report of the commission investigating the TMI accident was published on October 30, 1979.[14] The main conclusions of the report are outlined below:

• Fundamental changes to the organization, procedures, practices, and attitudes of government agencies (especially the Nuclear Regulatory Commission) and the nuclear industry are essential.

• Neither the utility nor the regulatory body paid adequate attention to previous incidents at nuclear plants; feedback from the plants seems to have been essentially ignored.

• Operator errors contributed significantly to the accident; so did defective and poorly designed instruments.

- The training of the nuclear reactor operators was deficient.
- The operator of the plant (i.e., the utility) "did not have sufficient knowledge, expertise and personnel to operate the plant or maintain it adequately".
- The emergency response by the Nuclear Regulatory Commission, the state, and the federal authorities "was dominated by an atmosphere of almost total confusion".
- Although the amount of radiation leaked to the atmosphere will probably have a negligible effect on the physical health of people living in the vicinity of the plant, many of these people were subjected to severe, albeit short-lived, mental stress.
- Much stricter safety requirements should be enforced for nuclear power stations operating in the vicinity of heavily populated areas.

The Royal Commission's Comments. The following comments are not based on first-hand knowledge of the TMI accident, but rather on the detailed findings of our own inquiry, as they relate to nuclear power, on the official report of the Ontario Government investigating team, and on several articles that have been published in the scientific literature. These comments have special relevance to our conclusions and recommendations, given in later sections of this chapter.

- The accident was caused by a combination of human error, equipment failure, and poor information display. Subsequently, the sequence of events that led to the ultimate serious accident was due to a combination of instrument and mechanical failure, and the failure of human operators to make the correct responses. The sophisticated automatic controls, predicated on computer operations, appear to have worked perfectly. Indeed, this is the basic reason why a major accident of very serious proportions did not occur.
- The failure of the Nuclear Regulatory Commission to ensure that operating rules were strictly adhered to was a contributory cause. A resident inspector might, *a priori*, have identified the specific fault conditions.
- Reactor 2 of the station appears to have been "dangerously out-of-control" for at least 24 hours, and probably 48 hours, after the initiation of the accident. This suggests a lack of effective communication and co-ordination between utility, regulating agency, and government personnel.
- In a presentation to the Ontario Hydro Board of Directors (April 9, 1979), a senior member of Ontario Hydro's staff stated:

> This was an extremely serious accident in terms of the amount of radioactivity released from the fuel pins. Also it was serious because it appears to have been caused mainly by operator action.

We concur with this view. Human operator performance was considered in the *Interim Report*. While we are not in a position to comment meaningfully on TMI operator training programmes, we believe nevertheless that it is simplistic to refer to "human errors". Brought into question is the selection of personnel and their training. We have noted, in particular, a recent comment by a former chairman of the General Advisory Committee to the U.S. Atomic Energy Commission:

> We were concerned lest reactor operations be left in the hands of operators with insufficient depth of knowledge and sense of responsibility and that unnecessary mistakes or carelessness might then occur, as apparently they did at Three Mile Island.[15]

- The evidence to date suggests that the health impact of the released radiation at the station was virtually negligible. Indeed, a United States Interagency Task Group has estimated that fewer than two fatal cancers and two non-fatal cancers would arise on a lifetime basis as a result of the accident, among the entire population residing within 50 miles of the TMI site (approximately 2,100,000 people). But note that the total number of deaths due to cancer that would normally occur over the lifetime of this population would be in the order of 325,000.[16] The consensus is that, although the release of radioactive material was extremely low, the public were given a distorted picture and this gave rise to serious psychological problems.
- The U.S. Nuclear Regulatory Commission, and also the Canadian Atomic Energy Control Board, rely to a great extent on computer analyses of postulated reactor accidents. It is obvious that the TMI accident sequence, and its consequences, were not adequately simulated or taken into account in the operation of existing plants. As a result, there were no specific guidelines for dealing with, for example, the hydrogen bubble situation. The lesson is, if such phenomena are predictable, that appropriate guidelines for dealing with a consequential accident should be developed and made available.
- Perhaps the most disturbing aspect of the accident related to the lack of co-ordination of the information released to the public. Indeed, the public relations aspects appeared to be in a chaotic state. Press and media received conflicting reports from different sources, e.g., the utility, the

Nuclear Regulatory Commission, and the State Governor's office. This clearly exacerbated the psychological impacts referred to above.

• Because of major differences in design philosophy between the light-water reactors of the TMI type and CANDU reactors, an identical accident sequence would be unlikely at a CANDU station. However, it is not inconceivable that a sequence of equipment, instrument and human malfunctions in the operation of a CANDU reactor could be postulated that would give rise to an accident of the same level of severity as that at TMI. It is for this reason that the lessons learned as a result of the accident are so important in the operation of nuclear reactors all over the world, regardless of their specific nature.

CANDU and TMI. Since the accident, there has been much public discussion of the comparative levels of safety of CANDU reactors and reactors of the TMI type (pressurized light-water reactors – PWR). Although judgements of this kind, in so far as they involve human operator behaviour, cannot be assessed quantitatively with confidence because the data are not available, Ontario Hydro has pointed out that there are nevertheless some unique characteristics of the CANDU reactor that appreciably enhance safety. For example, according to Ontario Hydro, the following attributes of the CANDU reactor would minimize, and probably prevent, an accident sequence such as occurred at TMI.

• The larger amount of water available in the "deaerator" and storage tank in the CANDU feedwater system would have prevented the flow of condensate from stopping suddenly as it did at TMI – water would continue to flow to the boilers to provide cooling at full reactor power for five minutes, or for 30 minutes after reactor shutdown.[17]

• In the TMI accident, cooling of the reactor was restored about eight minutes after the failure of the feedwater pumps to deliver water to the boilers. If a similar situation had occurred in a CANDU station, there would have been no interruption of cooling (via the steam generators), because the plants have a large reservoir of water available that would provide cooling for about 20 minutes. This is because there is a larger inventory of water in the CANDU recirculating boiler than in the PWR design.

• A high-pressure separate shut-down cooling system that can be directed into the reactor cooling system provides a second line of defence for the removal of reactor heat under conditions that applied to the accident.

• The "failed open" relief valve that caused a continuing loss of reactor cooling water, which was deposited on the floor of the containment building, is a situation that could not arise with CANDU. The CANDU system is designed to withstand 10 times as much pressure. The corresponding relief valve discharges the cooling water into a high-pressure vessel, which, when full, would terminate the discharge.

The above CANDU characteristics, Ontario Hydro says, considerably reduce the possibility of a TMI-type accident occurring, for example, at the Bruce or Pickering generating station. According to Ontario Hydro, there are other features of the CANDU system that should also be taken into account in any comparison between CANDU and PWR safety:

• The power density of the CANDU reactor core, per unit volume, is only one-tenth that of the TMI reactor.

• The heavy-water moderator, which has an independent cooling system, provides a massive heat sink that appreciably enhances public safety.

• CANDU nuclear stations have a large vacuum building that automatically becomes part of the reactor containment system when pressure builds up as it does in the event of an accident. The combination of containment system and vacuum building causes the pressure inside the containment to drop to below the surrounding atmospheric pressure, thereby minimizing the risk of leakage of radioactive gases to the environment.

The above information has been extracted from Ontario Hydro's presentation to the Select Committee on Ontario Hydro Affairs.[18] While accepting the utility's specific claims with respect to the CANDU nuclear system and its operational advantages over the TMI nuclear system, from a safety point of view, we have noted that several significant factors were not mentioned. First, there was no reference in the presentation to the fact that the emergency core cooling (ECC) system of a pressurized light-water reactor is essentially a high-pressure system, whereas the corresponding CANDU system relies on lower-pressure gravity feed of the emergency core coolant. The former is markedly more effective. Furthermore, questions have been justifiably raised concerning the efficacy of a potential CANDU pressurized ECC because of the pressure-tube configuration. Second, in the light of several Bruce A "significant event reports" (see below), we have concluded that, not only might certain aspects of the

TMI accident occur in a CANDU system, but some have in fact already occurred. Accordingly, the accident has more important lessons for Ontario Hydro, in the design and operation of CANDU reactors, than might be recognized on the basis of Hydro's presentation to the Select Committee. We have noted in particular that:

- A loss of feedwater supply, in one or more units of the Bruce A station, has occurred on several occasions.
- The auxiliary feedwater pumps have failed to start when required.
- An operator error has resulted in over-pressurization of the "bleed condenser" (this is analogous to the TMI "quenching tank"), with the loss of some heavy water to the vault.
- In connection with the above, there was operator failure to detect the leakage of heavy water.
- Excess concentrations of deuterium (heavy hydrogen) have occurred in CANDU systems (cf. the TMI "hydrogen bubble" phenomenon). Indeed, a deuterium explosion occurred at the Bruce Generating Station — fortunately at a time when the reactor concerned was shut down.
- Leaks have occurred in the steam generators of unit 2 at Bruce, and these have resulted in the leakage of some radioactive water outside the containment building.

Noteworthy is the fact that each of the above events is identical, or analogous, to a corresponding event in the TMI accident.

Reactor Operators. The TMI accident validated the view, expressed in the *Interim Report*, and supported independently by the Lewis Task Group, that operators in nuclear power stations have crucial roles in assuring public safety.[19] However, we have concluded that an acceptable level of risk in the operation of a CANDU station cannot be achieved by exclusive reliance on the human element. For example, all CANDU safety systems (excluding the containment system) are computer-controlled. Consequently, the reactor is shut down automatically when a range of monitoring signals (notably relating to neutron flux density in specific regions of the reactor core, pressure increases at various key locations, indication of loss of regulation, etc.) shows that a condition exists that might pose a threat to reactor personnel and to the public.

Operators may either make matters appreciably better during an accident situation or make them appreciably worse. The TMI accident was aggravated by operator error and poor design of control-room information display, both human errors. Accordingly, the question must be raised — how can the risk of operator error, associated with the operation and maintenance of CANDU reactors, be reduced to an absolute minimum, bearing in mind the special conditions that arise in the operation of a nuclear power station? Even a cursory examination of some of the Bruce A significant event reports (see next section) suggests that human error is still a significant factor in the operation of CANDU reactors, and it is fortunate that the results of human errors have been coped with effectively up to now.

A major conclusion of the Lewis study group was that there are insufficient relevant data relating to reactor operator performance and that, until a systematic programme for developing and assessing the data has been carried out, the uncertainties in the estimates of risk in nuclear reactor operation will remain. We believe that the data contained in Ontario Hydro's significant event reports, if adequately analysed by an independent group, would help appreciably in minimizing these uncertainties. Furthermore, such an analysis would provide an excellent basis for improving communications channels and management practices and, not least, developing better quality control procedures to identify defective components before they are installed. Each of these areas relates to the effectiveness of the human factor.

There is no doubt that the TMI accident has put reactor operator selection, training, and performance into better perspective, and that important changes in present practices will probably be made. For example, and this point was made by the Presidential Commission on TMI, prior to the TMI accident operator performance was based primarily on the economic performance of the plant rather than on safety. From this Commission's standpoint, we are gratified that the *Interim Report* was probably the first of its kind to stress the significance of the human factor in nuclear reactor safety, and we recall that there was considerable criticism of our findings in this regard prior to the TMI accident. Although we do not reiterate here our specific findings relating to the training of reactor operators and the purpose of reactor simulators, we stress their continuing significance.

The Bruce Generating Station Significant Event Reports. As a direct consequence of the TMI accident, several "unsolicited documents" in the form of Bruce A Significant Event Reports were forwarded to a member of the Ontario Legislature and subsequently passed to the Select Committee of the Ontario Legislature on Ontario Hydro Affairs. Copies of these reports were made available to the public in April

1979.[20] Because of time constraints, we have not been able to examine these documents in the depth they deserve. But it is clear that they offer a representative sample of abnormal events at the Bruce generating station that have safety implications. The reports show that some events similar to those that initiated the events at TMI have already occurred at the Bruce generating station. The fact that these events were handled effectively by the station staff, and that some subsequently gave rise to design changes, is encouraging. Nevertheless, we draw attention below to certain related factors, identified in the reports, which provide further evidence of the need for improved procedures and operational research. Of the seven event reports we received,[21] four (SER 79-9, 78-22, 77-50, 77-53) provide evidence of inadequate communications between the shift supervisor, operators, the chemical laboratory, etc. For instance, in connection with an event related to an increase of deuterium concentration in moderator cover gas (cf. the hydrogen bubble of TMI), we note that several errors contributed to the delay in taking action:

- A communications breakdown between operators and chemical technicians pointed out the need for great care in relaying accurate and complete information, irrespective of the pressures of other activities (SER 77-50).
- Human operator errors, some quite serious, were identified in three of the events (SER 77-50, 78-4, 78-22). In connection with the "increase of deuterium concentration" event mentioned above, for example, there was an error on the part of the Shift Supervisor in not investigating immediately a very high deuterium alarm (SER 77-50). In another event (SER 79-9) it is noted that "subsequent operations bent the spindle", and the "limit switch coupling was found broken". In this latter connection, the Station Manager commented: "The failure was compounded by the decision to attempt to shut the valve 'more' when it had already over-travelled."
- The Shift Supervisor and/or First Operator were identified in five of the events (SER 77-47, 77-50, 77-53, 78-22, 79-9). For example, in the situation that involved a "deuterium explosion", apparently triggered by a welder's torch, no sampling or purging of the cover gas had been undertaken for more than a week prior to the incident (SER 78-22). Further, we noted that although the explosion occurred at 11:20, the chemistry laboratory reported "a better follow-up could have been done if we knew of the problem before 14:30" (SER 78-22). We noted, also, in the reports such statements as "clear labelling needed" (SER 78-1), as well as "valve opened for no apparent reason" (SER 78-1).
- The Station Manager's comments recorded on the significant event forms in five of the seven cases were quite inadequate (SER 77-47, 77-53, 78-4, 78-22, 79-9). Indeed, in some cases (SER 78-4, 78-22) the comments did not even mention the recommendations and comments of, presumably, the Shift Supervisor; in other cases they appear to be irrelevant. It seems to the Commission extremely important that the senior officer's remarks regarding a significant event should provide management and design staff with cogent comments and recommendations. For example, there is no excuse for the "logging and turn-over" of details of an incident from one shift to the next shift not being undertaken (SER 79-9). A Shift Supervisor unaware of actions taken during the previous shift could be in a difficult position. In this specific event the Station Manager's comments did not even mention this serious lack of communication.
- Our impression is that abnormal events at the station are handled very largely through strict adherence to the "Operating Manual Procedures" (SER 78-4). And we regard this as good, and indeed as essential, practice. However, it is clear from several of the significant event reports (SER 77-50, 78-22) that certain situations are not covered in the Operating Manual — as was the case at TMI. This suggests the need for additional scientific and professional support in the operational teams.
- There is no reference to the Resident Inspector of the AECB in any of the reports. We believe that some indication of the awareness by the inspector of specific incidents and his comments would significantly improve existing practices.

Conclusions and Recommendations with Respect to Reactor Safety

In reaching the following conclusions and recommendations relating to the safety of CANDU reactors, we have been guided by the findings of the *Interim Report*, as well as by some of the Bruce significant event reports, by the Lewis report, and by official reports on the TMI accident.[22]

Fig. 5.1: p. 79
- The present structure of the Design and Development Division of Ontario Hydro (see Figure 5.1) shows that the nuclear safety section is only one of a large number of sections, and the implication

is that its importance is equivalent to such sections as "advanced nuclear concepts" and "nuclear analysis".

We recommend that:

5.2   A new division devoted exclusively to nuclear power safety, reporting directly to the Executive Vice-President (Operations) of Ontario Hydro should be established.

• This division, although independent of the group responsible for normal operational activities, should co-operate closely with it. In particular, the new division should be interdisciplinary and should include nuclear physicists, chemists, biologists, mechanical and electrical engineers, human factors specialists, and metallurgists. The normal activities of the new division should include: studies of the safety and performance of nuclear power stations and heavy-water plants, involving in-depth analysis of significant event reports; the development of event-tree and fault-tree analyses; and the designing of simulator-trainers, and research on human factors that relate to reactor safety.

We recommend that:

5.3   The new safety division recommended for Ontario Hydro should establish a small emergency task force, available 24 hours a day on an "on call" basis. This force should be one that could be transported expeditiously in an emergency, by road or helicopter or both, to any nuclear generating station in the province.[23]

5.4   A systematic attempt should be made by Ontario Hydro to look for patterns in operating and accident experience available from both CANDU and other reactor systems. These patterns should be fed back into the process of setting design, operating, and safety criteria.

In the light of the significant event reports, we recommend that:

5.5   Operational procedures and especially the reporting systems at CANDU stations should be critically assessed to improve communication.

5.6   The current CANDU control room and indicator design should be reviewed and assessed from a human factors perspective to ensure that the equipment will display clear signals on reactor status to the operator under both normal and accident conditions.

5.7   The educational requirements and training programmes for all nuclear supervisory, operational, and maintenance personnel should be critically reviewed.

• We believe that at least two senior members of a shift should possess a university degree, preferably in engineering science (the nuclear power option) and should as well have undertaken an appropriate "hands on" training programme, and should have at least two years of experience as operators. These members of the team should be capable of assessing unexpected events and taking appropriate action. It is clearly desirable to have personnel on site who understand the dynamics of nuclear reactor behaviour in some depth. Senior nuclear station operator personnel should be given salaries and status commensurate with their resoponsibilities. The skill and care with which they do their job may play a significant role in ensuring the safety and well-being of very large numbers of people.

We recommend that:

5.8   Provision should be made for the continuous updating and monitoring of the performance of all reactor operators and maintenance personnel; there should be much more imaginative use of simulators in this regard.

• To enhance the competence and, indeed, the vigilance of operating personnel, especially first operators, "retreats" should be organized at frequent intervals. These might take the form of two-day workshops or seminars on various aspects of reactor operation. They might be attended by several shift supervisors and reactor operators, members of the staff of the proposed safety division, and two or three academic persons.

We recommend that:

5.9   The Atomic Energy Control Board should establish a human factors group to ensure that human factors concepts and engineering become central elements in the safe design, construction, operation, and maintenance of Ontario's nuclear stations. Further, human factors concepts should be reflected in the licensing requirements for both nuclear stations and their key operating personnel.

• The importance of effective contingency planning to minimize the potential effects of a nuclear accident was clearly demonstrated at TMI. Despite the low probability of a serious nuclear accident, contingency planning must be viewed as an integral and central part of nuclear safety and the licensing of plants. While we are encouraged by the apparent effectiveness of the contingency planning at the railway accident in Mississauga (November 1979), it is uncertain whether the public response to a nuclear incident would be so orderly.

We therefore recommend that:

5.10 All aspects of contingency planning should be assessed in the light of the experience at Three Mile Island, and a comprehensive plan for each nuclear facility should be made publicly available. The public must be aware of these plans, which must be rehearsed regularly if they are to be credible. Special attention should be paid to preparing in advance for the sensitive and accurate handling of information during an accident.

• We do not endorse the recent recommendation of the Ontario Select Committee on Hydro Affairs that an in-depth study be undertaken to analyse the likelihood and consequences of a catastrophic accident such as a core melt-down at a nuclear power plant.[24] Our reasons are:

   • There is no evidence that a CANDU reactor safety study of this magnitude would be successful. Indeed, evidence to date as provided by the Rasmussen report and the Lewis report suggests that it would be virtually impossible for such a study to come up with a definitive conclusion. A basic reason for this, and it will be clear from the previous discussion, is that it is impossible to include in such a study the vagaries of human operator performance that affect nuclear safety. Note, in particular, the findings of the TMI investigations.

   • Even ignoring the human factor, the methodologies involved in such investigations are extremely complex. Apart from mathematicians, scientists, engineers, and computer specialists within the "nuclear establishment" (i.e., Ontario Hydro and AECL), there is very limited expertise in Canadian universities and industries in the methodology of "event-tree" and "fault-tree" analysis applied to reactor safety. How, therefore, could an independent task force be assembled?[25]

   • In the light of the Rasmussen/Lewis studies, it is highly probable that the methodology and results of such a CANDU study would be unacceptable to many Canadian scientists. Far from enlightening the general public, such dissension would probably create even more confusion than exists at present.

   • We believe that, instead of embarking on a very costly mission that is virtually doomed to failure, much higher priorities should be given to human operator studies, to the in-depth analysis of significant event reports, and to the safety studies at present in hand at the Whiteshell Nuclear Research Establishment and the Sheridan Park Power Projects Division of AECL.

Because of the safety record of Ontario's nuclear power stations — more than 60 reactor-years of operating experience, during which no member of the public has been killed or injured as a result of a nuclear accident — and because of the more than 1,600 reactor-years of experience, on a world-wide basis, during which there has been no major release of radioactivity, we reaffirm our earlier finding to the effect that CANDU reactors are safe within reasonable limits. However, in view of the additional information made available to the Commission since publication of the *Interim Report*, especially relating to the TMI accident, we cannot overemphasize the importance of continued vigilance. The fundamental consideration in connection with nuclear safety is attitude. The only attitude that will maintain public confidence is one that openly recognizes that nuclear plants are by their very nature potentially dangerous, and therefore recognizes the necessity of continually questioning whether the measures in place are sufficient to prevent major accidents.

Uranium Mining and Milling

Since publication of the *Interim Report*, new information relating to the low-level radiation associated with the mining and milling of uranium has become available. Although our hearings did not cover in detail the front end of the CANDU fuel cycle, we have nevertheless reached some general conclusions. First, we reaffirm our endorsement of the recommendations of the Royal Commission on the Health and Safety of Workers in Mines (the Ham Commission) that relate specifically to the health and safety of workers in the Elliot Lake uranium mines. However, in the words of Dr. E.P. Radford, Chairman of the BEIR III Committee:

It is regrettable that beyond the preliminary data presented by Ham in the Royal Commission report of 1976, no further follow-up data on the Ontario uranium miners have been made public.[26]

In view of the inadequacy of the data relating to the long-term effects of exposure to low-level radiation, it is clear that there is an urgent need to follow up all cases of known exposure, especially those in conditions that are reasonably well established. In this regard, Dr. Radford argued:

> It is my opinion that an adequate and continuing epidemiologic evaluation of this large group of Ontario miners offers one of the best opportunities to understand the many complexities of this important issue. Indeed, such a study is an obligation of government to the miners.

A further significant observation by Dr. Radford related to new evidence showing "that cigarette smoking is not as crucial a factor in the radiation-induced cancer as has been thought". Apparently, in the cases of miners exposed to excessive low-level radiation, the main effect of smoking is to reduce the latency period. Accordingly, cancer can be induced either by smoking or by working in the uranium mines; when the two are combined, the time until the onset of detectable cancer is reduced.

We understand that, as a result of the introduction of modern ventilation techniques, the radon daughter levels have been appreciably reduced in the Elliot Lake uranium mines, and this is encouraging. However, we strongly endorse Dr. Radford's view to the effect that there should be continuing epidemiologic evaluation of the Elliot Lake miners and that the data should be made public.

The subject of uranium tailings was addressed briefly in the *Interim Report*. However, because of the lengthy hearings conducted by the Environmental Assessment Board in connection with uranium mill tailings at Elliot Lake and, in view of our previous conclusion relating to an independent review of the problem, there is need for an update.

Although the additional radioactivity, released to the general environment, that can be attributed to mill tailings is small, the emanations are essentially never-ending. Consequently, significant health implications can be postulated, especially for people living or working in the vicinity of the tailings deposits. There are already approximately 100 million tonnes of tailings in the Elliot Lake area. In view of the expansion of Ontario's (and the world's) uranium industry, it is clear that major steps are needed to clean up abandoned mill-tailings areas and to ensure the isolation of functioning mill-tailings ponds and areas. This should be a high-priority operation.[27]

The principal hazards associated with uranium mine and mill tailings are:
- exhalation of radon-222 from tailings
- leachates from tailings and settling ponds containing acids, heavy metals, radium-226, and other radionuclides entering ground water and waterway systems
- movement of tailings containing radionuclides due to water and wind forces

There is insufficient quantitative knowledge at present concerning the pathways whereby radionuclides can reach man and the biosphere from tailings areas. While standards have been established concerning radiation exposure of persons exposed to tailings (500 millirem) and the content of radium in drinking water (10 pCi/L), no standards can be established for abandoned tailings until these pathways are better understood.

It is clear that future research to mitigate the environmental and health effects of tailings must place high priority on pathways analysis, which should reveal the extent of the problem. Concurrently, work must proceed on improved methods to stabilize existing tailings against movement, leaching, and exhalation. Equally important, however, is the need to find methods of removing more of the radioactive waste within the mill circuit before the tailings become exposed to the biosphere.

Since uranium tailings occur in Saskatchewan as well as in Ontario, and since other provinces, notably British Columbia and Newfoundland, also contain economic uranium deposits, the problem is national in scope. The existing 100 million tonnes of tailings in Canada are expected to triple over the next 20 years.

We endorse the present efforts of Energy, Mines and Resources Canada to establish a national programme for uranium mine and mill waste research. While each individual mine will require site-specific treatment of its tailings, some problems are common to all mines. Generic areas that would form the core of the national programme include:
- pathways analysis involving the development of pathways models
- modelling the source of radioactive substances in tailings, including comparisons with existing tailings
- measurement of the environmental effects of existing tailings
- understanding the chemistry of radionuclides in tailings
- conducting general research on disposal techniques that can be applied to all types of tailings

It is clear that Ontario should underwrite its share of this work, in addition to the site-specific activities being pursued by the uranium-mining companies operating in the province.

It is not known with any degree of certainty whether any person has contracted cancer as a result of radiation exposure due to mill tailings; an in-depth study of the whole problem is clearly essential. Indeed, this has been recognized by the Ontario Ministry of the Environment and especially by the Environmental Assessment Board (EAB), which conducted public hearings in the Elliot Lake area. The *Final Report on a Public Hearing by the Environmental Assessment Board into the Expansion of the Uranium Mines in the Elliot Lake Area* was published by the Ministry of the Environment in May 1979, about eight months after the publication of the *Interim Report*. The EAB's report deals with the health and environmental implications of the mill-tailings ponds in the area and their potential extension, and considers the concerns we raised in September 1978. We endorse the findings and recommendations of the EAB with respect to uranium tailings and draw particular attention to the following:

- With respect to the long-term impact of tailings, current knowledge is limited. It is evident to the Board that considerable effort and time is required before solutions to the long-term aspects of waste management can be found.[28]
- The Board recommended that the companies immediately commence studies to determine the extent and quantity and impact of radium deposited downstream from discharges of tailings effluent, and to determine ways and means of removing the deposits if this should be required (Recommendation 10-7).
- The Board found that the long-term impermeability of tailings basins – specifically dams containing the tailings – cannot be guaranteed (Recommendation 10-9).
- The Board found little evidence to give it confidence in the use of synthetic membranes, asphalt, cement, or chemical means to cover tailings areas to inhibit water infiltration during the long term (Recommendation 10-19).
- The Board found that the mill tailings have the greatest potential impact on the natural environment of all the activities related to the mines' expansion (Recommendation 10-30).
- The Board recommended that intensive studies be undertaken by the mining companies with the assistance of CANMET (Canadian Centre for Mineral and Energy Technology) into ways of stabilizing pyrite in the tailings (Recommendation 10-34).
- The Board recommended that all research programmes in the area of waste management be required to produce results within three to five years from May 1979 (Recommendation 10-37).
- The Board recommended that no new tailings areas be approved at least until the results of research projected for a period of three to five years from May 1979 can be evaluated (Recommendation 10-38).
- The Board recommended that there be a public hearing in 1984 into the developed waste management techniques and that, if necessary, funding mechanisms for intervenors be established (Recommendation 10-41).
- The Board recommended that careful consideration be given by the province and the AECB to requiring the monitoring of effluents from tailings basins during the post-operations period – at least until trends are established (Recommendation 11-12).

Conclusions and Recommendations with Respect to Uranium Mining and Milling

Because of the uncertainties surrounding the health consequences of the exposure of humans to chronic doses of low-level radiation, it is important to ensure the minimization of risks (resulting from the ingestion and inhalation of radiation) to uranium mine and mill workers and to the general public.

We recommend that:

5.11 Continuing epidemiologic evaluation of Elliot Lake miners and uranium mill workers should be undertaken. The public should be informed of the progress of these studies.

5.12 Ontario should contribute its share to any national programme for uranium mine and mill waste research.

5.13 Measures should be taken to ensure that the costs of long-term tailings monitoring, management, and R&D are reflected in the cost of uranium fuel rather than becoming a general charge to the Ontario taxpayer, not least because most of the uranium is currently being exported (over 90 per cent).

5.14 The future expansion of the nuclear power programme in Ontario, and in particular the uranium mining and milling portion of the fuel cycle, should be contingent on demonstrated progress in research and development with respect to both the short- and the long-term aspects of the low-level uranium tailings waste disposal problem, as judged by the provincial and federal regulatory agencies and the people of Ontario, especially those who would be most directly affected by uranium mining operations. It would be unacceptable to continue to generate these wastes in the absence of clear progress to minimize their impact on future generations.

The Management and Disposal of Spent Nuclear Fuel

It is important to distinguish between non-reprocessed and reprocessed spent fuel, and to recognize that high-level radioactive wastes, resulting from nuclear weapon production as well as from nuclear power stations, are accumulating in many countries.[29] There is clearly an urgent need to develop ultimate disposal facilities to ensure that these wastes are isolated from the world's ecosystems. The *Interim Report* considered in some detail the nature of the spent fuel management and ultimate disposal problem, and particularly the steps being undertaken by Canada to deal with it.

It is improbable that Ontario Hydro will be forced to resort to plutonium-239 reprocessing of spent uranium fuel (to enhance the security of nuclear fuel supplies) or to the thorium/uranium-233 reprocessing fuel cycle before the year 2010. Therefore, the immediate concern facing Ontario is the storage of spent fuel in on-site water-filled bays at nuclear power stations. It may also be possible to store spent fuel at a station site in retrievable form for a period of between 50 and 100 years in large concrete silos.[30]

On the assumption that there will be no reprocessing of spent fuel (and consequently no accumulation of high-level radioactive liquid wastes) before the end of this century, we restrict the discussion to the management and potential disposal of non-reprocessed spent fuel.[31] In fact, a simple alternative to reprocessing is to dispose of the spent fuel in the form in which it is removed from the CANDU reactor, i.e., as durable ceramic pellets of uranium dioxide (and a very small proportion of plutonium dioxide) sealed in non-corrosive tubes of zirconium alloy.

In the *Interim Report* we made extensive reference to the independent studies conducted by the Hare Study Group and Professor Uffen;[32] this material is still relevant. The purpose of this section is to update the Commission's earlier findings.

The Ontario Hydro and Canadian Programme

We have concluded that the current practice of storing spent CANDU fuel bundles in special water-filled bays at Ontario's nuclear power stations (i.e., Douglas Point, Pickering, and Bruce) is both safe and effective. The radioactivity, and consequently the thermal energy, of the spent fuel bundles diminishes rapidly during the first few years of storage and continues to diminish, less rapidly, thereafter. We have concluded, further, that on-site storage facilities should be available at all nuclear stations in order to accommodate a station's spent fuel until a permanent disposal facility is available, probably between 2000 and 2010. Consequently, the on-site facilities will be required to store the accumulated spent fuel resulting from the operation of Ontario's nuclear generating stations for the next 20 to 30 years. The present and projected situation is shown in Figure 5.2 and Table 5.1 which show, respectively, the spent fuel generation rates and the storage capacity of the Pickering, the Bruce, and potentially the Darlington generating stations. Note, in particular, that the Pickering generating station has enough storage capacity in two main bays and an auxiliary bay to accommodate station operation until 1996. As well, Bruce A and Bruce B can store all their spent fuel until the year 2000 and 2007, respectively. Darlington will have storage capacity adequate for about 25 years of station operation. Ontario Hydro's original policy was to provide sufficient storage for six years of station operation combined with the storage required to "dump" all the fuel bundles in one reactor core. However, the existing policy, which will apply to all new stations, is to build storage capacity for up to 20 years of station operation (presumably auxiliary bays could be built to extend this period). It has been estimated that the total accumulated volume of spent fuel in Ontario, to the year 2000, will be approximately 9,000 m^3 (i.e., $120\,m \times 25\,m \times 3\,m$).

Fig. 5.2: p. 80

Ultimately the spent fuel stored in on-site bays, or on-site silos, at the nuclear stations will have to be transported to a permanent disposal facility or, in certain circumstances (not foreseen for at least 30 to 40 years), to a spent fuel reprocessing facility.[33] Ontario Hydro is responsible for the development of the high-level radioactive waste transportation technology. As at present envisaged, the spent fuel would be transported in large flasks, each weighing 50 tonnes and capable of handling 3.5 tonnes of spent fuel bundles.

We have concluded that, even in its present state of development, the technical problems for the safe

Table 5.1 On-site Storage Capacity of Ontario Hydro's Nuclear Generating Stations (1979-2007)

	Storage capacity (bundles)	Bundles generated annually (80% ACF)	Bundles to 2000	Capacity full in
Pickering				
A	84,000	13,300	360,000	1979
B	160,000	13,300	225,000	1995
Auxiliary	234,000	–	–	1997
Bruce				
A	390,000	23,000	530,000	1994
B	407,000	25,500	380,000	2001
Darlington	486,000	25,500	310,000	2007

Source: RCEPP

transportation of spent fuel from nuclear power stations to a permanent disposal facility (or to a reprocessing facility) are not intractable.

Canada's major spent nuclear fuel management, research, development, and demonstration programme is essentially a co-operative effort involving AECL, the Canadian Geological Survey, and Ontario Hydro.[34] The programme is co-ordinated by the Whiteshell Nuclear Research Establishment, Pinawa, Manitoba. Although the majority of the basic laboratory and testing facilities are located at Whiteshell, research and development activities in connection with the handling of high-level radioactive material are being undertaken in universities, industries, and research institutions across Canada.

Members of the Commission and staff visited the Whiteshell Establishment on July 25-26, 1979, to obtain a first-hand briefing relating to the spent fuel management programme. We understand that three phases are envisaged – concept verification, site selection, and the construction and operation of a demonstration disposal facility. The target date for completion of the demonstration facility is the year 2000.[35]

The main briefing during the Commission's visit concerned the concept verification phase. We were given an opportunity to assess progress to date by visits to the laboratories doing research on geotechnical problems, the immobilization of spent fuel and radioactive wastes, the environment, and the sorption properties of "buffer" and "back-fill" materials. The briefing and laboratory visits helped materially to put into perspective the risks associated with the management and disposal of spent CANDU fuel.

In parallel with the programmes of other countries, the Canadian effort is concentrating on the ultimate disposal of spent fuel (non-reprocessed or reprocessed) in deep underground stable geologic formations. (Techniques for immobilizing both reprocessing wastes and the normal spent fuel bundles are, in fact, in process at Whiteshell – both options are being kept open.) It is anticipated, and there is international consensus on this point, that high-level radioactive materials can be isolated in this way from the earth's ecosystem during the long period necessary to ensure that they are comparatively harmless; after about 700 years, the remaining radioactivity is about the same as that of uranium and thorium deposits in the earth. Clearly, the basic requirement is to ensure that radioactive material is not allowed to penetrate into the biosphere through ground-water movements, glacial action, or accidents.

The first objective is a study of the geological, hydrogeological, physical, and chemical characteristics of hard-rock formations in the Canadian Shield; these are known as "plutons", in recognition of their volcanic origin. To determine the suitability of specific hard-rock formations, it is necessary to drill to a depth of about 1,000 m and undertake extensive laboratory testing (for fractures, porosity, ion-exchange characteristics, etc.). Drilling programmes will be undertaken in Ontario during the next two years. It cannot be emphasized too strongly that, without comprehensive research based on deep drilling of the igneous rocks of the Canadian Shield, the long-term programme of nuclear spent fuel disposal cannot succeed.

If hard-rock disposal of the spent fuel proves non-viable, other alternatives such as salt domes, the ocean bed, and shales may prove to be effective. It is important to note that, because the question of radioactive waste management and disposal has international connotations, other countries are involved in the study of these alternatives. Canada is kept in touch with these alternative developments.

As a direct result of the hard-rock drilling programmes and the subsequent laboratory experimentation, it will be possible to deduce the "containment characteristics" of various rock formations, in several locations. But, because of the very long periods involved in the decay of some of the actinides, it is impractical to study the potential migration of radionuclides experimentally through cracks and fissures. The alternative is to undertake computer model studies, known as "pathway analyses", and to develop associated validation tests.

To ensure isolation from the biosphere, not only must the geologic containment system provide an effective barrier to prevent the escape of radionuclides and actinides (e.g., through ground-water movement), but other barriers are also essential. In the first place, the spent fuel must be immobilized in the sense of effective local containment. This containment must resist leaching by water and must maintain a suitable stable form for several hundred years, during which period most of the radioactive products of fission will have decayed to comparatively low levels. The first containment barrier must also be capable of withstanding bombardment by high-energy radiation and the associated high temperatures (i.e., in the order of 200°C). However, if it is assumed that the spent fuel has been stored on site in water-filled bays, the level of radioactivity and the corresponding temperature will have moderated appreciably.

The fuel immobilization methods at present under consideration, on the assumption that there will be no reprocessing of spent fuel and hence no high-level radioactive liquid wastes, are based on the containment of the intact fuel bundles. For example, the design of simple primary (thick- and thin-walled) containment systems consisting of a single corrosion-resistant outer shell capable of containing the spent fuel bundles for about 500 years (the high-toxicity phase) is under way. For the very long term, in the order of several thousand years, after which toxicity remains virtually constant, it is proposed that the simple containment system will be extended and multi-barrier systems will be developed. In these, the fuel bundles will be matrixed, probably with lead, and the matrix will be surrounded by both sorptive and reactant particulates that will retard the escape of radionuclides. Alternative spent-fuel forms might also rely on vitrification or ceramic containment techniques; these techniques have been developed extensively in Canada and several other countries. The outer shell will probably be fabricated in corrosion-resistant ceramic or copper.

Still another formidable barrier is required to absorb any radioactive wastes that may penetrate the aforementioned containment. It will be completely surrounded by many tonnes of a mixture of bentonite clay and quartz sand. The sheer mass and physical and chemical characteristics of the various containment barriers will determine for how many years they will remain stable.

To ensure full co-operation and consultation in nuclear spent-fuel management, the governments of Canada and Ontario have established a co-ordinating committee, under the chairmanship of Dr. S.R. Hatcher, Vice-president, AECL. Representatives of Energy, Mines and Resources Canada, the Ontario Ministry of Energy, and Ontario Hydro serve on the committee. Furthermore, a technical advisory committee has been appointed to review the programme and to advise AECL. Membership of this committee, under the chairmanship of Professor L. Schemilt of McMaster University, includes representatives of the Canadian Association of Physics, the Canadian Federation of Biological Sciences, the Canadian Geoscience Council, the Canadian Institute of Mining and Metallurgy, the Chemical Institute of Canada, and the Engineering Institute of Canada. This committee, with a membership of 10 distinguished Canadian scientists and engineers, will monitor the spent nuclear fuel management programme on behalf of the people of Canada. We have confidence in their ability and integrity, and we believe the people of Ontario will be able to trust their technical judgement. This process will constitute an important form of peer scientific and technical review. However, a more broadly based and public review will ultimately be needed if the results of the nuclear waste management programme are to be credible and acceptable to the people of Ontario.

While the science and technology of managing and disposing of high-level radioactive wastes appears to be advancing purposefully, the social and political aspects of the problem are much less tractable, not least because they have been virtually ignored. Indeed, we believe that these problems transcend science and technology and that the debate will be conducted with increasing emphasis on politics and ethics. Unless consensus is achieved on a broad basis, the resulting uncertainties, delays, and cost escalation will impact negatively on the scientific programmes.

It is important that appropriate mechanisms be put in place to ensure a meaningful dialogue between the critics of nuclear power and the proponents. It must be obvious to all that the waste disposal problem will not vanish. The problem must be resolved within the next two or three decades. It will require co-operation among individuals of all shades of opinion. In particular, it will call for a degree of reasonableness and compromise on the part of both critics and advocates of nuclear power.

We have concluded that the social, educational, and psychological aspects of the nuclear waste management issue are even more significant than the scientific, engineering, and economic aspects.

Conclusions and Recommendations with Respect to the Management and Disposal of Spent Fuel

- The reprocessing of spent fuel will not be required in Ontario before 2010 at the earliest. If uranium resource constraints require that commercial reprocessing and advanced fuel cycles be considered, this will become a matter for public discussion in order to ascertain whether the risks and benefits of this technology are acceptable to the government and people of Ontario. Ontario Hydro is interested in exploring the possibility of using slightly enriched uranium fuel (1.5 per cent uranium-235). If this proves feasible, it would result in a more efficient use of finite uranium fuel resources and would postpone the day when reprocessing would be economical or necessary.

- The interim storage of spent fuel in on-site water-filled bays, and the possible option of on-site storage in concrete silos, is safe and effective.

We recommend that:

5.15 All existing and planned Ontario Hydro nuclear stations should be retrofitted or designed for the interim storage on site of their spent fuel for the next 30 years by which time a disposal facility should be available.

- We have concluded that, even in its present state of development, the technical problems for the safe transportation of spent fuel from nuclear power stations to a permanent disposal facility (or to a reprocessing facility) are not intractable.

- We are satisfied that the existing scientific research and testing programme looking to a demonstration facility for the safe and permanent disposal of Ontario Hydro's spent nuclear fuel shows promise.

- It is absolutely essential to ensure public confidence in the waste management programme. However, the socio-political implications of managing the wastes have not been adequately assessed. Because we are dealing with wastes that will be potentially hazardous for at least 500 to 1,000 years, socially acceptable steps on a continuing basis must be taken to ensure that the long-term risks are minimal. The social, health, psychological, and political aspects may be more important, for arriving at an acceptable solution, than the geological, ecological, and engineering aspects of waste management.

- The technical and social aspects of the problem should be dealt with in parallel. The waste-disposal issue provides an opportunity for inter-disciplinary study that was not possible during the early stages of the nuclear power programme, when the technology was the sole concern; few people were then aware of the potential environmental, health, and social concerns.

- Field studies and tests (such as those being undertaken by AECL) are an essential first step in the identification of geological structures and patterns that meet the technical criteria. However, the criteria must be established on the basis of socio-political as well as technical consensus.

- The technical advisory committee that has been established to review AECL's R&D programme constitutes an important form of peer technical and scientific review.

Accordingly, we recommend that:

5.16 An independent "nuclear waste social advisory committee" should be established to ensure that broad social, political, and ethical issues are addressed. This committee should be chaired by an eminent Canadian social scientist.[36]

- We reaffirm the *Interim Report* conclusion that the future evolution of the nuclear power programme in Ontario must be predicated on demonstrated progress in the waste management R&D programme. It would be unacceptable to continue to generate these wastes in the absence of clear progress to minimize or eliminate their impact on future generations through the availability of a technically credible and socially acceptable nuclear waste disposal facility.

Because we are concerned that this vital R&D effort not be rushed, and since we have confidence in Ontario Hydro's ability to store spent fuel safely on site for an interim period (to between the years 2000 and 2010) until a final nuclear waste repository will likely be available, we have revised our Interim Report view on this matter. We recommend that:

5.17   If progress in high-level nuclear waste disposal R&D, in both the technical sense and the social sense, is not satisfactory by at least 1990, as judged by the technical and social advisory committees, the provincial and federal regulatory agencies, and the people of Ontario — especially in those communities that would be directly affected by a nuclear waste disposal facility — a moratorium should be declared on additional nuclear power stations.

The Economics of Nuclear Power

It is salutary to reflect on the cost of the TMI accident, even excluding the not inconsiderable social costs. According to one estimate, the cost to bring the reactor back into service will probably exceed $500 million, and, assuming a three-year clean-up, refurbishing, and recommissioning period, the cost of coal-generated energy to replace the nuclear energy is expected to be significant.

The purpose of this section is to update Chapter 7 of the *Interim Report*, with special reference to the coal-nuclear choice, and to Canada's nuclear industry.[37]

Assuming, for example, a 3.5 per cent annual compounded load growth, Ontario's major nuclear generating stations in service by the year 2000 will include Pickering A and B, Bruce A and B, and Darlington. Depending on whether or not there is much greater reliance on imported power (especially from Manitoba), and on the extent to which co-generation plants are introduced, the maximum additional nuclear power requirement would be another "Darlington-type" generating station. It is assumed that nuclear power stations operate in a base-load mode.[38] While it is conceivable that the additional nuclear generating station could be replaced with an equivalent coal-fired station, the possibility of this is remote.

A recent cost comparison of a CANDU nuclear station and a coal-fired station indicates that the general conclusions of the *Interim Report* relating to the costs of new nuclear and new coal stations still apply.[39]

The sensitivity of the nuclear-coal cost comparison to alternate assumptions, especially when three sample annual capacity factors (ACF) are assumed, is shown in Table 5.2. The cost advantage of nuclear power increases as the ACF increases and, correspondingly, the "time to break even" (i.e., the time from a 1987 in-service date for the "accumulated present worth" of nuclear and fossil to break even) decreases markedly as the ACF increases. Assuming a 40 per cent ACF, the nuclear advantage is reduced to 8 per cent, especially if the price of coal remains as forecast. Note, however, the marked sensitivity of the "nuclear advantage" to changes in the annual discount rate, as well as to the assumed fossil-fuel price escalation. Furthermore, the nuclear-coal cost comparisons, on a life-cycle basis, vary dramatically with the capital cost of nuclear power. The corresponding accumulated future discounted cash flows, based on years beyond the assumed 1987 in-service date, are shown in Figure 5.3, for ACFs of 40 per cent and 80 per cent. These graphs also demonstrate, convincingly, the cost advantage of nuclear power, especially at high ACFs, and the cost advantage of United States coal compared with Canadian coal. Fig. 5.3: p. 81

However, these data should not be interpreted too literally because several unknown factors (especially social costs) have not been included in the comparative costing. Some uncertainties are:
- the costs of stricter environmental and safety requirements, especially on a life-cycle basis
- the cost of managing and ultimately disposing of nuclear spent fuel
- the cost of fuel, especially coal, over a 30-40 year operational life
- the interest accumulated during the construction phase of a generating station (the longer the lead time the greater the uncertainty)
- the cost associated with unexpected problems such as the pressure-tube retubing required at Pickering A and Bruce A

Notwithstanding the above uncertainties, we believe that comparative cost studies of CANDU and coal-fired generation should be continued in the light of changing circumstances; an up-to-date assessment of comparative life-cycle costs will be of value to decision-makers.[40]

Because heavy-water production is an important component of the economics of nuclear power, we summarized in the *Interim Report* the potential of Bruce Heavy Water Plants A, B, and D, and the supply situation to the year 2000. In the light of the downward revision in the load forecast, and a likely maximum nuclear component to 2000 of one additional 4 × 850 MW generating station (after the

Table 5.2 Summary of Sensitivity Analysis

Parameter varied from reference case[a]	Value of Parameter	At 80% ACF		At 60% ACF		At 40% ACF		Lifetime ACF[d] (%)
		Nuclear advantage[b] (%)	Time to break even[c] (years)	Nuclear advantage[b] (%)	Time to break even[c] (years)	Nuclear advantage[b] (%)	Time to break even[c] (years)	
Annual discount rate	8.5%	60	7	39	9	16	16	29
	9.75%	49	7	30	11	8	20	33
	11.0%	39	8	21	12	2	27	39
Fossil-fuel mix	100% U.S. bituminous	46	8	27	11	7	22	35
	90% U.S. + 10% Cdn. bituminous	49	7	30	11	8	20	33
	50% U.S. + 50% Cdn. bituminous	61	6	39	9	15	16	29
	100% Cdn. bituminous	78	5	53	7	25	13	25
Fossil-fuel price escalation	+20%	76	6	51	9	24	14	26
	As forecast	49	7	30	11	8	20	33
	−20%	28	9	13	15	−4	>30	44
Nuclear fuel price escalation	+20%	39	8	23	12	4	24	37
	As forecast	49	7	30	11	8	20	33
	−20%	58	7	36	10	12	18	30
Fossil capital cost	+20%	55	6	36	8	15	14	28
	Best estimate	49	7	30	11	8	20	33
	−20%	43	9	23	14	1	28	39
Nuclear capital cost	+20%	36	10	18	16	−2	>30	42
	Best estimate	49	7	30	11	8	20	33
	−20%	65	5	44	7	22	11	23
In-service date	Nuclear delayed one year (fossil on schedule)	41	10	25	13	7	23	35
	Both stations on schedule	49	7	30	11	8	20	33
	Fossil delayed one year (nuclear on schedule)	53	7	32	10	9	20	31

Notes:

a) The reference case in the sensitivity analysis is always the second line of data in each of the seven parameter studies. In each study only one parameter is varied from the reference case.

b) "Nuclear advantage" is the margin by which the accumulated present worth of the nuclear alternative is more economic at year 30.

c) "Time to break even" is the time from 1987 in-service for the accumulated present worth for fossil and nuclear to break even.

d) "Lifetime ACF" is the annual capacity factor at which the accumulated present worth for fossil and nuclear break even. This ACF is the same in each of the 30 years.

Source: Ontario Hydro, "Cost Comparison of 4 × 750 MW fossil-fuelled and 4 × 850 MW CANDU Nuclear Generating Stations", Report No. 584SP, January 1979.

Darlington generating station), Ontario Hydro's heavy water requirements would be as shown in Figure 3.6, in Volume 2 of this Report. Note that, by the year 2000, Bruce Heavy Water Plants A and B together will be able to supply a 30,000 MW nuclear programme.

During the nuclear power public hearings the question of economy of scale was raised. We now reaffirm the position first stated in the *Interim Report* that concerns the potential development of a 1,250 MW CANDU reactor. The reasons we have given previously are now reinforced by the Commission's view that the load growth will be even smaller than we previously assumed.

Accordingly, we recommend that:

5.18   **No further development of the 1,250 MW CANDU reactor, even in the concept stage, should be** undertaken by Ontario Hydro. Any additional nuclear base-load power stations in the post-Darlington period should be based on 850 MW CANDU reactors. We believe that such standardization will facilitate reactor safety as well as optimizing the annual capacity factors of these stations.

The future of Canada's nuclear industry, as we first intimated in the *Interim Report*, remains a cause for concern. At the time the *Interim Report* was prepared, the "minimum order level" for economic viability of the Canadian nuclear-components industry was assumed to be 1,200 MW per year (Figure 8.2 in the *Interim Report*). For the period 1980 to 2000, orders totalling 24,000 MW (i.e., 28 × 850 MW or seven Darlington-type power stations) would therefore be required. Unless there is a marked increase in foreign exports of CANDU, it is impossible that this minimum order level will be achieved. Two major reports on Canada's energy future have been published since the publication of the *Interim Report*. These reports suggest that the nuclear component of electric power generation in Canada by the year 2000 may total in the order of 60,000 to 70,000 MW.[41] We believe these projections are misleading to government, and especially to the Canadian nuclear industry, because they are predicated on obsolete or incredible Ontario Hydro and AECL forecasts.[42] Advancing the in-service date of a committed Ontario nuclear station by several years before the power is actually needed in Ontario may offer some relief to the industry if United States export markets are available, but the situation is clearly becoming very serious.

We recommend that:

5.19   The Ontario government should advise the federal government that Ontario's requirements will be insufficient to ensure an order level of one reactor per year and, therefore, that the maintaining of CANDU as a viable option for the future suggests a need for urgent federal initiatives to fill the order gap. Our estimate of the likely total installed nuclear capacity in Ontario to the year 2000 is in the order of 17,500 MW; this means one additional 3,400 MW four-reactor nuclear station after Darlington, and it could be a high estimate, depending on, for example, actual load growth, success with conservation, co-generation, and potential imports of hydroelectric energy from Manitoba or Quebec. If the industry wishes to survive, it must begin to search for opportunities to diversify.

Uranium Supplies

In the *Interim Report* we expressed concern about the uranium requirements associated with the massive programme then envisaged by Ontario Hydro's (LRF48A) system expansion plans. These requirements appeared to exceed the maximum production capacity likely to be available to Ontario by the end of the century. Major developments have since occurred that effect these concerns.

• Because of the 1979 and 1980 downward revisions in Ontario Hydro's load forecast (now 3.4 per cent annual growth to 2000), the availability of uranium will not be as critical a factor in Ontario Hydro's nuclear power programme as was envisaged in the *Interim Report*.

• The nuclear power growth patterns in the western industrial nations indicate a marked reduction in growth rate. On a global basis this will relieve the pressure on uranium supplies to some extent. On the other hand, the spent fuel reprocessing programmes envisaged four or five years ago have been appreciably slowed down, and this will to some extent tend to sustain pressure on demand.

• To the year 2000, the period of major concern to the Commission, Ontario Hydro's annual needs for uranium will not exceed 2,500 tonnes if one additional Darlington-sized station is built in the 1990s (3.5 per cent load growth to 2000) and nuclear units operate at 80 per cent ACF. Consumption may be as low as 1,800 tonnes if no more nuclear capacity beyond Darlington (the current commitment) is built before 2000 (3.0 per cent load growth) and existing units operate at an average ACF of 70 per cent, as a result of the retubing required at Pickering A and Bruce A.

• Ontario Hydro's existing massive uranium contracts could provide life-time fuelling at an 80 per cent ACF for all committed capacity plus one additional Darlington-size station required before 2000 for a load growth of 3.5 per cent per annum (see Figure 5.4).

Fig. 5.4: p. 82

• Given current load forecasts and existing contracts, the demand for additional sizeable quantities or uranium for domestic or Ontario use will not manifest itself before 2000. With no immediate domestic markets available, producers will continue to seek to export. (In 1977, Canada exported 91 per cent of its uranium production.) There will be a tendency for lower-cost, higher-grade deposits to be exported over the next two decades. It therefore becomes a question of whether lower-grade ore will be economically recoverable at a later date. Policies will have to be developed that protect domestic utilities such as Ontario Hydro that might wish to exercise the nuclear option after 2000 while still allowing Canadian producers access to a lucrative international market. As we indicated in the *Interim Report*, uranium is likely to be subject to an unpredictable and highly complex set of forces that will be amenable to only minor control by the government of Ontario.

Table 5.3 Uranium Resources Recoverable, Canada

Category	Thousands of tonnes
Measured	80
Indicated	155
Inferred	302
Total	537
Adjusted reserve	415

Notes: Resources are no longer broken down by province, to protect the confidentiality of the holdings of specific companies. Estimates are of reserves recoverable at up to $175/kgU.
Source: "1978 Assesssment of Canada's Uranium Supply and Demand", Energy, Mines and Resources Canada, June 1979.

Because of concern over the availability of a secure, stable supply of uranium, several countries that lack indigenous supplies (i.e., France, Germany, Japan) are actively and rapidly developing fast-breeder reactors. The Science Council of Canada in a recent report entitled "Roads to Energy Self-Reliance"[43] recommended embarking upon a major CANDU advanced uranium/thorium reactor R&D programme at a total cost of $1.75 billion. Does such a proposed programme, especially in the light of reduced load forecasts, fit into any Ontario Hydro future planning concepts?

> *We recommend that:*

> 5.20   Although it is important to keep open the thorium fuel cycle option by engaging in an R&D programme, a firm decision to go ahead with a major demonstration and/or commercial programme should be delayed at least until 1990, and then made only if it is acceptable to the public after appropriate dialogue and study concerning the full implications and impacts of such a project.

Indeed, only after the future of Canada's nuclear industry has been clarified can the thorium fuel cycle proposal be put into an adequate perspective. There is no possibility that Ontario Hydro will be reliant on a second-generation CANDU fuel cycle before the year 2010, at the very earliest. Therefore, at this time, we do not endorse a thorium fuel cycle demonstration project with the concomitant spent-fuel-reprocessing facility.

> *We recommend that:*

> 5.21   Nuclear power should no longer receive the lion's share of energy R&D funding, and R&D priorities in the nuclear field should be focused primarily on the human factor in reactor safety, on the management and disposal of wastes at the front and back ends of the fuel cycle, and on the decommissioning of nuclear facilities.

Socio-Political Issues

Governments are faced with a major dilemma. They must balance the economic merits of nuclear power including, in the case of Ontario, the availability of a high level of professional expertise, the maintenance of a highly skilled work force, the availability of indigenous nuclear fuel, and the existence of a large number of jobs, against the potential health and environmental hazards of nuclear power, the concern about the ultimate disposal of high-level radioactive wastes, the containment of mill tailings, and the concern expressed by many people relating to the ethical implications of nuclear power.[44]

Value judgements of a particularly complex kind are involved in nuclear power decisions. In particular, we have concluded, an assessment of the acceptability of the risks and benefits of nuclear power and decisions about these matters must include an assessment of the social, ethical, and political implications of its use. Nor can these decisions ignore the risks associated with alternative generating technologies. For example, as will be discussed in Chapter 9, the health and environmental consequences of burning fossil fuels are by no means trivial. Because the ultimate decisions are the responsibility of government, and government responds to public opinion, the need for a higher level of understanding of the nuclear issue by the public is an educational issue of central significance.

Is a higher level of scientific literacy among the public at large essential as a prerequisite for public participation in decisions relating to nuclear power? We are not convinced that it is. We have noted, for example, especially during the public hearings, that scientific literacy *per se* has by no means prevented the deep and bitter divisions that exist amongst highly qualified and respected scientists with respect to virtually all aspects of the nuclear fuel cycle. Furthermore, the divisions between scientists on the nuclear issue have given rise to confusion in the minds of many people who have felt justified in summing the matter up by saying that "the best-informed appear to be the most confused". If, as some claim, the nuclear power controversy is more quasi-religious than technological, then, in the words of R.L. Meehan: "Exposure and examination of the ideological aspects of the issue, using both traditional liberal arts and contemporary social science techniques, might do more to restore rationality than widespread improvement of scientific literacy."[45]

The "Mind-Set" Syndrome

The report of the Presidential Commission on the TMI accident makes repeated reference to the mind-set of the nuclear industry.

> After many years of operation of nuclear power plants, with no evidence that any member of the general public has been hurt, the belief that nuclear power plants are sufficiently safe grew into a conviction. One must recognize this to understand why many key steps that could have prevented the accident at Three Mile Island were not taken. The Commission is convinced that this attitude must be changed to one that says nuclear power is by its very nature potentially dangerous, and, therefore, one must continually question whether the safeguards already in place are sufficient to prevent major accidents. A comprehensive system is required in which equipment and human beings are treated with equal importance.[46]

This syndrome, we believe, applies in some degree to Ontario Hydro. Indeed, we drew attention to one manifestation of it in the *Interim Report* (page 81) when we said that "there is some circumstantial evidence that criticism [of operating procedures]... is not always welcomed by management personnel". Further, as we have already noted in connection with the Bruce significant event reports, there does not appear to be a systematic search for generic warning signals that may presage nuclear reactor abnormalities.

An underlying social problem of some concern is exemplified by the cases of two former Ontario Hydro employees, Messrs. Taves and Kaponerides, who felt their concerns with respect to safety were not being taken seriously by the utility,[47] and other similar cases in the United States. A professional scientist, engineer, or reactor operator is committed to protect the health and welfare of society at large, but he also has a sense of loyalty to his employer. When fundamentally important societal issues such as the safety of nuclear power plants and the disposal of spent nuclear fuel are at stake, it is clearly desirable to ensure that some credence be given to the dissenting opinions of perceptive employees. Dissent on the part of an employee should not be equated with disloyalty. Indeed, dissent is vital to ensure the vigilance necessary for the safe operation of nuclear facilities.

In some respects, the question of "professional dissent" is inextricably related to the issue of the rights of the individual. If employees' freedom of speech is to be protected, especially in such sensitive areas as nuclear reactor safety, some new ground rules are required. Ewing, for example, has suggested the concept of a new Bill of Rights for employees.[48] We believe this is worthy of detailed study, not only from the point of view of protecting an employee's human rights but also to protect the general public. Furthermore, we have concluded, such a step might enhance the confidence of the public in technologies characterized by very-low-probability accidents but with very high consequence levels; such technologies include nuclear power and airline operations. In a real sense, each employee should act as a "regulator" in his or her own right. There must, of course, be protection for the management of the utility, who are required to comply with regulatory criteria and procedures. How might both employee and management be protected in cases of dissent such as those mentioned above? The issue is of fundamental concern.[49]

Accordingly, we recommend that:

5.22 Procedures should be established to ensure fair handling of bona fide cases of professional dissent. Procedures should include the following concepts:

- Concerns should be expressed in writing and considered by a special review group consisting of representatives of management, professional engineering staff, and at least one outside expert.
- The review group should obtain evidence from the dissenting staff member's colleagues.
- The review group should assess management's response to the concerns.

Nuclear Decisions

The decision-making process, in the context of the whole inquiry, is considered in Volume 8, and in Chapter 12 of this volume. However, because decisions that relate to nuclear power are of a special kind, we summarize below our main conclusions.

• The central issue in Ontario's electric power system, as perceived by the people, is the role nuclear power should play. Many people appear largely to discount the scientific and technological data and information relating, for example, to the safety of nuclear power stations and the ultimate disposal of high-level radioactive wastes.

• Novel and imaginative ways are needed of involving the public in decisions relating to nuclear power, and, indeed, in energy and environmental problems, in general. At the same time, understandable information relating to these decisions must be available. Further, and most important, because the complexity of the health, social, and political dimensions of nuclear power are at least as important as the technology, the information base should be biased towards the socio-political implications.

• There should be a determination on the part of government to ensure that decision-making is more open, and, unless national and public security implications dictate otherwise, all information relating to nuclear power systems should be made available to the public.

• Ontario Hydro and the government should ensure, to provide a sound basis for decision-making, that future hearings relating to nuclear power have a primary commitment to the candid exploration of the issues. To facilitate this openness, legitimate public interest groups concerned with the major nuclear issues should be supported financially.

• To inspire more confidence, especially in those who are most affected by nuclear power decisions, quantitative and qualitative assumptions should be open to public scrutiny.

• It is essential that the AECB, before finalizing licensing criteria, especially concerning nuclear safety, should hold public hearings. Only in this way will public trust and confidence in the regulatory body be enhanced.

Regulation of Nuclear Power

With declining load growth and the resulting diminishing prospects for a large nuclear programme, it is our belief that the licensing of new facilities will not be a major aspect of nuclear regulation in Ontario during the next decade. Instead, attention should be focused increasingly on the management of existing plants, and on compliance with regulations. Vigilance by the regulatory authorities will be essential, and greater emphasis will have to be placed on the role of the human operator in reactor safety. It will be essential, for effective compliance and public confidence, that the Atomic Energy Control Board continue to widen the gap between itself and the utilities it regulates. However, events at TMI have strengthened our conviction that the burden of proof for demonstrating the health and safety of its nuclear power plants should be placed squarely on the utility.

In the *Interim Report*, we endorsed the AECB's responsibility to regulate nuclear developments for the purpose of protecting health, the environment, and safety. Further investigation leads us to reaffirm that this allocation of responsibility is appropriate, if for no other reason than that there appears to be no evidence that the health and safety of the citizens of Ontario have been jeopardized by the presence of the federal regulatory agency. However, we also indicated that co-operation between the federal and provincial authorities is needed if undue delay, expense, and uncertainty are to be avoided. The AECB's involvement with the Select Committee of the Ontario Legislature and their discussions with the Ministry of the Environment about the appropriate agreements for co-operation are most encouraging. We believe that this type of co-operation is essential for the future. In the *Interim Report*, we indicated that, starting in 1978, the AECB began to make rules and guidelines for all aspects of the fuel cycle. This is a major undertaking and one that must be done with considerable thought. The slow-down in the nuclear power programme suggests that there may be adequate time to prepare guidelines and rules in a manner that is understandable to the public.

We recommend that:

5.23   **Standard-setting for the nuclear fuel cycle should be done in an open manner, including opportunities for public participation in the process.**

We believe that earlier and more active participation than was permitted during the study by the Interorganizational Working Group should be undertaken in the future.

The importance of compliance and monitoring cannot be underestimated. As much attention as has been directed in the past to the assessment of proposals for development should also be applied to the development of programmes for the monitoring and supervision of performance standards. A full regulatory process that includes an extensive programme of compliance could make a significant contribution to the assessment of future projects as well as aiding in interim enforcement. Greater attention to compliance should include greater attention to the role and responsibility of the resident inspector.

We recommend that:

5.24 **The role of the AECB on-site resident inspector should be strengthened and the reports of the inspector should be made public.**

Both the public and the government have a right to know what actions have been taken to ensure the safety of nuclear facilities. We reaffirm our conclusion that the principle of "openness" in the regulatory process is essential to public confidence. Public participation, including access to all relevant information and documents, should be recognized and implemented as quickly as possible.

The regulatory process has concentrated to date, justifiably, on the issue of nuclear reactor safety. Consequently, the independent advisory committees established by the AECB are essentially science- and technology-based.

While we support fully the concept of the AECB independent scientific advisory committees, in view of the compelling social and environmental implications of nuclear power, we recommend that:

5.25 **Advisory committees based on the social sciences should be established by the Atomic Energy Control Board.**

We believe this to be particularly important in connection with the nuclear proliferation issue, the siting of nuclear power stations, and the social questions relating to the management of high-level radioactive wastes. The role of the social advisory committees should be largely educational, and they should pay special attention to the question of ensuring adequate communication between the AECB and the public.

Because Ontario Hydro and the AECL (and indeed all electricity utilities that operate nuclear power stations and nuclear research establishments) rely extensively on the utilization of sophisticated computer codes and fault-tree methodology in reactor safety analysis and pathway analysis techniques in radioactive waste disposal research, there is clearly a strong case for the AECB to strengthen considerably its mathematical and computer section. Special attention should be paid to probability theory and statistical analysis.[50]

To ensure that the people of Ontario can have confidence in the process that regulates nuclear power, we recommend that:

5.26 **Appropriate steps should be taken to guarantee that the AECB has adequate human and financial resources. The AECB, or its eventual successor, must not become a victim of government spending restraints.**

5.27 **The Government of Canada should ensure the separation of the promotional and regulatory aspects of nuclear power by drafting appropriate legislation to replace the Atomic Energy Control Act as a matter of the highest priority. This would ensure that the AECB and AECL would report to separate ministers, reflecting their very different roles, thereby avoiding public confusion and possible conflicts of interest of the sort that have in the past strained public confidence in the regulatory process.**

5.28 **The Atomic Energy Control Board should expand its membership to include a broad representation of the general public as well as members of the scientific and technical community.**

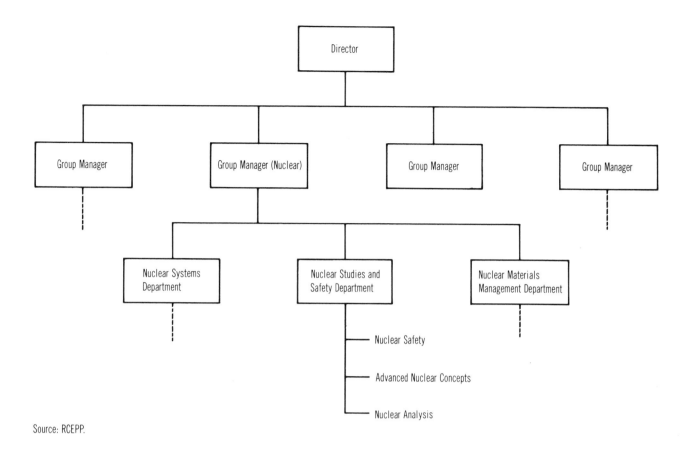

Source: RCEPP.

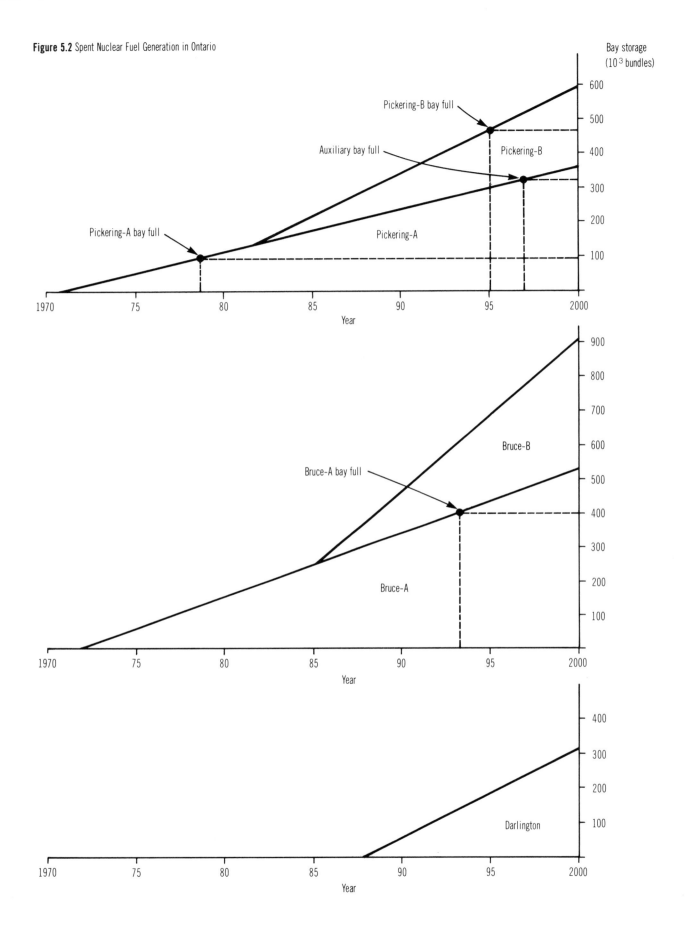

Figure 5.2 Spent Nuclear Fuel Generation in Ontario

Sources: Ontario Hydro and RCEPP.

Figure 5.3 Accumulated Discounted Cash Flow versus Years from 1987 In-Service Date for Three Types of Station (9.75% Discount Rate)

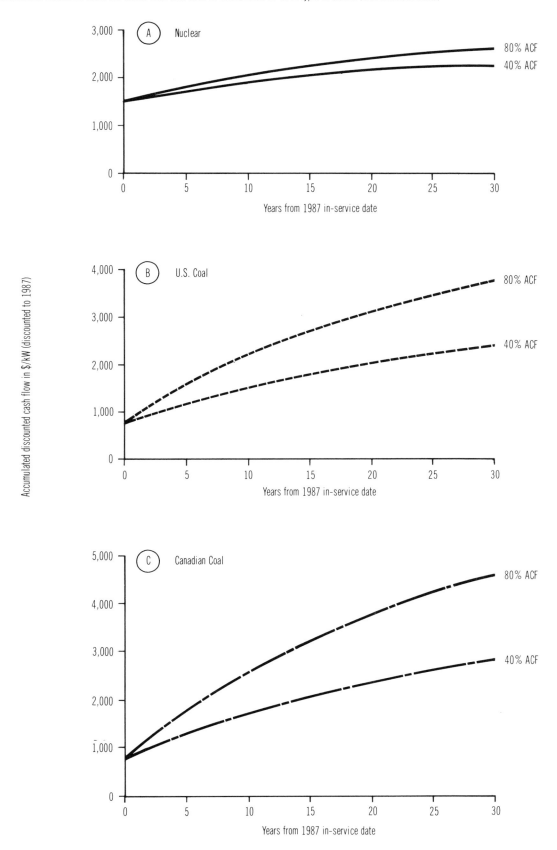

Source: "Cost Comparison of 4 x 750 MW Fossil-Fuelled and 4 x 850 MW CANDU Nuclear Generating Stations", Ontario Hydro, Report No. 584 SP, January 1979.

Figure 5.4 Uranium Supply and Demand (Ontario Hydro)

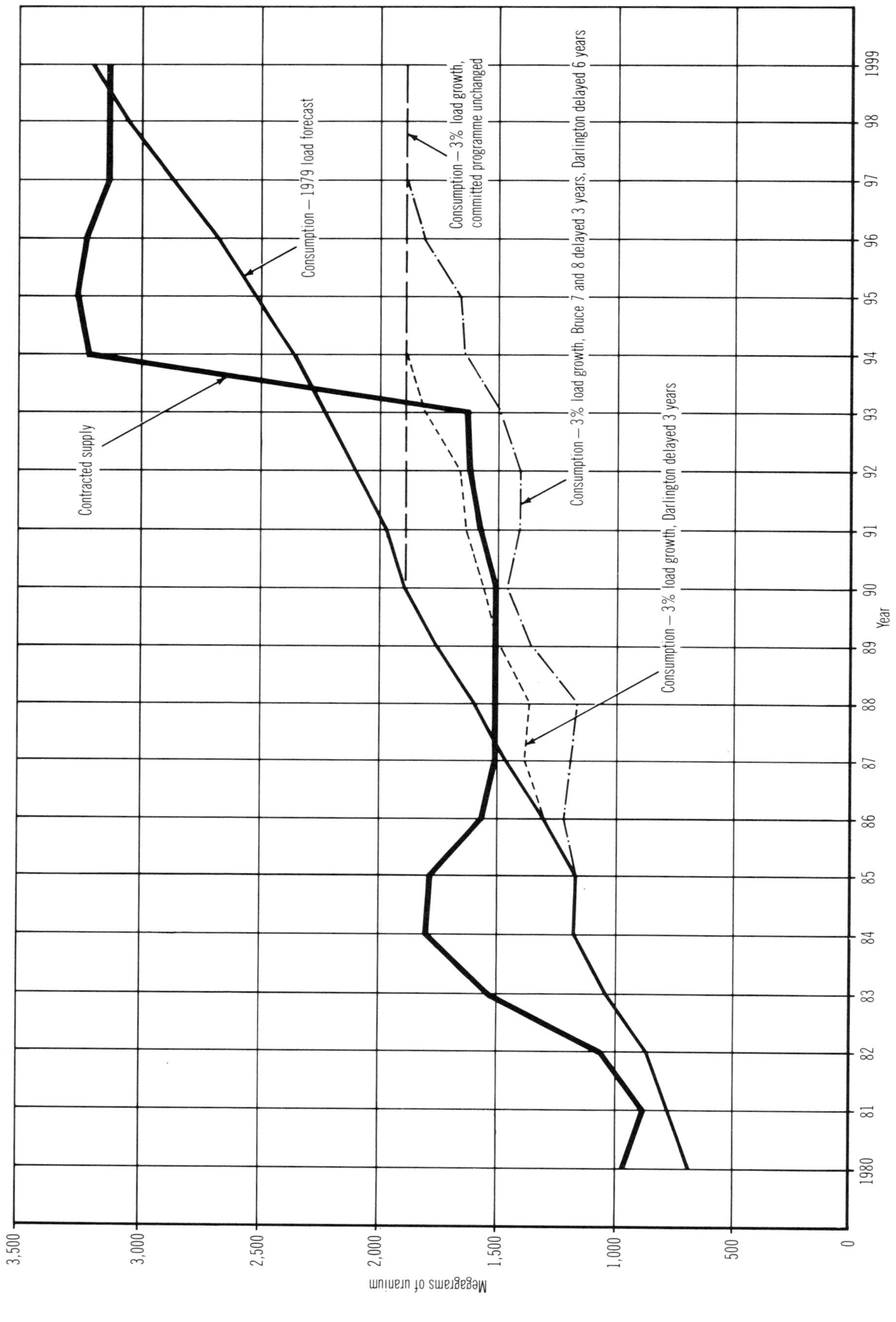

Consumption — 1979 load forecast

Contracted supply

Consumption — 3% load growth, committed programme unchanged

Consumption — 3% load growth, Bruce 7 and 8 delayed 3 years, Darlington delayed 6 years

Consumption — 3% load growth, Darlington delayed 3 years

Consumption — 3% load growth, Darlington delayed 3 years

Megagrams of uranium

3,500
3,000
2,500
2,000
1,500
1,000
500
0

Year

1980 81 82 83 84 85 86 87 88 89 90 91 92 93 94 95 96 97 98 1999

Sources: RCEPP and Ontario Hydro.

CHAPTER SIX

Bulk Power Transmission

Technical Considerations

The single most important reason why Ontario's electric power system is an integrated system, which necessitates bulk power transmission, is to ensure high reliability of supply. Although only a comparatively small proportion of electric power outages have been due to failures in the bulk power system, it is important to note that transmission-line failures may lead to rejection of load on a regional scale. Hence the importance, in planning bulk power transmission systems, of ensuring that adequate redundancy is built into the system in order that failure of a single circuit will not necessarily lead to a regional black-out.

Although reserve generating capacity (capacity that is available over and above the peak power requirements) can be expressed in terms of thousands of megawatts, the concept of "reserve capability" in a bulk power transmission system is much more difficult to quantify. Indeed, the only way of assessing the overall reliability of the system is by simulating a variety of fault conditions, using a large-scale digital computer, and determining whether or not the system remains stable and achieves an acceptable steady state. If, for example, fault conditions due to hurricanes, lightning strikes, etc., give rise to unstable behaviour in the network, whereby surges of electric power may necessitate the virtually instantaneous rejection of major generating units, serious mechanical and electrical component failures may arise. In such circumstances, a major base-load nuclear generating station may be put out of service for extended periods.

The choice of bulk power transmission voltage levels is determined essentially by the amount of power to be transmitted, the efficiency and economics of transmission, and the system configuration.[1] Although extra-high-voltage (EHV) lines are more expensive than lower-voltage lines, their power-carrying capability is substantially higher. Consequently, the number of EHV transmission lines needed between a generating station and the load, and the overall costs, are lower. Under varying circumstances, it may take from four to seven 230 kV lines to equal the power transfer capability of a single 500 kV line, whereas the cost of a 500 kV line may be only twice that of a 230 kV line. There are other issues associated with voltage levels as well. The higher the voltage used, the wider the transmission corridor required and the higher the towers. However, generally, the greater the generating capacity of an electric power system, the greater the need for EHV bulk power transmission (e.g., 500 kV). In the case of Ontario Hydro, the bulk power transmission voltages have been successively 115 kV, 230 kV, and 500 kV, as the system has evolved. At present, while the 115 kV and 230 kV circuits comprise more than 80 per cent of the bulk power transmission circuit miles, the 500 kV circuits are becoming increasingly predominant and are planned for future construction. In other words, as Ontario Hydro's transmission system has grown, in step with the increase in the capacity of new generating stations, there has been an increasing trend towards the use of 500 kV lines for bulk transmission.

Although a small proportion of Ontario's bulk power transmission circuits are underground – notably in the vicinity of densely populated urban areas – more than 99 per cent of the system consists of overhead power lines. Not only are overhead lines appreciably less expensive than underground cable,[2] but they have the inherent advantage of being able to utilize the excellent electrical insulating properties of air to dissipate the thermal energy generated in the lines.

An important additional advantage of overhead lines is the fact that they can be upgraded, whereas, because increased conductor temperatures are unacceptable in underground cables, such upgrading is not feasible for them. Overhead lines can be upgraded by the installation of new conductors with increased power-carrying capacity, by accepting overhead conductor temperatures in excess of established practice, by raising the height of towers, and by the installation of static capacitors at transformer stations. Although these practices result in somewhat lower reliability and may push the bulk power transmission system to its limits, they nevertheless provide a degree of flexibility in system planning. Such upgrading steps, which are at present being undertaken by Ontario Hydro in certain sections of the bulk power transmission system in southwestern Ontario and eastern Ontario, are particularly useful as stop-gap measures pending the installation of additional higher capacity (for example, 500 kV) circuits. Note, in particular, that as the generating capability of the system is increased to deal with growth in the load, it is necessary, in parallel, to increase the load-carrying capability of the bulk power transmission system.

We are encouraged by Ontario Hydro's efforts to date to upgrade existing transmission facilities and thus improve the use of existing rights of way. We believe, however, that this should not be a short-term undertaking but rather a continuing programme aimed at optimizing the use of existing rights of way.

Accordingly, we recommend that:

6.1 Ontario Hydro should continue to undertake research and explore all alternatives that will permit the upgrading of existing transmission facilities and lead to optimizing the use of existing rights of way. Evidence of this research should routinely form part of Ontario Hydro's submission for approval of the acquisition of a new transmission corridor and/or the siting of a new transmission line.

In the normal operation of the system, the bulk power transmission network must have load-carrying capabilities such that synchronism between all the synchronous generators in the system is maintained in spite of major shifts in the load and forced outages of some transmission lines and/or generators. An integrated electric power system such as Ontario Hydro is designed for a high level of stability and, incidentally, for compatibility with the systems of neighbouring utilities. It may not be generally appreciated that Ontario's electric power system copes on a day-to-day basis with a wide variety of faults and disturbances (e.g., those associated with load changes, equipment outages, and weather conditions) and that the monitoring and control systems associated with each generating unit must react within a fraction of a second in order to maintain system stability in spite of such disturbances.[3]

Reference to system stability brings up also the potential of DC bulk power transmission. All the early electric power systems used DC. The advantages of such systems are that they are inherently stable, they are economical in the use of conductors, smaller towers are possible, there is undoubted economy in land utilization, and the aesthetic impact is less.[4] The main disadvantage of DC bulk power is its lack of flexibility (i.e., the inability to change voltage levels by transformation).

As power systems across the world grew in size, transmission voltages increased accordingly, and this inflexibility of DC, coupled with the infeasibility of generating and utilizing electricity at very high voltages, led to a virtual extinction of DC power systems. Recently, however, there has been renewed interest, not in DC generation or utilization, but in high-voltage DC transmission because of the advantages cited earlier. There are a number of high-voltage transmission systems operating around the world. The Nelson River system of Manitoba Hydro is a notable example.

The terminal equipment for conversion of AC to DC at the generating station and from DC to AC at distribution transformer stations is still costly although, because of the advent of solid state conversion devices, the comparative costs may decrease in the future. It is noteworthy that DC transmission is necessary to provide an asynchronous tie-line between two incompatible systems such as those of Ontario Hydro and Hydro-Québec.[5] A case can be made, also, for employing DC bulk power transmission when the transmission distances are in excess of 650-800 km.

Accordingly, we recommend that:

6.2 Given the advances in converter technology that suggest that high-voltage direct current (HVDC) transmission has now become economically attractive for distances in excess of 650-800 km, Ontario Hydro should carefully re-examine the advantages of HVDC for the proposed east-west interconnection and study its application for the line connecting the proposed Onakawana Generating Station with load centres in southern Ontario.

With respect to the technical characteristics and performance of Ontario Hydro's bulk power transmission system, we have concluded that:
- The transition from 230 kV circuits to 500 kV circuits for most of the new bulk power transmission lines is fully justified on grounds of system efficiency and minimum cost. Except in special circumstances, for example, in the design of tie-lines between isolated systems, DC bulk power transmission is not a viable alternative for Ontario in spite of some of its undoubted advantages.
- With the exception of limited requirements for underground cables in certain urban areas, the bulk power transmission system should continue to be based on overhead transmission lines.
- Although, during recent years, the additions to Ontario Hydro's generating station complement have been large-capacity nuclear and coal-fired stations, the power transmission system must handle, in addition, a broad spectrum of generating unit capacities. To enhance system resilience it will be desirable in the future to consider the incorporation of comparatively small-capacity

generating units (e.g., hydroelectric units, co-generation units, and eventually, perhaps by early in the next century, units based on solar photovoltaics and fuel cells). From the standpoint of increased efficiency of utilization of fuel, economic advantage, and enhanced system resilience, important trade-offs must be taken into account between small and large units. The question was raised in the Commission's *Report on the Need for Additional Bulk Power Facilities in Southwestern Ontario.*

The Siting and Routing Process

Ideally, the siting of generating stations, switching yards, and transformer stations, and the routing of bulk power transmission lines should be regarded as a single systemic planning problem. But, inevitably, when an electric power system such as Ontario Hydro evolves with an exponential rate of growth, the planning process itself must be flexible, and a range of alternatives will be on the drawing board at any particular point in time. The detailed site location and route selection processes, for example, are complex tasks that have geographic, demographic, economic, environmental, political, and complex technological dimensions. Noteworthy, too, are the provisions for the protection of man and his environment. Before Ontario Hydro's proposals can be put into effect they are subject to such government legislation as the Planning Act, the Expropriation Act, and the Environmental Assessment Act. For example, several tribunals may be involved in:

- an assessment of the need for the additional facilities
- an assessment and evaluation of alternative methods or routes for meeting the technical and social requirements
- an environmental impact assessment, guided by the Environmental Assessment Act's definition of "environment"

The central problem in the routing of high-voltage transmission lines is the identification of alternative strips of terrain, or "bands", each of which satisfies environmental, economic, and technical criteria. It is particularly important to note, moreover, that the final choice of a band must be based not only on the immediate requirement but also on the location of potential generating facilities and associated transmission lines. This is only good planning practice. There may, however, be important social consequences. For example, the question has been raised by regional communities that may be affected by a bulk power transmission line as to the influence the projected line will have on future generating station site selection and on urban development. In fact, there is concern that some routings of a second line out of the Bruce Nuclear Generating Station could encourage the development of similar large stations, as well as industrial growth, in areas noted for their excellent farmland.

While other linear systems such as expressways, oil and gas pipelines, and even railways, have comparatively low profiles, a 500 kV two-circuit transmission line incorporating towers 50 m high can be seen for a distance of several kilometres. The land beneath the power lines cannot be used for expressways or railway lines, but agricultural use can be made of it, and in the United Kingdom and several European countries, homes, farmhouses, and other low buildings are built under high-voltage transmission lines. The latter practice is not permitted in Ontario.

The basic concepts and procedures involved in route selection may be summarized as follows:
- determining the origin and destination of the transmission corridor
- defining the associated area to be studied in depth, i.e., the band[6]
- drawing up a detailed inventory of environmental factors, including the natural features and topographical characteristics
- plotting these environmental factors on a series of maps and choosing the criteria for the selection of the route (The route-selection criteria may be established on the basis of appropriate weighting of the various environmental factors.)
- public participation in the route-selection process

We have concluded that Ontario Hydro's procedures in the above respects, with the exception of public participation, are effective and we have no specific comments. Concerning the role of the public in the route-selection process, however, we have concluded that it is not adequately defined and does not appear to be as effective as it should be. (We will deal with public participation in such decision-making processes in Chapter 12.) Several guidelines have emerged; perhaps the most important is that the public should be brought into the selection process at the earliest possible stage and that all relevant information should be made available. It is noteworthy that the public participation aspects of transmission line routing are proving to be the most intransigent of the problems confronting the Route and Site Selection Division of Ontario Hydro.

Accordingly, given that transmission routing will be the major undertaking of Ontario Hydro in the next decade, we recommend that:

6.3   Ontario Hydro should utilize even more imaginative approaches to public involvement in transmission routing. In particular, we believe the utility should leave more of the initiative in the public participation process to affected citizens, permitting those who will be most immediately impacted and involved to select alternate routes and to designate the preferred route; independence will be essential. The chairman of an appropriate citizens study committee should be selected by the citizens. Ontario Hydro should clearly state its criteria for routing, and this information with any other required by the committee should be readily provided by the utility. While the time period for study should be established by the utility, the procedures should be established by the study committee.

Health and Safety

During the Commission's hearings, especially in the agricultural counties of southwestern Ontario, concern was expressed repeatedly about the possible deleterious health and environmental effects due to high-voltage transmission lines. Increasing the voltage of bulk power transmission lines from 230 kV to 500 kV, and the ever-increasing number of circuit kilometres required as the power system grows, has exacerbated the concern. High-voltage transmission lines, under certain circumstances, give rise to the production of ozone, to audible noise, to interference of radio signals, to electric shocks, and, of most serious concern, to possible biological effects as a result of exposure to comparatively high intensity electric fields.[7]

Especially when the level of humidity is high, a high-voltage transmission line can generate a so-called "corona" effect, which is characterized by a glow; the corona discharge is usually most pronounced at points of attachment between transmission line and insulator. It is well known, moreover, that corona discharges are accompanied by audible noise and can give rise to radio and television interference. Further, corona discharges produce traces of ozone in the vicinity of the discharge. During the last few years, however, as a result of improvements in conductor and insulator designs, the occurrence of corona discharges has been considerably reduced and this trend is expected to continue. The level of production of ozone arising from corona discharges is now accepted as being negligible and not comparable to the appreciably higher levels of ozone due to other causes (see Chapter 9).

Of a quite different nature, and by no means fully understood at present, is the possible biological impact of lengthy exposure of living organisms to the electric fields resulting from high-voltage transmission lines. It is well known that the behaviour of all living organisms, and particularly the electrical activity in cell membranes and motoneurons, depends to some extent on inter-cellular electrical potentials. Consequently, it is conceivable that if an external alternating field is superimposed on the naturally occurring fields, undesirable biological effects may occur. It is noteworthy, for instance, that living organisms are susceptible to damage by very-high-frequency electromagnetic radiation (e.g., X-rays and gamma radiation), but, because the frequency of the electromagnetic radiation due to high voltage AC transmission is very low, we would expect the biological effects due to associated electromagnetic radiation to be negligible.

Is there any experimental evidence of such effects? One of the major difficulties in carrying out experimental work of this kind is the setting up of controlled experimental conditions. It is extremely difficult, for example, to differentiate between a range of potential causes and their effects. However, some health disturbances have been reported among Russian switchyard workers exposed to high-voltage fields for long periods, and among a few workers in Spain, and several countries have launched research projects to study the problem.[8] The U.S.S.R. and Spanish studies were not based on comparisons between "exposed" and "controlled" groups, and they have been criticized on the grounds that statistical analysis of the data was not possible. Furthermore, the observations (both studies were undertaken with a comparatively small number of workers) were not confirmed by similar studies that were undertaken in other countries, notably Sweden and the U.S.

In 1975, the Department of Preventive Medicine and Biostatistics, University of Toronto, was commissioned by Ontario Hydro and the Canadian Electrical Association to undertake an in-depth study of the health impacts, if any, of the exposure to high voltage transmission lines of a group of linemen with an average of 11 years of service. The study employed an "exposed" group consisting of 30 maintenance workers and a control group of another 30 workers of matched ages, who had not been exposed to high-voltage fields. A pair of workers (one exposed and one controlled) attended each week for comprehensive medical examinations (which took two days to complete). The men were examined by consultants in psychiatry, internal medicine, neurology, and psychology, and underwent electrocardiographic and electroencephalographic examinations as well as extensive biochemical and blood studies. The medical assessments of the workers were undertaken by consultants attached to the teaching hospitals of the University of Toronto. The detailed findings of this comprehensive study, which has taken four years to complete, are now available in a final report that has been submitted to Ontario Hydro.[9] The conclusions of the University of Toronto study, which is probably the most detailed study undertaken to date on the health effects of exposure to extra-high-voltage power lines, were summarized as follows by the investigators:

> The conclusion of this study is that among 30 men exposed to extra high voltage and high voltage power line and switchyard environments, no difference in health status could be determined when compared with 30 age-matched controls drawn from maintenance workers in the same utility. The findings are in keeping with those of a Swedish study reported in 1978 and of the study of 10 United States linemen reported earlier. The findings are at variance with the Soviet and Spanish studies of switchyard workers. It appears that in none of the studies have men exposed only to the high voltage environment of transmission lines alone been found to have unusual medical complaints or findings. The complaints and medical findings in the Soviet and Spanish switchyard workers appeared to be mild in nature and the part played in their causation by the electric field, in contrast to other possible causative agents, remains to be defined.

During 1977, the Commission undertook a survey of 12 major high voltage laboratories (located in Holland, Italy, Sweden, the United Kingdom, and the United States) to ascertain the extent to which laboratory technicians and other workers exposed to electric and magnetic fields had reported occupationally related sickness; in addition, we requested information on comparative absenteeism and family statistics. The responses, without exception, indicated that exposure of workers to appreciably higher-than-average electric and magnetic fields, in some cases over many years, had not apparently caused harmful health effects; on a statistical basis exposed and non-exposed workers were indistinguishable with respect to absenteeism and family size. However, it is important to stress that, while this information is encouraging, it has no scientific validity.

The Commission, recognizing the obvious importance of this subject, has obtained information also from several laboratory-oriented investigations, notably the work of Dr. R.O. Becker and Dr. A.A. Marino at the University of Syracuse, on the effects of induced electric currents both on bone and on the central nervous system, undertaken with laboratory animals.[10] Because of the extreme complexity of the electrical phenomena associated with biological cells, and the difficulty of isolating specific causes of certain biological impacts, we hesitate to accept the findings of these experiments as evidence of the potentially harmful effects of transmission lines on man. In particular, too many uncertainties are involved in extrapolating the results of experiments with laboratory animals, trees, plants, etc., and applying them to human beings.

On the basis of the broad range of epidemiological studies that have been undertaken, relating to the impact of high-intensity electric fields on human beings, we have concluded that, while there may be minor health disorders of a transitory nature, there is no evidence of harmful health impacts on either linemen or switchyard workers exposed to 500 kV lines and equipment. However, we support the conclusion of the University of Toronto research team to the effect that if further epidemiological studies are required they should be undertaken on an international basis and the research team should include investigators from Spain and the Soviet Union.

Is it safe to operate farm machinery under high-voltage transmission lines? This question was raised many times during the Commission's public meetings and hearings. We have concluded that the answer is in the affirmative if all precautions relating to the adequate grounding, and in some cases insulation of the agricultural appliances, are taken – this is particularly important in respect of spraying equipment. It is well known, for example, that inadequate grounding of metallic fences and objects located close to transmission lines can give rise to shocks when these objects are touched by a human or an animal.

We have concluded that the majority of scientific evidence to date indicates that normal exposure of people to the electrical fields in the vicinity of high-voltage transmission lines is not harmful.

Accordingly, we recommend that:

6.4 Ontario Hydro should take all possible steps to ensure the safety and convenience of all persons working in the vicinity of extra-high-voltage transmission lines.

6.5 Ontario Hydro should continue to plan the integrated electric power system on the basis of 500 kV and 230 kV transmission lines.

Environmental Implications

The siting and routing of a specific bulk power transmission line is essentially a problem of land use and environmental impact assessment. The complexity of the problem is exemplified when we reflect on the comparative environmental impacts of routing a transmission corridor through:

- a centuries-old natural marsh
- Class 1 or 2 farmland
- a well-established forest consisting largely of high-quality trees (e.g., marketable hardwoods)

Clearly, a quantitative assessment of the respective environmental impacts of the above is impossible. Nevertheless, in addition to the economic component, the impact on the ecology must be taken into account. The choice must inevitably involve value judgements based on qualitative factors that are often very difficult to define. In contrast, decisions relating aspects of design, routing, and construction of a high-voltage transmission line are comparatively straightforward.

Although the land-use and environmental impacts of the siting and routing of generation and transmission systems are introduced in Chapters 8 and 9, in this section certain environmental considerations are introduced that relate specifically to the design and emplacement of bulk power lines. Since the early phases of our inquiry we have been aware of the psychological impact of transmission lines as well as their health and safety impacts. Indeed, we opened our Issue Paper No. 4 on transmission and distribution with the sentence:

Undoubtedly, the most visible aspect of Ontario's electric power system is the network of transmission lines which criss-cross many square miles of the province.

It has been asserted, and there is some supporting evidence, that when power lines cross private property, especially farmland, there is an inevitable reduction in the value of the property.[11] The reasons for this appear to be:

- The conventional self-supporting lattice towers, which in the case of 500 kV double-circuit towers may be 50 m high, and correspondingly lower for lines of lower voltage, have, in the view of the majority of people attending our hearings, a negative aesthetic impact. In particular, it was suggested that, in cases where more than one transmission line must be located in an area, a single multi-line transmission corridor should be constructed to minimize the aesthetic impact as well as the costs.
- Transmission-line construction inevitably necessitates a degree of soil compaction, and it may take several years for the soil to recover. During the last few years, Ontario Hydro has succeeded in minimizing this problem by new line-stringing procedures, but there is no doubt that this problem, together with that of ensuring ready access to the power lines, inconveniences some farmers.
- Especially with the advent of very large mobile agricultural machinery (up to about 20 m in length and 5 m in height) there is clearly inconvenience to farmers undertaking operations in the vicinity of high-voltage transmission towers because of the difficulty of by-passing the tower bases.
- The difficulty of eradicating weeds growing in tower base areas was also mentioned as a problem for some farmers.

On the other hand, in the case of some suburban residential subdivisions, it has been argued that the presence of a high-voltage line ensures for some home-owners a degree of privacy and a more open environment. Psychologically, a power line may have a more positive impact than several adjacent rows of houses.

Taking the above impacts into account, we have concluded that further development is required in the design of transmission towers that are more compatible with the land uses and environments into which they must be fitted. An example of this is the newly designed lattice steel tower for use in orchards and vineyards which has a 10-foot clear elevation under the centre part of the tower to permit most types of equipment to pass underneath.

The steel single shaft tower should, we believe, be the benchmark against which other designs are evaluated. We acknowledge the higher cost of these towers, but, essentially for aesthetic reasons, for reasons of convenience, and for reasons of greater acceptability, we believe that in the long run the additional cost will be largely offset by reducing delays that now arise from lengthy public hearings. The indications are that the single-shaft tower is more acceptable to a public conscious of the natural environment than the alternative designs. We have noted, in particular, that some United States utilities, for example the Wisconsin Electric Power Company, are abandoning lattice towers in favour of single-shaft towers largely for the reasons outlined above.

Accordingly, we recommend that:

6.6   Ontario Hydro should work with the appropriate farm organizations and the Ministry of Agriculture on the design of an appropriate single-pole and/or lattice tower for use in cultivated fields.

Technological Innovation

The spectacular advances in electric power generation technologies, and especially in nuclear power, usually overshadow significant though less spectacular advances in high-voltage transmission, transformation, and switching. Some examples of the new concepts and technologies at present either in place or in process of development are:

• Widespread use of computer analysis to determine the characteristics of bulk power transmission networks and to assess the technical feasibility of alternative transmission line configurations. Computer analysis is also being used increasingly to facilitate the assessment of the economics, as well as to a limited extent the environmental impact, of alternative transmission routes.

• With the introduction of sulphur hexafluoride (SF_6), a gas with exceptionally good insulating properties, the possibility of using underground cables is increased. Although SF_6-insulated underground cables cost appreciably more than equivalent overhead transmission lines, it is anticipated that these cables may become competitive by the end of the century. We have already mentioned their respective advantages and disadvantages. Further, research programmes aimed at the development of so-called cryogenic cables (based on the principle that the electrical resistance of some metals and alloys is virtually zero at very low temperatures – in the order of -260°C) have been in hand for more than a decade. The chief virtue of these cables is that they virtually eliminate transmission losses, and they offer a considerable economic incentive for that reason alone. However, the cost of installing and maintaining cryogenic cables precludes their large-scale commercial utilization probably until the beginning of the 21st century.

• As well as the potential of SF_6 as an insulator for underground cables, this gas is being used increasingly in transformer and switching stations, especially in Europe. Ontario Hydro is installing new gas-insulated transformer stations at Clareville and Milton. Because SF_6 has at least a 10 to 1 advantage over air in its insulating properties, it is now possible to design new stations with major reductions in land use and hence in visual environmental impact. Again, because of the reduction in scale of gas-insulated switch-gear and other components, and the reduced separation distances, it is possible to house most of the components in buildings rather than in the open air as with air-insulated stations.

• Advances in the design of conductors and insulators to improve their mechanical and electrical characteristics are ongoing and are aimed at minimizing line outages, especially as a result of adverse weather conditions.

• Due to the recent developments in the reliability of semi-conductor rectifiers, increasing attention is being focused on DC transmission. When the cost of terminal stations for rectifying and modulating the electric power decreases, as it probably will, DC transmission will become increasingly attractive. Its main advantages are that fewer conductors are required compared with three-phase AC transmission, towers that are smaller and shorter and on narrower rights of way will suffice, stability problems are eliminated, and the safety hazards are probably less serious. Both B.C. Hydro and Manitoba Hydro are utilizing DC transmission successfully, as also are several European countries, notably West Germany and Sweden. We have concluded that research and development in all the areas mentioned above, with the possible exception of cryogenic underground cables, should be pursued by Ontario Hydro.

Not yet mentioned is the possibility of "transporting electric power" in the form of liquid hydrogen. When water is electrolysed it can be dissociated into hydrogen and oxygen. It has been suggested to the Commission by several intervenors that high-voltage transmission lines might eventually be replaced by pipelines carrying liquid fuel based on hydrogen generated at large-scale electrolysis plants. While recognizing the conceptual merits of such a hydrogen-based economy (indeed, one in which electricity might eventually replace all non-renewable fossil fuels, through the hydrogen cycle, as the primary energy source), we believe such a development is unlikely, on a large commercial scale, for at least 30 to 50 years.

The Southwestern and Eastern Ontario Regional Hearings: Supplementary Findings and Recommendations

During March and April 1979, the Commission conducted a series of public hearings in connection with the need for additional bulk power facilities in southwestern and eastern Ontario. These were in response to Paragraph 4 of the Terms of Reference of the Commission, which was amended and supplemented under Order-in-Council No. 2065/78 dated July 12, 1978. The amendments required the Commission to study the need for additional bulk power facilities in specified geographical areas. The reports relating to the southwestern Ontario and eastern Ontario studies were published on June 13 and July 23, 1979, respectively. We stress that those reports should be read in conjunction with this Report, the reason being that only with the availability of the latter can the need for additional bulk power facilities in the specified regions be put into the perspective of the whole inquiry. Because of the excess generating capacity available in the system during 1979, and expected to continue to be available in varying degrees for at least the next eight or 10 years due to the reduced load growth, the regional studies were concerned essentially with the strengthening of the bulk power transmission system in several respects.

At the time of preparing the regional reports, the Commission had almost completed its studies on the projected growth in electric load in the province to the year 2000. It was clear that Ontario Hydro's forecast rates of load growth, especially in southwestern Ontario, were higher than the Commission's growth projections for the system as a whole. Indeed, because the timing of the need for additional bulk power transmission facilities, especially in southwestern Ontario, was based essentially on the utility's load forecasts, the Commission concluded that delays of a least two or three years beyond Ontario Hydro's proposed in-service dates for the new facilities would not unduly prejudice the security of electricity supply to the region.[12] Further, several important transmission lines are in the process of being upgraded and these will ameliorate the security of supply to such cities as London, Kitchener, and Waterloo.

The Commission's southwestern Ontario regional report stressed, in particular, that studies of future configurations for the 500 kV bulk power network should take into account the possibility that dual-purpose generating plants (with combined thermal and electric power outputs, perhaps based on technologies such as fluidized-bed combustion) might be constructed in the vicinity of key load centres. A major reason for this suggestion is that such steps might minimize the need for bulk power transmission in a certain region and thus facilitate the conservation of prime foodlands. Furthermore, because we foresee serious social as well as environmental and economic problems associated with the possible construction of a second 500 kV line from the Bruce Generating Station following any route that crosses the prime foodlands of Ontario, all other alternatives, even if there are apparent economic penalties, should be explored fully before further consideration is given to such a proposal.

Notwithstanding our continuing concern for the protection of critically important agricultural land in southwestern Ontario, we are convinced, from the point of view of the effectiveness of the total system (see Chapter 7), that it is indefensible for power to be "bottled up" at the Bruce Generating Station. Indeed, this was recognized during the southwestern Ontario hearings; a review of the record reveals that many participants considered load and generation rejection to be very undesirable. For example, Pat Daunt, a member of the Food Land Steering Committee, stated:

> I would hate to be responsible for anything that prevents you incorporating . . . eight units [at Bruce G.S.]. I think there is a real problem there, and I know common sense tells me you have, and you are, looking for a solution to it.[13]

Accordingly, we recommend that:

6.7   **The farming community with the collaboration of Ontario Hydro should develop, as soon as possible, alternative routes for a second 500 kV transmission line from the Bruce Generating Station that will have minimal and acceptable impact on Class 1 and Class 2 agricultural land. Ontario Hydro should provide the necessary funding.**

6.8   **In order to facilitate the co-operation of the farming communities, Ontario Hydro should not site a thermal generating station in the vicinity of Goderich or Kincardine, or indeed on the eastern shoreline of Lake Huron south of the Bruce Generating Station, before the year 2000. Ontario Hydro should make a public statement to this effect as soon as possible.**

In the southwestern Ontario report we pointed out that, because of limitations on the capability of transmission circuits into Buchanan Transformer Station, Ontario Hydro's capability to move power to United States border points, assuming normal security criteria on its system, would drop from 1,600 MW in 1979 to 900 MW by 1986, if loads in southwestern Ontario grew at the 4.3 per cent rate predicted in the 1979 load forecast. Under the most severe design contingency, export capability could drop to zero by about 1990.

Consequently, we concluded that, if no capability to export power is maintained and loads grow at Hydro's assumed rate, new transmission facilities would not be needed until 1990. In this case there would be a two-to-three-year breathing space before transmission plans would have to be presented for environmental review and approval. We contended that Hydro could make good use of this period to study the prospects for industrial co-generation in the industrial centres of the southwest. These studies, we understand, are under way. However, on the basis of the Commission's overall review of the total system, we have concluded that it is not in the best interests of the people of the province to allow capability of exporting surplus power to United States utilities to dwindle or to reduce the level of potential emergency support from these utilities.

As for the need for additional bulk power facilities in eastern Ontario, the Commission concluded that the studies Ontario Hydro is making in respect of the strengthening of three major transmission lines should be continued. We noted, in particular, that the electric power supply to Ottawa, even assuming only a moderate rate of growth of load, may be in jeopardy by 1983. While the upgrading of the appropriate transmission lines will provide adequate security of supply for the next few years, we agreed that more permanent strengthening, by the construction of new facilities, will be needed before 1990. We further concluded that a strengthening of the 230 kV bulk power transmission network in eastern Ontario would also enhance the possibility of considerably strengthened interconnections between Ontario and Quebec and between Ontario and New York State. Interconnections are considered in more detail in Chapter 7.

The Ontario Government has confirmed the Commission's finding of a need for additional bulk power facilities in eastern Ontario and has, through the Minister of Energy, requested Ontario Hydro to prepare for the next step in the approval process. We understand, further, that Ontario Hydro is preparing a system plan for the reinforcement of its transmission system in eastern Ontario, together with environmental assessments of the specific projects under consideration. We believe that the approval process should proceed as expeditiously as possible because, as stated in the eastern Ontario report, "at any rate of load growth exceeding 3 per cent per annum, the normal delivery capability of the [Ottawa area] system will be exceeded before new facilities can be provided... there are real problems with the supply of power to the Ottawa area."[14]

CHAPTER SEVEN
The Total Electric Power System

The systems concept is ubiquitous in nature and in the man-made environment. The solar system, the earth's ecosystem, the human nervous system, the electric power system, the telephone system, etc., exemplify the holistic nature of systems insofar as their behaviour is predicated on interactions between components and on internal and external interdependencies. Systems are responsive to both their internal and their external environments. They are purposive, or goal-oriented, in that their total behaviour is aimed at achieving a specific purpose or purposes. Especially in the case of man-made systems, of which the electric power system is a particularly important example, it is necessary to review from time to time the purpose of the system, the extent to which that purpose is being achieved, and the public's perceptions of the system's performance in a continuously changing environment. It is noteworthy that an important role of this Commission has been to assess the public's perceptions of Ontario Hydro and to assess the degree to which the utility is achieving its purpose as set out in its mandate.

Although Ontario's electric power system is usually identified exclusively with Ontario Hydro, we must bear in mind that important contributions to the system are made by several privately owned utilities, notably the Great Lakes Power Corporation and Canadian Niagara Power, as well as by many industries – these amount to a small percentage of the total. Further, for planning and administrative purposes, Ontario Hydro's system is conveniently split into two systems, joined by a tie-line. The "East System", by far the larger, serves the geographic regions of southern, eastern, and most of northwestern Ontario, while the "West System" services western and northwestern regions of the province. The demarcation line between the systems corresponds roughly to a north-south line passing through the community of Wawa, north of Sault Ste. Marie. The East System bulk power system, discussed in Chapter 6, is an integrated grid system, while, in contrast, the West System is essentially a linear system. In this Report, however, because we are concerned with planning concepts rather than with detailed planning processes, it is desirable to consider Ontario's electric power system as a single entity.

The major components of Ontario Hydro's electric power system are people, both employees and consumers, generating stations of various kinds, bulk power transmission lines, transformer and switching stations, distribution networks, and a complex information processing and control system. The power system can be represented schematically as shown in Figure 2.2, Volume 2. The diagram shows the various stages of transmission and transformation that exist between the generating stations and the customers. The large direct users are large industrial establishments such as Dow Chemical, Imperial Oil, Union Carbide, and others. The distribution of power to other customers in Ontario is the responsibility of public utility commissions; however, most rural areas are served directly by Ontario Hydro.

High technology, in every sense of the term, is manifest in the electric power system – technology that incorporates some very powerful tools, such as the tool of nuclear power. Although this technology is obviously central, it must always be borne in mind that the social, environmental, and political implications also have profound significance. Technology, *per se*, has been defined as the organization of knowledge for the achievement of practical purposes. In the case of Ontario's electric power system, this definition suggests that the system far transcends the machines, the energy conversions, the energy flows, and the vast multiplicity of end uses of electricity in the province. There is the implication that electric power technology has a pervasive influence on virtually all our institutions and on many of our values. Indeed, this pervasive influence on our lives and our culture is an increasingly intangible aspect of electric power planning, and it is the basic reason why the input of the general public to major power decisions is so important. Not surprisingly, it is the uncertainties that underpin the planning process manifest, for example, in the increasingly long lead times required to put major generating facilities into service and in the growth in demand over this period, that exacerbate not only the complexity of the planning process but also public concern. Electric power planning must, therefore, have strong technical components and just as strong social and environmental components. In this chapter we provide a technical framework for the essentially social and environmental aspects of electric power planning that are introduced in subsequent chapters.

Operational Concepts and Characteristics

Load Patterns

The load on a large electric power system, such as Ontario Hydro, varies from second to second; at any specific instant it is essentially a random quantity. However, for reasons related to societal life-style and seasonal weather changes, we can readily identify characteristic daily load patterns (the load peaks during certain hours of the day and drops to lower levels during the night and during weekends) as well as seasonal load patterns (in Ontario, for example, as is the case with all Canadian provinces, the loads during winter are higher than during other seasons). The hourly, daily, and seasonal variations of the load are shown in Figure 2.4, Volume 2. The aggregate generating capacity, together with adequate bulk power transmission and ancillary equipment capacity, should be sufficient to meet the annual peak load, and in addition there should be a level of reserve capacity to cope with both anticipated and unpredictable generating station and line outages.

Load Management and Control

Load management is related to conservation in that the more efficient the utilization of generating facilities, the more efficient the utilization of primary fuels, and consequently the more effective the conservation of non-renewable resources of energy.

The primary purpose of load management is to flatten the load profile and thereby reduce the magnitude of peak power requirements while increasing the load during off-peak periods. If the utilization of peaking generating facilities, for example, oil- and gas-fired generation, can be reduced with a consequent reduction in capital costs and in the use of high-cost fuels, then clearly there is an economic advantage. Load management can also reduce the need for cycling on large base- and intermediate-load stations.[1] In the long term, load management will increase the share of high-capacity-factor generation such as nuclear and base-load hydraulic. In the shorter term, however, it may delay the need for some base-load capacity.

The methods of load mangement may be classified broadly into three categories: pricing schemes and incentives, load control, and load reduction through voltage reduction, customer appeals, load-shedding, and rationing. Pricing schemes such as time-of-use rates can play a significant role in the long term in shifting peak loads to off-peak hours. The marginal cost of supplying an additional kilowatt hour is, in general, higher during periods of high demand than during periods of lower demand. Thus, if the pricing scheme is based on this concept, there will be incentives to the customer to move some of his consumption from peak hours to off-peak hours. Marginal cost-pricing schemes have received considerable attention among North American utilities in the last decade.[2] One pioneer is the Long Island Lighting Company (LILCO), whose time-of-day rates (a form of marginal cost-pricing) for commercial customers with demand higher than 750 kW have been in effect since February 1977. LILCO is also planning to expand this application to residential customers who consume more than 45,000 kW·h per year. Ontario Hydro is studying marginal cost-pricing for its system and the study was reviewed by the Ontario Energy Board. We believe that time-of-use pricing is a desirable approach for Ontario, although the implementation of time-of-day rates can be deferred some years for reasons discussed in Chapter 11.

One form of load management through rate incentives has been practised in Ontario for many years in the form of the interruptible service offered by Ontario Hydro to its large industrial users. This service, which is offered at reduced rates, can be interrupted in system emergencies or for reasons of economy. The total interruptible load in Ontario Hydro's system is about 750 MW.

Load control refers to direct control of customer loads by the utility, for example, the control of residential water heaters. It has been used by many municipal utilities in Ontario to a limited extent. Load control, like interruptible service, can be exercised for economy as well as in emergencies. The residential hot-water heater is an example of economic load control, whereas decorative lighting and residential clothes-dryers are appropriate targets for emergency load control. An indication of the potential for load control may be obtained by noting that, at present, 7 to 8 per cent of electricity used in Ontario is consumed by residential and commercial hot-water heaters.

The third approach to load management is used in extreme emergency situations. The order in which it is imposed is, voltage reduction, customer appeals, load-shedding, and load-rationing. Voltage reduction and customer appeals were used by Ontario Hydro during the 1976-7 winter. Load-shedding is practised routinely in many developing countries. Load-rationing was used in the United Kingdom in 1976 and in California in 1974 and 1976.

In Ontario, experience indicates that, provided advance notice is given, there are essentially no negative customer impacts as a result of voltage reduction and that a 6 per cent voltage reduction can lower peak demand by about 2.7 per cent. Consequently, this appears to be an effective load-management technique and one that does not jeopardize system reliability.

The experience during the 1976-7 winter also showed that consumer appeals during a serious emergency are effective — a 250 MW load reduction was effected by consumer appeals through the mass media. This is in addition to load reduction through industrial appeals, which are more predictable primarily because of the close contacts between Ontario Hydro and its large industrial customers. This potential is estimated to be between 400 MW and 600 MW. Accordingly, the total load reduction possible through consumer appeals represents about 5 per cent of the system demand. However, this potential can only be realized on an emergency, short-time basis; much less would be available on an extended or regular basis.

An example of a load-shedding procedure is the one installed by Ontario Hydro in the Ottawa area. This will shed up to 300 MW, in order to avoid a voltage collapse under a transmission contingency, especially during the period of upgrading work on critical transmission circuits.[3]

We believe that load management will be facilitated when the "digital (chip) watt-hour meter" (see Chapters 10 and 11) is introduced into the system. Indeed, this new approach to electricity monitoring and metering will not only ensure more effective load management but will provide a degree of controlability of electricity use both at the consumer level and at the system control level.[4]

However, it is important to note that load management saves only capacity — not energy. Accordingly, in view of the existing excess reserves (in generating capacity) and the availability of hydraulic peaking capacity, there is little to be gained by the large-scale development of load management in the short run (i.e., to the late 1980s). But later on, if surplus nuclear power becomes available in off-peak periods, the long-run potential of load management will be impressive.

> *Accordingly, we recommend that:*
>
> **7.1   Ontario Hydro, working with the municipal electricity utilities, should give high priority to completing the load-management experiments now under way so that the technical problems, cost, and public acceptability of alternate systems can be assessed.**

System Reliability

The reliability of an electric power system may be defined, briefly, as the system's ability to meet the demand for power while maintaining acceptable frequency and voltage levels. An electric power system is composed of generating stations, bulk power transmission lines, and a distribution network, and the reliability of supply to the customer depends on the reliability of each of these subsystems.[5]

The reliability of a generation system is to a large extent predicated on the availability of individual generating units. Since availability is affected by both planned and forced shut-downs, or deratings of units, generation reliability cannot be stated quantitatively except in terms of probabilities. The most widely used reliability index is the "loss-of-load probability" (LOLP) index. In simple terms, this is the expected number of events over a specific period (say, a year) during which the load on the system will exceed the available generating units. The generation reliability increases with the amount of reserve generating capacity that can be called upon when one or more generating units go out of service. This reserve capacity is usually referred to as "reserve margin", and it represents the difference between the generating capacity and the peak demand, expressed as a percentage of the peak demand.[6]

Until quite recently, Ontario Hydro's policy was to plan generating capacity to meet an LOLP of about 1 in 2,400 in the month of December. This corresponds to a loss of load on one working day in 10 years (on the basis of 240 working days per year).

In response to recommendation III-19 of the Ontario Legislature's Select Committee on Ontario Hydro Affairs in June 1976, Ontario Hydro developed a reliability programme based on the "frequency and duration" method for its System Expansion Programme Review. This programme reduces the reliability target to 10 "system minutes" of unsupplied energy per year and the required generation reserve by 5-7 per cent (from the present 30 per cent to 23-25 per cent). Hydro is no longer using the LOLP of 1 in 2,400 but has not yet formally adopted a new reliability planning criterion. Due to the present large reserve margin, the adoption of a new reliability criterion is not urgent. In the interim, Hydro is using a reserve margin of 25 per cent in planning capacity requirements.

Whereas the availability of a typical large thermal generating unit is in the order of 77 per cent of the time, the availability of Ontario Hydro's bulk power transmission system is generally much higher – in fact, about 99 per cent. A second aspect of reliability, one that is not very significant in the evaluation of generating system reliability but critical in transmission reliability, is "security". The security of a bulk power transmission system is a measure not only of its ability to withstand major disturbances (caused, for example, by the sudden loss of a transmission line) but also to settle down subsequently to an acceptable operating state. The stability of a power system is discussed in the next section.

Historically, only a relatively small proportion of electric power outages have been due to failures in the bulk power system (generation and transmission). However, failure of the bulk power system due to inadequate generating capacity, or to a transmission-line breakdown, may lead to a large-scale rejection of load on a regional scale, while failure of a component or line in the distribution system normally gives rise to a local outage only. A major regional electric power outage is generally the result of a failure of the system to respond rapidly enough to sudden major changes in load or to line outages caused by a storm or by lightning.

While it is comparatively easy to determine the cost of providing electric power, it is much more difficult to assess the economic consequences of interruptions of electric power to specific customers. Perceptions of the cost of a power interruption differ widely among various classes of customers. Ontario Hydro, for example, has recently carried out a customer survey and published the estimated costs of electric power outages of varying duration in respect of selected classes of customers (see Figure 7.1). Fig. 7.1: p. 110 Note the conclusion that the agricultural and industrial sectors of Ontario's economy are the most sensitive to electric power outages.

System Stability

Stated simply, power system stability depends upon the maintenance of synchronism between all the synchronous generators in the system, in spite of a range of major and minor disturbance manifested, for instance, in equipment failures and load variations.

During normal operation, the power system is subjected continually to minor disturbances caused by a multiplicity of load changes (e.g., when consumers switch power on and off); these disturbances are handled automatically by the individual generator excitation systems. Under normal operating conditions when all generators are operating in synchronism, no power is being transferred from one generator to another – the mechanical output of the turbine virtually matches the electrical output of the generator, the speed of which remains constant. However, when the system is subjected to a major fault (i.e., loss of a bulk power transmission circuit), the generators on line react within a fraction of a second and imbalances occur between generators. In this state, the power flows across the system change continuously as the generators try to regain synchronism. If these oscillations are gradually damped out – which normally occurs within a few seconds – the system remains stable. On the other hand, if the imbalances escalate, the system rapidly becomes unstable and synchronism between generators is lost within a few seconds.

To minimize the possibility of such instability, high-speed response circuits are necessary. Suffice it to add that the study of the transient behaviour of a synchronous generator operating in an electric power system is a highly complex subject. For instance, transient stability depends on both mechanical and electrical properties of the system as well as on the nature of the disturbance that gave rise to the transient behaviour in the first place.

If it proves impossible for the control and regulatory circuits to stabilize the system under certain serious disturbances, or if disturbances result in severe overloading of transmission circuits, the only design option available to stabilize the system is "generation rejection", sometimes accompanied by "load rejection". The latter, as the term implies, would give rise to highly undesirable regional blackouts or brown-outs. Generation rejection, also undesirable, is based on a generating unit being tripped (i.e., removed from the system) extremely rapidly with concomitant reduction in the power available to be transmitted; system stability is thereby facilitated. Until comparatively recently, in Ontario's power system, the generation rejection technique was restricted to hydraulic units, which are more rugged than thermal units. But with the growth of the thermal generating component it has become necessary, when the situation demands it, to trip these units as well.

Although the control and regulating systems are designed to withstand massive mechanical and electrical shocks due to generation trips, these have been known to fail, and the consequences can be serious, i.e., they can cause mechanical and electrical damage necessitating major maintenance work and a consequent high probability that the unit will be unavailable over extended periods. Generation rejection is especially undesirable in the case of nuclear units, because if the governors do not work perfectly the reactors may "poison out", with consequent heavy economic penalties until they can be brought up to power again.

Major advances have been made in control technology and generator design, but, because the bulk power transmission network is the means whereby power is transferred through the system (e.g., from generator to load and from generator to generator), it is the network's security that is crucial in ensuring the stability of the total system. It is for this reason that Ontario Hydro is embarking on a programme of implementing certain stop-gap measures to increase the load-carrying capacity of certain critical circuits in eastern and southwestern Ontario. These measures include the restringing of critical transmission lines to provide higher current-carrying capacity, operation of the lines at higher line temperatures (up to 150°C from 90°C at present), and the installation of large amounts of static capacitance. However, it is important to note that the use of such engineering solutions to enhance stability, which pushes transmission technology to the limits, also gives rise to a reduction in system reliability.

Control of the Total System

The impressive performance of Ontario's electric power system, as indicated previously, is due in large measure to the fact that the system is an integrated system of generation, transmission, and distribution. In common with all other major electric power systems, Ontario Hydro's system depends on a large number of remote monitoring instruments, telemetry systems to convey the information to control points, human operators, and, of most recent origin, a central computer, installed at the Richview Control Station. Because the electric power system is hierarchical in structure, control is necessary at all levels. For example, the power output, the rotational speed that is synonymous with frequency, and the output voltage of each turbogenerator are regulated automatically; the values of the controlled variables are transmitted to the central control station.

It is essential for the supply to match the demand instantaneously. To carry this out expeditiously, as well as economically, the load must be apportioned among the available generating units. For example, when the load increases, additional kilowatts must be supplied in the most efficient and hence the most economical way; similarly, when the load decreases, the least efficient generating unit in service is taken off line. In the case of a large complex system such as Ontario Hydro's, this task would be virtually impossible without a central control station equipped with large capacity computers. The computers have the following major roles:

- First, there is the "display role", in which the computer, using a large-scale total system display, together with detailed regional cathode-ray-tube displays, can display continuously, on demand, the status of all major generating units, switching stations, and tie-lines interconnecting one area with another. These are dynamic displays that are continuously updated on a real time basis.
- Associated with the operation of the electric power system, a vast amount of data is generated. The computer assembles, processes, and carries out logical operations on these data at all levels of the system. In this way, the human operators are supplied with integrated status information that provides them with insights relating to system performance in an on-going way. This "distilled" information is vitally important in the assessment of the security of the system.
- The central control computer has a significant role to play in decision-making. For example, in the minute-to-minute operation of a complex power system, decisions relating to the bringing into and taking out of service of generating stations have to be made. The allocation of generating units at a particular time must be predicated on the minimum cost criterion, but without at the same time prejudicing the security of supply. Indeed, at all times, security margins must be maintained from both the generating standpoint and the transmission-line capacity standpoint. Further, the assessment of the security of the system on a minute-by-minute basis must take into account the system's ability to withstand sudden disturbances such as a major transmission-line outage.
- An increasingly important role of the central computer, in controlling the power system as a whole, is to check the telemetered data from remote locations for consistency (e.g., real and reactive power flows in lines, the voltages at critical points, and the status of circuit-breakers). In effect,

these variables can be utilized to set up a model of the power system in the central computer. This is a valuable adjunct to computer control. It not only enhances system security but is valuable also in the training and updating of human operators.

• The primary purpose of the central control system (the Richview Control station) is to facilitate the economic allocation of generating capacity. Because of the complexity of the computations and the short time available, it is obvious that human operators, without the computer, cannot undertake this crucial activity. Consequently, it is a truism that the central control facilities, by minimizing the total cost of producing power, probably save Ontario's consumers many millions of dollars each year.

The central console of the control centre, by providing information relating to the status of Ontario Hydro's major system components, also facilitates communication with neighbouring utility control centres. Noteworthy are the communication channels between the Richview Control Centre and the control centres of Hydro-Québec, New York state utilities, Michigan utilities, and Manitoba Hydro. This ensures co-ordination in the operation of Ontario's system with the operation of contiguous systems and is of particular significance during emergency conditions. Accordingly, not only pre-planned maintenance outages are taken into account in the planning of power exchanges; unpredictable failure of equipment, especially of a large facility, can also be compensated for with minimum risk to the consumer.

We strongly endorse Ontario Hydro's present operational and control practices and congratulate the utility on having developed probably one of the most effective systems at present in operation anywhere in the world.

Systems Planning Concepts

The planning of an electric power system, although based on high technology, combines both engineering and art. The engineering aspects, backed by scientific principles, relate to the design of major facilities such as generating stations, transmission lines, and switch gear and control systems, as well as of the system as a whole. System design relies increasingly on large-scale mathematical models implemented through highly sophisticated computer programmes. But in spite of this powerful array of design tools and methodologies, the art implications of systems design remain paramount. It is particularly important to note that the art of planning a major electric power system can be largely identified with the process of integrating societal and individual values with scientific facts. Only human imagination and foresight can deal with the subjective aspects of transmission-line routing, or make effective decisions when the main factors are social and environmental in nature. Further, in the operation of a large electric power system, states of emergency, perhaps caused by the breakdown of a large generating station or a double-circuit 500 kV transmission line, inevitably necessitate human judgement and decisions by the control station operators.

With the increasing complexity of society and in view of current world problems relating to the future supply of energy, it should not be surprising that the planning of Ontario's electric power system is being undertaken in an environment of increasing uncertainty. Reflect, for example, on the uncertainty of the future demand for electricity, the uncertainty about the lead times that will be required to complete major facilities already committed (these will probably be increasingly influenced by the time required for public discussion and debate), the uncertainty relating to potential environmental threats due particularly to electric power generation, and, not least, the uncertainty relating to Ontario's, Canada's, and the world's economic growth. It is natural that the load-forecasting process, the ultimate basis for electric power planning, is becoming increasingly difficult. We recall that the basic issues relating to load forecasting and system planning are:

• Are the assumptions relating to future electricity load growth in Ontario acceptable to a broad range of trans-disciplinary specialists and to the general public?

• How adequate are the data used in load forecasting and, in particular, how effectively have electricity end-use patterns been developed and utilized?

• Is the province's projected electricity supply for, say, the period 1990-2000 predicated on a mix of generation technologies that will provide sufficient resilience in the system as a whole?

• Is the planned bulk power transmission system adequate to deal with the most probable scenarios? And will the system cope adequately with probable increases in interconnection capabilities between Ontario and Quebec, Ontario and Manitoba, and Ontario and the U.S.?

Generation Planning Concepts[7]

We agree with Ontario Hydro that a satisfactory programme for the generation of electricity in a large-scale electric power system involves the following steps:

- determine the requirements for new generating sources
- determine the restrictions in the manner in which the requirements can be met
- determine the feasible alternatives, i.e., those that meet the requirements and conform to the restrictions
- compare the feasible alternatives by weighing up and trading off their advantages and disadvantages
- identify the best alternative, when all factors are considered

Furthermore, we concur with the basic criteria upon which the design, construction, and operation of generating facilities are predicated. The first of these is safety — safety to the public and to employees of Ontario Hydro is the primary objective. The second is the objective of reliability of power supply to customers. And the third criterion is that electricity be generated at the lowest feasible cost.

Ontario Hydro's total installed generating capacity and the respective contributions of hydroelectric, nuclear, and fossil-fuelled generation and firm purchases, are shown in Table 7.1. Further, in Table 7.2 we show the mix of generating capacity as well as the mix of electric energy generated in 1978. This table exemplifies the important distinction between the mix of capacity of various types of plants and the mix of energy, or the various fuels used, to generate electricity. Note also that, in 1978, nuclear with a 20 per cent share of capacity generated more than 30 per cent of the energy whereas oil and gas with a 14.5 per cent share of capacity generated only 4 per cent of the energy. This illustrates the concept of various modes of operation of generating plants, i.e., base-load, intermediate-load, peaking, and reserve.

Table 7.1 Ontario Hydro's Installed Generating Capacity

| | Dependable peak resources[a] (January 1980) | |
	(megawatts[b])	(% of total)
Hydraulic	6,407	26.2
Nuclear	5,248	21.4
Coal-steam	9,337	38.1
Oil-steam	2,232	9.1
Natural gas-steam	585	2.4
Combustion turbine	471	1.9
Firm purchases	209	0.9
Total	24,489	100.0

Notes:

a) Based on Ontario Hydro's forecast of dependable peak resources in January 1980 with currently installed facilities. Note that the output of hydraulic and combustion turbine plants varies from one season to another.

b) Arithmetical sum of the peak resources of the East System and the West System.

Source: Ontario Hydro Power Resources, Report No. 790201.

Table 7.2 Generating Capacity and Energy Mix: Ontario Hydro System — 1978[a]

Resource type	Capacity mix[b] (%)	Energy mix (%)
Hydraulic	28.2	37.4
Nuclear	19.9	30.3
Coal	37.4	28.3
Natural gas	2.6	2.2
Oil	11.9	1.8
Total	100.0	100.0

Notes:

a) Does not include purchases.

b) Based on December dependable peak resources.

Sources: Ontario Hydro Power Resources Report — 790201; Ontario Hydro Annual Report 1978.

The existing mix of generating facilities, as shown in Table 7.1, underlines the fact that Ontario Hydro's existing system is both flexible and resilient insofar as generation is concerned. It will be noted that approximately a third of the province's electric energy is produced by hydroelectric stations, a third by nuclear stations, and a third by fossil-fuelled stations. This represents a well-balanced mix of the generating resources currently available. For example, if problems should arise in any one generating resource, there could be more reliance on the other generating resources and on interconnections, to reduce the severity of the impact. Coal-fired stations could be operated at higher capacity factors if a shortage occurred in hydraulic or nuclear generation. Assistance through the interconnections could be sought during fossil-fuel shortages.

From the standpoint of operating flexibility in the system, that is, the ability of the system to adapt to daily, weekly, and seasonal load variations, the operating characteristics of various generating stations are very important. We note the following operating characteristics of the generating units in the Ontario Hydro system:

- Hydraulic units provide excellent load-following capability.
- Fossil-steam generating units are suitable for all modes of operation – from base load to peaking.
- CANDU nuclear units are best suited for relatively continuous operation, that is, for supplying base loads. These units can be operated at lower capacity factors by reducing their output overnight by as much as 50 per cent and by weekend shut-downs. Existing CANDU units, however, are not suitable for load-following.

We also note that the present Ontario Hydro system has good operating flexibility, largely because the operating characteristics of generating units effectively match the system load profiles.

With respect to planning of generation in the period up to 1995, Ontario Hydro set out its position relating to the basic principles that underpin its generation policies in a report on the planning of the East System.[8] With minor changes, these principles still apply, and we believe it would be useful to reproduce the appropriate ones and to add the Commission's comments.

(a) the major portion of the base load electric generation in Ontario can be provided most economically and most reliably by the installation of CANDU nuclear stations. Reserves and supplies of uranium in Canada should be adequate for such stations constructed beyond 1990, provided exploration and development are actively pursued over the next 10 years and export limitations are ensured;

(b) primary fuel reliance should be placed on uranium, provided the related capital requirements can be met. Any large additional consumption of fossil fuels should be in coal, although this should be limited because of concerns related to supply, cost, and air quality. Further major commitments to use of oil and gas should be avoided, if possible, due to their relative scarcity and cost;

(c) most new nuclear and fossil-steam generating stations should be large central power stations located adjacent to major bodies of water. However, smaller power stations with multipurposes such as electric generation, steam production for district heating or industrial purposes, and refuse burning may become economic in certain locations; some of these may be located inland;

(d) none of the new technological alternatives currently being discussed in the public domain (solar power, wind power, geothermal power, nuclear fusion, etc.) are likely to have been sufficiently developed as economic and reliable generating sources to form a significant component of the Ontario power system; wind power may have applications for electricity generation in remote communities; solar energy will be used primarily for heating rather than electricity generation;

(e) to meet the need for reserve, peak load, and intermediate load generating capacity, and to replace fossil-steam generating units which have come to the end of their useful life, different combinations of additional hydraulic and fossil-steam capacity and energy storage schemes may be developed;

(f) the only major sources of hydraulic energy remaining for development in the province are on rivers emptying into James Bay and Hudson Bay. One possibility is the development of the Albany River. This could involve 15 power dams and several major river diversions. The development of this and other hydraulic projects is likely to be affected by economic, social, and environmental considerations, and provincial policy with respect to the development of renewable resources;

(g)...

(h) power and energy purchases from neighbouring utilities should continue to be investigated, and arranged when they are economic or required to enable the electric load in Ontario to be met;

(i) as it may not be possible or desirable to install additional heavy-water production capacity at the Bruce Nuclear Power Development subsequent to BHWP [Bruce Heavy Water Plant] A, B, C, and D, it would be desirable and prudent for Ontario to make provision for heavy-water production at another site.

The Commission endorses most of these basic principles that relate to generation planning with the following comparatively minor reservations:

- While agreeing in principle with (f), we are conscious of the fact that the economic, social, and environmental considerations with regard to the potential, albeit unlikely, development of the

Albany River system for hydraulic power purposes will be considered in depth by the Ontario Royal Commission on the Northern Environment. At the time when Ontario Hydro's 1976 report was in preparation, this Royal Commission had not been established.

• Taking into account the considerably reduced nuclear programme at present envisaged by Ontario Hydro, to the end of the century, we have concluded that the heavy-water production capacity at present committed at the Bruce Heavy Water Plants A, B, and D will be more than adequate to provide the necessary heavy-water inventory to the year 2000 and beyond (see Chapter 5). Accordingly, we suggest that item (i) relating to the possible development of heavy-water production at another site is no longer appropriate.

While the conventional policy of building large nuclear and coal-fired stations may be desirable from the standpoints of achieving economy of scale and optimizing the use of land in siting power stations, it may not be the best policy to ensure the flexibility of the planning process. We note, for example, that as the long-range demand for electric power becomes increasingly uncertain a case can be made for planning a number of smaller generating plants rather than a single large plant. These small plants could be brought into service sequentially and, because of the appreciably smaller lead times involved in obtaining approvals, and in the construction process, a higher degree of planning flexibility would be injected. On the other hand, a large central facility might not be ready when it is needed, to provide additional generating capacity, essentially because of the uncertainties relating to licensing and environmental hearing processes and other unpredictable delays. Or, alternatively, a plant might be completed several years before all of its capacity is required. Either of these situations could seriously reduce the substantial cost advantages that nuclear power enjoys, at any rate currently, over fossil-fuelled plants.

Translated into the future electric power requirements of Ontario beyond 1980, this suggests that serious consideration should be given to alternative sources of base-load generation that will not only provide a capability to "fine tune" the planning process but also preserve as high a degree as possible of diversity of resource base. In Table 7.3, we present nine scenarios for generating capacity additions to the year 2000, beyond the Darlington G.S. For each of the three load-growth projections (3 per cent, 3.5 per cent, and 4 per cent average annual growth in primary peak demand) we provide three supply options, distinguished by the degree to which nuclear base-load generation is replaced by the alternatives. The most promising alternatives to the year 2000 appear to be industrial and non-industrial co-generation,[9] the development of the Onakawana lignite reserves, and the purchase of firm hydroelectric power on a medium-term basis from Manitoba and Quebec. For example, under the low-nuclear scenario, for a 3.5 per cent load growth projection the 3,400 MW from alternative sources could include 1,400 MW from co-generation, 1,000 MW from Onakawana lignite, and 1,000 MW from firm imports. In Table 7.4 we also provide the mix of total generating capacity in the year 2000, under the nine scenarios.

Table 7.3 Supply Scenarios – Additional Generating Capacity beyond Darlington Generating Station, in megawatts

Load growth rate	3%	3.5%	4%
High-nuclear scenario			
Nuclear	0	3,400	6,800
Hydraulic	600	850	1,100
Other	0	0	0
Medium-nuclear scenario			
Nuclear	0	1,700	5,100
Hydraulic	600	850	1,100
Other	0	1,700	1,700
Low-nuclear scenario			
Nuclear	0	0	3,400
Hydraulic	600	850	1,100
Other	0	3,400	3,400

Source: RCEPP.

Table 7.4 Total Capacity in the Year 2000, in megawatts

Load growth rate	3%	3.5%	4%
High-nuclear scenario			
Nuclear	13,860	17,260	20,660
Hydraulic	7,115	7,365	7,615
Coal	10,304	10,304	10,304
Oil	3,947	3,947	3,947
Gas	592	592	592
Other	0	0	0
Total	35,818	39,468	43,118
Medium-nuclear scenario			
Nuclear	13,860	15,560	18,960
Hydraulic	7,115	7,365	7,615
Coal	10,304	10,304	10,304
Oil	3,947	3,947	3,947
Gas	592	592	592
Other	0	1,700	1,700
Total	35,818	39,468	43,118
Low-nuclear scenario			
Nuclear	13,860	13,860	17,260
Hydraulic	7,115	7,365	7,615
Coal	10,304	10,304	10,304
Oil	3,947	3,947	3,947
Gas	592	592	592
Other	0	3,400	3,400
Total	35,818	39,468	43,118

Source: RCEPP.

Notwithstanding the possibility that the alternative generation technologies will prove to be more costly than nuclear power, we believe that, because of potential lower social and environmental costs, Ontario Hydro should give serious consideration to such alternatives.

Accordingly, we recommend that:

7.2   An in-depth study of the Commission's supply scenarios should be undertaken and the findings should be used as a basis for future planning of the electric power system.

Only after detailed analysis of the alternative scenarios we have presented will it be possible for the government to evaluate the respective merits and disadvantages of each scenario and make the necessary decisions. We assume that the input of Ontario Hydro and of the public to the decision-making process, as considered in Chapter 12, will be key factors.

No consideration of energy mix (or indeed of load management) would be complete without reference to energy storage. Because of the fluctuating demands for power and the need to optimize the utilization of base-load generating stations, especially nuclear stations and hydraulic stations, the possibility of storing base-load generated energy during off-peak periods is attractive. It is also recognized that hydraulic pumped storage is already in a mature state of development in both Ontario and abroad, especially in the United States.[10]

If electric energy could be generated during times of low demand and stored for use during times of high demand, there would be at least two major advantages. First, there would be increased ability to satisfy peak power demands utilizing reasonably inexpensive fuels. Second, storage units enhance the overall economics of power-plant operation through optimum utilization of efficient base-load units.

In Ontario, the Delphi Point and Matabitchuan above-ground hydraulic pumped-storage sites have been of interest over the last decade. Although the scope of other conventional pumped-storage facilities in Ontario is limited, there is the possibility of utilizing underground pumped storage in which the lower reservoir is located in an excavated cavern. In one configuration, studied by Acres Consulting Services Ltd., Lake Ontario would provide the upper reservoir; power would be generated during the day when water flowed from the lake to the underground cavern, and during the night this water would be pumped back into the lake using surplus electricity.

We fully endorse the hydraulic pumped-storage concept. The pumped-storage facility at Niagara is, of course, ideally situated and of considerable benefit to Ontario's electricity consumers. However, on the basis of our projections of the average rate of load growth to the end of this century, we believe that the economic potential of hydraulic pumped storage (above-ground or underground) in Ontario Hydro's system is limited. Ontario Hydro's load-management programme and its decision to examine in detail the remaining potential of peaking and intermediate-load hydraulic generation in the province must be taken into account here. Both of these alternatives achieve the same objective as storage, and thus the potential for storage will be determined by the success of load management and the economics of peaking and intermediate-load hydraulic generation.

It is worth noting, however, in connection with energy storage, that major research programmes to develop high-performance electric batteries are in hand. Such batteries, which may not be available commercially at competitive prices for at least 20 years, may have an important role to play in the development of small-scale storage units with capacities in the range of 100-500 MW·h. We do not advocate Ontario Hydro's participation in such research at this time.

Another energy storage concept that will probably come to fruition on a reasonably large scale before the end of the century is the electrical road vehicle. Most of the major automobile manufacturers are undertaking research in this field (e.g., on lithium-iron disulphide, zinc-nickel oxide, and sodium-sulphur batteries). The new batteries have appreciably higher performance capabilities than the conventional lead-acid battery. It has been estimated that more than 2 million electrically powered vehicles may be on the roads of North America within the next 25 years (verbal communication from General Motors and media reports). The power implications of this energy storage medium (resulting from the fact that batteries would be recharged during off-peak periods) may become significant in improving overall system performance by the late 1980s.

Interconnections

Interconnections have many potential advantages in the areas of improved system reliability and reduced costs. These advantages have consistently outweighed the disadvantages.

This quotation from Ontario Hydro's report 573 SP (page 9.1) on "Planning of the Ontario Hydro East System" sums up the utility's position with respect to interconnections between Ontario and its neighbours, that is, Quebec, Manitoba, and the United States. The Commission fully endorses this conclusion, and believes that existing interconnections between Ontario and neighbouring utilities should be strengthened.[11] The economic implications of interconnections are considered in Chapter 11; in this section we are concerned with the planning implications.

Although Ontario Hydro has used its interconnections, until comparatively recently, primarily for importing power, during the last three years the export of power, especially to the United States, has increased appreciably. We are encouraged to know that Ontario Hydro proposes to continue its use of interconnections and "to continue its ongoing studies on the expansion of interconnection capacity, and to profit where possible from reserve savings, firm purchases, co-ordinated developments and operations".[12] Below, we summarize the advantages and disadvantages of interconnections.

Advantages
• System generating reliability may be increased through interconnections without adding reserve capacity. Conversely, a reduction in reserve generation capacity may be possible while retaining the same level of reliability.
• Particularly with respect to interconnections between Ontario and the United States utilities, the important question of summer-winter diversity arises, and it strengthens the case for interconnections. For example, Ontario Hydro is a winter-peaking utility and as a result more reserve generating capacity is available in summer than in winter. The majority of United States electric utilities are summer-peaking. Consequently, some opportunities may exist for exchange of power on a seasonal basis. Unfortunately, however, the capabilities of United States utilities to export power to Ontario during the winter months is very limited, especially on anything approaching a firm basis. Furthermore, because of Ontario's present and projected excess generating capacity position, the need to import power during the winter is minimal. However, during the summer months of recent years Ontario Hydro has been in a position to export power to United States utilities, and it will continue to be for several years. The limiting factor in a few years' time may be the capacity of Ontario Hydro's bulk power transmission system, although steps are being taken to strengthen the network, particularly in southwestern Ontario and southeastern Ontario; furthermore, the tie-lines at border points will need upgrading.
• Especially when long-term plans are being formulated, it may be desirable and to the benefit of

both utilities for firm purchases of power to be arranged as well as interruptible power exchanges. This is particularly appropriate in connection with the potential interconnections between Ontario and Manitoba in the west, and between Ontario and Quebec in the east. The point is that both Manitoba and Quebec have large-scale hydroelectric developments in hand and being planned, and we understand that each of these provinces may have surplus generating capacity during the 1990s. It makes a great deal of sense for Ontario to consider importing electric power under long-term firm contracts to take effect in the 1990s.[13]

• Because a utility may have surplus generating capacity, capable of producing electric energy at lower cost than a neighbouring utility, a sale of the power can be beneficial to both utilities. Interconnections provide a facility for economy transactions that may be profitable for both utilities. Ontario Hydro's net export revenues – the gross revenue less the estimated incremental cost to the utility of generating and transmitting the energy – are shown in Table 7.5.

Table 7.5 Economic Benefits of Sales over the Interconnections – Ontario Hydro

Year	Energy sales (TW·h)	Total revenue (10^6\$)	Net revenue (10^6\$)
1973	5.4	61	32
1974	5.9	101	55
1975	2.0	42	20
1976	4.1	87	37
1977	8.4	206	88
1978	10.4	284	119
1979[a]	12.0	353	157

Note a) Estimates Jan. – Oct. 1979: 9.4 TW·h, 275.4 (total 10^6\$), 122.4 (net 10^6\$).
Source: Ontario Hydro.

• Interconnections tend to reduce the severity of major contingencies caused, for example, by extreme weather conditions, common mode outages of generation and transmission equipment, strikes, and fuel shortages. This was clearly demonstrated during the extremely cold winter of 1976-7 when, combined with mechanical problems at the Nanticoke G.S. and low-water conditions in northern Ontario, Ontario Hydro was forced to buy large quantities of power from U.S. utilities to ease the situation in Ontario.

• Interconnections enhance the frequency stability of an individual electric power system. Assuming that adequate facilities are in place, the frequency stability of interconnected systems is related to the total "mechanical inertia" of the aggregated systems.[14]

Disadvantages

• Each of the interconnected systems loses a degree of autonomy, both in planning and in operation.

• If the main purpose of the interconnection is to reduce generating reserves, there is increased reliance on the neighbouring utility in order to maintain the required level of reliability. It is important to note that failure to achieve adequate reliability levels on the part of one utility affects the other utility, which also suffers a lowering of reliability standards.

• If there is a change in government policies with respect to the exporting of electric energy, for example, from Ontario to the United States or vice versa, this might detract from the advantages of interconnections. However, bearing in mind the good relationship that has been established between neighbouring governments and utilities, we consider that this possibility is extremely remote.

• If voltage reduction is considered as a method of load management, as it is on very rare occasions, interconnected operation would militate against this procedure.

Interprovincial Interconnections

During the last few years the Inter-Provincial Advisory Council on Energy (IPACE), with membership consisting of the provincial deputy ministers of energy, has been studying the problem of interconnections within Canada. This is, of course, in addition to bilateral initiatives between the utilities directly concerned. In general, the Council has expressed approval of increased inter-provincial electricity exchanges; in particular, a proposed Ontario-Quebec HVDC tie-line has been given support.[15]

Below, we consider, first, the question of Ontario-Quebec interconnections and then Ontario-Manitoba interconnections.

Ontario-Quebec Interconnections. We have concluded that the present situation relating to Ontario-Quebec interconnections is unsatisfactory. We agree with Ontario Hydro that the existing interconnection capabilities militate against mutually beneficial interchanges of electric power and energy. We consider that strengthened interconnections, probably in the form of an HVDC tie-line, between Ontario and Quebec would eventually prove beneficial to both provinces, from the standpoints of operational economy and ability to deal with system emergencies. The fact that Ontario's power system is increasingly based on thermal generation (63 per cent of total energy generated by Ontario Hydro in 1978) while Quebec's system is predominantly hydraulic and will probably remain so for several decades is a powerful reason for strengthening the interconnections.[16]

A study is at present being undertaken by Ontario Hydro in connection with the possibility of establishing an interchange capability with Hydro-Québec of 1,000 MW in the first place and 2,000 MW eventually. We strongly endorse these efforts to forge a strong interconnection between Canada's two largest provinces. It should be noted that increasing the capacity of the tie-line will necessitate strengthening the bulk power transmission network in eastern Ontario. This subject was addressed in our *Report on the Need for Additional Bulk Power Facilities in Eastern Ontario* in July 1979.

Accordingly, we recommend that:

7.3 The studies aimed at strengthening the electricity interchange capability with Quebec should be expedited, and in particular they should be extended to ensure close collaboration between Ontario Hydro and Hydro-Québec in the future planning of their respective systems for the mutual benefit of both provinces.

Ontario-Manitoba Interconnections Manitoba Hydro and Ontario Hydro's West System have been interconnected since 1956. In October 1972, firm power deliveries from Manitoba Hydro's Whiteshell hydroelectric generating station to Ontario were put into effect. With the increased potential for hydraulic power associated with the Churchill and Nelson rivers (and especially the lower reaches of the Nelson River), Manitoba Hydro, looking to the end of the century, could generate an additional 5,000-6,000 MW of hydroelectric power. At present, the 1,100 MW Limestone generating station on the Nelson River has been "mothballed" pending the arrangement of contracts with several states of the United States and perhaps with Ontario Hydro.

We believe that Ontario Hydro should capitalize as much as praticable on the Manitoba hydraulic resources that remain to be developed. This would call for, first, a high-capacity DC transmission system to strengthen appreciably the existing interconnections, and, second, considerable strengthening of the ties between Ontario Hydro's West and East systems.

There are obvious advantages to accelerating the development of Canada's remaining renewable hydraulic potential, particularly on a river like the Nelson that is now subject to regulation from existing dams so that additional hydraulic development ought not to have significant environmental consequences. But it is clear that, by the time it is delivered to Ontario load centres, lower Nelson power will not be cheap. Manitoba Hydro estimates the capital cost of developing Conawapa, the next station downstream from Limestone, plus the associated high-voltage transmission to Winnipeg at about $2,800 to $3,000 per kilowatt (1990 dollars) or about 50 mills/kW·h. This compares with Ontario Hydro's estimate of about 45 mills/kW·h for CANDU nuclear at that time (plus necessary transmission costs).[17]

From this province's standpoint, the possibility of strengthening interconnections with Manitoba, backed by an all-hydraulic electic power system, appears to be desirable, not least because such a development would reduce Ontario's heavy reliance on nuclear power, especially by the late 1990s. There would then be a more equitable balance between hydraulic, coal-fired, and nuclear generating facilities and this we believe would improve total system security, especially in the event of an emergency at a nuclear power plant in Ontario or anywhere in the world, which might require a shut-down of some or all of Ontario's nuclear capacity.

Accordingly, we recommend that:

7.4 Ontario Hydro should co-operate with Manitoba Hydro in studies aimed at strengthening electricity interconnections and the purchase of substantial blocks of hydraulic power from the lower Nelson River; there should be closer collaboration between the two utilities in the future planning of their respective systems for the mutual benefit of the two provinces.

Ontario-United States Interconnections

The need for strengthening interconnections between Canada and the United States was recognized in a recent report on electricity exchanges between the two nations prepared jointly by Energy, Mines and Resources Canada and the United States Department of Energy. The report[18] emphasizes the importance of such exchanges through the following resolution:

> The United States and Canada have resolved to explore the potential benefits of increased international electricity transactions and have identified recommendations which can be acted upon by the governments and operating electric utilities.

The report concludes that:

> ... increased mutual benefits may be realized from increased United States/Canada electricity exchanges. There appear to be several specific actions that electric utilities and the governmental regulatory bodies in both nations could take to ensure that such benefits are maximized for individual consumers in both countries.

The Commission agrees in principle with the conclusions and recommendations of the Canada/U.S. study as they apply to Ontario. The advantages of interconnections between Ontario and Michigan, and Ontario and New York cannot be overemphasized. At present, in addition to the agreements with the Great Lakes Power Corporation, Hydro-Québec, and Manitoba Hydro, Ontario Hydro has formal interconnection agreements with the Niagara Mohawk Power Corporation, the Power Authority of the State of New York, the Detroit Edison Company, the Consumers Power Company, and the Minnesota Pulp and Paper Company in the U.S.

We have noted a unique characteristic of the interconnections between Ontario, Michigan, and New York. Because power circulates around Lakes Erie and Ontario (the circulating power level sometimes reaches 600 MW) it has been shown that, although the circulating power complicates the operation of interconnections, the transmission losses in the area are reduced.

Repeatedly during the last 25 years, Ontario Hydro has benefitted, especially in emergency situations, from its ties with the United States utilities and vice versa. Quite apart from the potential economic benefits of the interconnections, the enhanced security provided is significant.

Accordingly, for social and economic reasons, we recommend that:

7.5   The interconnections between Ontario Hydro and neighbouring utilities in the United States should be strengthened.

Fuel Requirements and Supply

Ontario's electric power system depends upon an assured primary fuel supply — primary in the sense that the production of electricity uses primary energy sources such as coal, uranium, oil, and gas, as well as the kinetic energy of falling water. Security of fuel supply is crucial to a reliable supply of electricity to Ontario consumers.

Based on our supply scenarios introduced in Tables 7.3 and 7.4, the forecasts of Ontario Hydro's annual electricity generation from various sources in the year 2000 are shown in Table 7.6. To put the utility's projected generation from each source into perspective, we show in Table 7.7 the energy mix of various sources. Finally, in Table 7.8, we show the annual consumption of various fuels in conventional units.[19] The forecast fuel consumption corresponding to generation from the "other" category is given in terms of milligrams of uranium equivalent. However, depending on the primary sources for generation in the "other" category, the forecast fuel consumption of coal, oil, and natural gas will be modified.

Table 7.6 Total Generation in the Year 2000, in gigawatt hours

Load growth rate	3%	3.5%	4%
High-nuclear scenario			
Nuclear[a]	91.1	113.4	135.7
Hydraulic	36.0	36.5	37.1
Coal	52.9	50.6	50.5
Oil	2.3	2.3	2.3
Gas	0.5	0.5	0.5
Other[a]	0	0	0
Total	182.8	203.3	226.1
Medium-nuclear scenario			
Nuclear[a]	91.1	102.2	124.5
Hydraulic	36.0	36.5	37.1
Coal	52.9	50.6	50.5
Oil	2.3	2.3	2.3
Gas	0.5	0.5	0.5
Other[a]	0	11.2	11.2
Total	182.8	203.3	226.1
Low-nuclear scenario			
Nuclear[a]	91.1	91.1	113.4
Hydraulic	36.0	36.5	37.1
Coal	52.9	50.6	50.5
Oil	2.3	2.3	2.3
Gas	0.5	0.5	0.5
Other[a]	0	22.3	22.3
Total	182.8	203.3	226.1

Note a) Assumed at 75 per cent ACF.
Source: RCEPP.

Table 7.7 Energy Mix in the Year 2000 (percentages)

Load growth rate	3%	3.5%	4%
High-nuclear scenario			
Nuclear	49.8	55.8	60.0
Hydraulic	19.7	18.0	16.4
Coal	28.9	24.9	22.4
Oil	1.3	1.1	1.0
Gas	0.3	0.2	0.2
Other	0.0	0.0	0.0
Total	100.0	100.0	100.0
Medium-nuclear scenario			
Nuclear	49.8	50.3	55.1
Hydraulic	19.7	18.0	16.4
Coal	28.9	24.9	22.4
Oil	1.3	1.1	1.0
Gas	0.3	0.2	0.2
Other	0.0	5.5	4.9
Total	100.0	100.0	100.0
Low-nuclear scenario			
Nuclear	49.8	44.8	50.1
Hydraulic	19.7	18.0	16.4
Coal	28.9	24.9	22.4
Oil	1.3	1.1	1.0
Gas	0.3	0.2	0.2
Other	0.0	11.0	9.9
Total	100.0	100.0	100.0

Source: RCEPP.

Table 7.8 Annual Fuel Consumption in the Year 2000

Load growth rate	3%	3.5%	4%
High-nuclear scenario			
Uranium (megagrams)	1,800	2,240	2,680
Coal (10^6 tonnes U.S. equivalent)	18.8	17.9	17.9
Oil (10^6 barrels)	3.6	3.6	3.6
Gas (10^9 cubic feet)	5.0	5.0	5.0
Other (Mg of U equivalent)	0.0	0.0	0.0
Medium-nuclear scenario			
Uranium (megagrams)	1,800	2,020	2,460
Coal (10^6 tonnes U.S. equivalent)	18.8	17.9	17.9
Oil (10^6 barrels)	3.6	3.6	3.6
Gas (10^9 cubic feet)	5.0	5.0	5.0
Other (Mg of U equivalent)	0.0	220.0	220.0
Low-nuclear scenario			
Uranium (megagrams)	1,800	1,800	2,240
Coal (10^6 tonnes U.S. equivalent)	18.8	17.9	17.9
Oil (10^6 barrels)	3.6	3.6	3.6
Gas (10^9 cubic feet)	5.0	5.0	5.0
Other (Mg of U equivalent)	0.0	440.0	440.0

Notes: Oil consumption corresponds to an ACF of 7 per cent for the total 3,947 MW of oil-fired capacity. While this number appears reasonable, a more accurate estimate can only be obtained by using simulation. Gas consumption corresponds to the Ontario Hydro 1979 forecast. The "other" category can be a combination of hydraulic, coal, and other fossil-fuel resources and therefore is expressed as megagrams of uranium, i.e., the base-load capacity displaced. Source: RCEPP.

It is clear from Table 7.7 that the share of nuclear generation rises significantly from an existing value of about 30 per cent to between 50 and 60 per cent in the year 2000, depending on the scenario chosen, and the share of fossil-fuelled generation drops somewhat from the existing one-third, whereas the share of hydraulic energy declines considerably from the present 37 per cent to between 16 and 20 per cent. Replacement of some nuclear-generated base-load energy by the "other" category can reduce the share of nuclear only by about 10 per cent.

Ontario Hydro relies heavily on United States coal supplies from Pennsylvania and West Virginia; at present, the utility has under contract about 10 million tonnes per year. These contracts expire within the next six years but it is expected that they could be renewed if required. A contract for 82 million tonnes of coal, over a 30-year period, was also negotiated with the United States Steel Corporation in 1976.

Although coal from western Canadian markets is not at present competitive with U.S. coal, because of high transportation costs, Ontario Hydro has recently negotiated contracts to purchase 2.5 million tonnes of Alberta and British Columbia bituminous coal and 0.9 million tonnes of Saskatchewan lignite annually, primarily to enhance the security of its coal supply. This coal is to be used at Ontario Hydro's West System coal-fired units (Thunder Bay and Atikokan) as well as in its Nanticoke G.S. Ontario Hydro has invested approximately $79 million in this project to cover the cost of the railroad equipment and a blending terminal at Nanticoke to blend the western Canadian bituminous coal with U.S. coal.

The extent to which Ontario Hydro can capitalize on the massive coal reserves of western Canada remains problematical. From the standpoint of energy conservation and environmental impact, on a nation-wide scale, the cost of transporting several millions of tonnes of coal from western Canada to Ontario may compare unfavourably with the transmission of hydraulically-generated electricity from Manitoba to Ontario. Considerations of this kind, especially on a long-term basis, must be borne in mind in systems planning.

The Onakawana lignite deposits in northern Ontario, with proven reserves in the order of 190 million tonnes, constitute Ontario's only significant coal resource. This is estimated to be sufficient to fuel a 1,000 MW thermal station at base-load capacity factors over 30 years. A study of the potential of this resource for electric power production was recently completed by Shawinigan Steag and Onakawana Development Limited, which owns the mineral rights to the lignite deposits between the Mattagami and Abitibi rivers. The study proposes that Shawinigan Steag finance and build a 1,020 MW station and that Ontario Hydro operate the station and purchase its power. Ontario Hydro is currently engaged in a major review of the proposal and expects to have an environmental assessment statement completed by late 1980. If the results of this review are encouraging, Ontario Hydro may consider the proposal as an alternative to the next large thermal generating station to be added to its system. Preliminary results of Ontario Hydro's review indicate that a 1,000 MW station could be feasible, operating initially as a base-load station but gradually moving to intermediate-load applications over its 30-year life. From the viewpoint of cost economics alone, the proposed Onakawana station does not compare favourably with a CANDU nuclear plant. However, although Onakawana may not be a substitute for a nuclear station, it may delay the need for such a station after Darlington. Furthermore, when compared with a thermal station fuelled with U.S. or western Canadian coal, Onakawana appears to be an acceptable alternative, taking into account its positive impact on the provincial economy.

Ontario Hydro's residual oil-burning generation will continue to be an important peaking facility to the end of the century. Residual fuel oil supplies appear to be reasonably assured. For example, the Corporation has a contract with Petrosar Limited of Sarnia to supply 7.3 million barrels of low-sulphur residual oil per year to 1992 using western Canadian crude oil as the feedstock. Due to the current and forecast surplus in generating capacity, Ontario Hydro expects to reduce the oil supplies to half the contract amount.

The Richard L. Hearn G.S. (400 MW fuelled solely with gas and 800 MW that can be fuelled with gas or coal) burns natural gas supplied by Consumers' Gas Limited. Ontario Hydro had contracted with this company to supply 49 billion cubic feet per year until November 1981. Again due to reduced demand, Ontario Hydro is reducing projected natural gas consumption to 10 billion cubic feet per year to 1988 and 5 billion cubic feet thereafter. Consequently, there do not appear to be any problems with natural gas supplies for Ontario Hydro's requirements to the year 2000 and probably beyond.

Our overall conclusion concerning the security of fuel supplies at likely load growth rates for Ontario's electric power system is that, in spite of some uncertainties relating to the long-term future security of supplies of coal from the United States, we believe the province to be in an extremely sound position, with sufficient flexibility to deal with changing electric load patterns in the future.

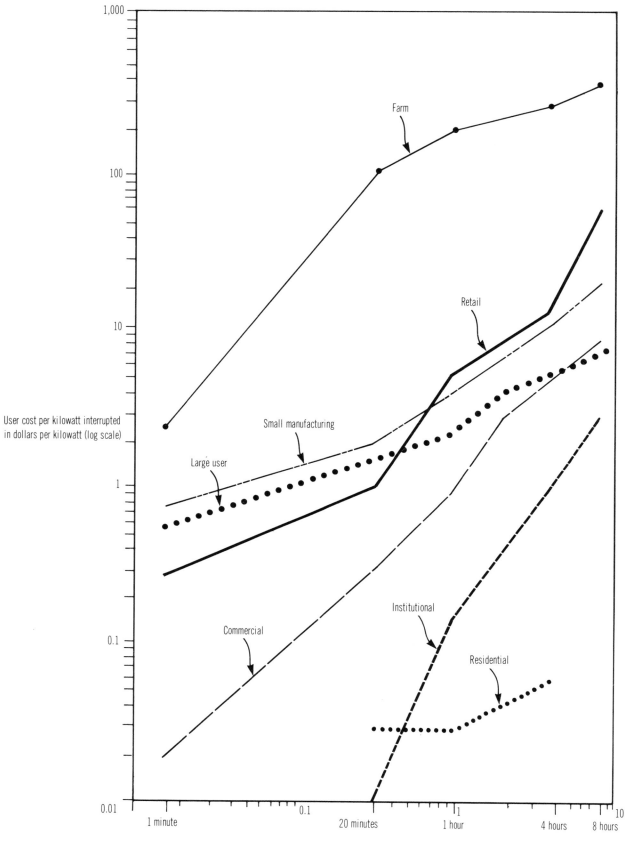

Figure 7.1 User Estimates of Interruption Costs

1,000

100

10

1

User cost per kilowatt interrupted
in dollars per kilowatt (log scale)

0.1

0.01

Farm

Retail

Small manufacturing

Large user

Commercial

Institutional

Residential

1 minute   0.1   20 minutes   1 hour   4 hours   8 hours   10

Duration of interruption in hours (log scale)

Sources: Ontario Hydro and RCEPP SW Ontario Report.

CHAPTER EIGHT

Land Use

Man uses land for agriculture, habitation, industry, transportation, recreation, and a variety of services such as pipelines, electric power generation and transmission, and water and sewage systems. It is obvious that these uses[1] should, to the extent possible, be planned holistically, and not individually, because of the multiplicity of interactions and interdependencies that characterize societal systems and related technologies.

Ontario's population is distributed unevenly. In particular, two major geographical regions can be identified. The northern part of the province is sparsely settled, essentially because of the severity of the winter, and the low level of agricultural productivity of the soil; however, this region is rich in mineral and forestry resources. In contrast, the southern half of the province is relatively densely populated because of its temperate climate, its large areas of good agricultural land (and hence its suitability for growing a wide range of crops and for rearing cattle, hogs, and poultry), and its proximity to the Great Lakes. The use of land adjacent to the Great Lakes and the St. Lawrence River is particularly important for Ontario. This land is unique and competition for its use, not surprisingly, is considerable. Note, in particular, the siting of:

- urban communities and industrial developments
- port and docking facilities
- recreational facilities, such as marinas
- power generating stations

Not least because the original settlements in Ontario were established on prime agricultural land, the process of urbanization has utilized large areas of Class 1 and Class 2 land; this encroachment continues.[2] Unfortunately, as far as agriculture is concerned, urbanization is an irreversible process.

During the Commission's hearings we were told, on many occasions, that the proper use for land should take into account social, environmental, ethical, and aesthetic, as well as economic factors. Clearly, within or closely adjacent to urban areas, complex social and political issues arise because of disparate land uses for agriculture, urban development, and industry. This is inevitable, particularly as pressure mounts to ensure that Class 1 and Class 2 agricultural lands are preserved. Trade-offs are unavoidable and some of the hardest decisions will relate to the siting of electric power facilities. Whether it is a dual-purpose generating plant (close to a populated area) or an additional transmission line from a remote generating plant, some encroachment on agricultural land may be expected, even if, in the future, urban development is directed to areas of low agricultural productivity.

Although the associated land-use decisions can be based on quantifiable data, it cannot be overemphasized that most of the land-use concerns, certainly those that have been brought to the attention of the Commission, are necessarily of a qualitative nature and involve value judgements.[3]

In common with other industrialized regions and nations, Ontario has experienced a marked change in the distribution of its population during the last 30 years. In 1951, 73 per cent of the population was classified as urban, but by 1971 this percentage had increased to 82 per cent, and the expectation is that by the year 2000 almost 90 per cent of the province's population will be located in urban areas.

In this chapter, we will consider the development of the power system in the light of growing demands on the province's finite land and water resources.

Ontario Hydro – Land Use

The choice of suitable locations for electric power facilities involves a large number of factors. We have concluded that site selection should be based on the following considerations.

Public Health and Welfare
- These are critical factors in determining the location of bulk power facilities. In particular, the maintenance or improvement of life-styles should be assured.
- Both air- and water-quality standards, as established provincially, should be met (see Chapter 9).
- Recreation areas, notably provincial parks and conservation areas, should be preserved.
- Although Ontario is comparatively stable seismologically, the siting of nuclear power stations should take the potential threat of earthquakes into account.

Protection of Food Lands

• Wherever possible, large thermal generating stations should be sited on marginal agricultural land.

• Coal-fired stations should be sited to assure minimal environmental impact on crops from the effluents.

• Routing of bulk power transmission lines should have minimal impact on food lands and, in general, on the practice of agriculture.

Protection of the Ecology

• Marshes, river banks, and lakeshores (often referred to as wetlands) are regarded as ecologically sensitive and biologically productive habitats for wildlife. This should be taken into account in siting thermal generating stations and routing transmission lines.

• The ecological impact of thermal energy discharges (e.g., into the Great Lakes) should be minimized.

• Hydroelectric installations have been shown to be major threats to the ecology of a region; the costs (including environmental costs) and benefits must be carefully weighed.[4]

In the course of the inquiry, we directed considerable attention to the impact of electric power facilities on agricultural land. Our investigation has led us to conclude that the losses of good farmland attributable to electric power generation, transmission, and ancillary facilities (and these constitute the major component of Ontario Hydro's land use) are small in comparison with the land-use implications of modern urbanized society. Table 8.1 puts the issue into perspective,[5] but it does not illustrate the concern of the agricultural community about the routing of bulk power transmission lines and the important influence this may have on the siting of future generating stations and, most important, on the siting of new towns and industrial parks. We received evidence during the inquiry that has led us to conclude that, in the case of generating station sites, the concerns of the farm community are well founded. This, we believe, is a fundamental reason why transmission-line routing must be evaluated on a systems basis, in order that the impact and the interrelationships can be more clearly evaluated.

Table 8.1 Comparative Land Areas Required for Ontario Hydro's Bulk Power Facilities, Farmland, Airports, etc.

Provinces of Ontario, total area (excluding lakes): 88,064,000 hectares
Ontario's farm land: 4,800,000 hectares
Area occupied by Bruce Nuclear Power Development, including controlled area where population is regulated: 9,200 hectares
Ontario Hydro bulk power transmission system: 80,000 hectares
Ontario Government land bank for the Pickering Township Development (including Pickering Airport): 34,000 hectares
Toronto International Airport: 1,755 hectares

Source: "Land Use", Ontario Hydro submission to the RCEPP, April 1976.

Notwithstanding Hydro's stated goal of minimizing the effect of its facilities on agricultural lands, we recognize that the problem, especially as it relates to the routing of bulk power transmission lines, is very complex. The topography of the area and the location of farm buildings, villages, towns, prime farmland, roads, rights of way, aircraft landing strips, airports, prime forest lands, and marshes and other ecologically sensitive areas must all be brought into the route-selection process, and routing criteria must be developed and applied.

Combined Energy Centres

In an attempt to spur the decentralization of industry out of the Golden Horseshoe (Oshawa-Toronto-Hamilton-Niagara Falls), an interministerial committee headed by the Ministry of Industry and Tourism recommended in 1976 that consideration be given to creating combined energy centres. The idea of such a centre is to co-locate Ontario Hydro thermal stations with industries with suitable locational requirements, and subsequently to plan a new community nearby.

Candidate industries would include those with large steam requirements, large energy users, users of reject heat (greenhouses, aquaculture), and plants that could provide or interchange materials, labour, or services with the other plants at the site.

Careful town planning would take into account the recreational advantages of a lakefront location and the possibility of heating the community from the generating station. Such centres would also minimize encroachment on good farmland.

Ontario Hydro pointed out that there has been no tendency for industries to locate near its power stations, partly because its policies of uniform geographic pricing of electric power offered no incentive to do so.

Norman Pearson discussed the joint development of industry and power stations in a study made for the Commission, and described in Volume 7. He concluded that, at locations that are actual or potential key transportation routes — accessible to Great Lakes shipping routes or within the Detroit-Toronto-Montreal corridor, Ontario Hydro plants could help trigger industrial development, provided that other necessary location factors were either present or could be developed economically. The Nanticoke case is often cited as proof of the potential synergistic effect of a large power plant. Today, a major refinery and steel mill share with Hydro's 2,200 MW coal-fired generating station part of a 12,000 ha industrial complex. There has been some joint development of the transportation infrastructure, cooling water and water supply facilities, harbour and dockage facilities, and community infrastructure. However, it is likely that the decision of the other industries to locate at this prime industrial site was simply a result of their search for the same combination of location factors that attracted Ontario Hydro.

We believe that the combined energy centre concept deserves a trial, but feel that the Bruce site is a more desirable location than the Golden Horseshoe.

The Bruce Agripark project seems now to be on its way under the auspices of the Ontario Energy Corporation. The communities of Kincardine and Port Elgin face a wrenching adjustment as the labour force at the Bruce site drops from 8,500 workers in 1978 to some 3,300 operating staff by 1987, losing in the process a construction payroll of some $2 million per week, and its multiplier.

Part of the problem has doubtless been insufficient advance planning by all levels of government and by Ontario Hydro for the "demobilization phase" of the Bruce project. It is important to mitigate future adverse socio-economic effects in the area, and to capitalize on an infrastructure already in place.

> *Accordingly, we recommend that:*
>
> **8.1**   Ontario Hydro and the Ontario government should build on developments already taking place at the Bruce site to test further the concept of a combined energy centre as described in the Ministry of Industry and Tourism's 1976 report.
>
> **8.2**   Ontario Hydro should accept financial responsibility for the debenture debt load of municipalities in the vicinity of the Bruce Generating Station that is over and above what would have been incurred in the absence of the Ontario Hydro projects.

Land Banking

To facilitate electric power planning by attempting to reduce the long lead times (in the range of 12-15 years) associated with the planning and construction of large thermal generating stations, Ontario Hydro has embarked on a "land-banking" programme. The concept involves the acquisition of land well in advance of its being required as a site for a generating station, and "keeping it in inventory". In this way, the utility has argued, several preliminary steps (comparatively inexpensive procedures) could be undertaken well in advance of any construction activities. For example, Hydro has argued, hearings of the Environmental Assessment Board and other public participation activities could be brought forward. Indeed, it is suggested that land banking might involve the consideration of several alternative potential sites from economic, environmental, and social standpoints. A current example is the land-banking programme in connection with a possible thermal generating station in the Georgian Bay North Channel area.

The Commission does not endorse land banking, for the following reasons:
• Because of the marked drop in the load forecast, coupled with a sizeable excess generating capacity, the need for a new site (additional to existing and designated generating station sites already under Hydro control) before the year 2000 appears unlikely.
• We do not believe that Environmental Assessment Board hearings could be undertaken with respect to a "land-banked" site. The site would have to be designated specifically as a fossil or nuclear site and the plans would have to be carried forward to an appropriate level. Accordingly, the lead time might not be decreased.
• Evidence to date suggests that a potential generating site (even though the probability of its being developed within 20 years is small) is regarded by local people as a potential nuclear station.

This has caused considerable concern and uncertainty, for example, among residents in the North Channel area.

- It is difficult, we believe, to justify land banking from an economic standpoint. How many potential sites should be banked? And at what cost to the consumer?

Accordingly, we recommend that:

8.3   Ontario Hydro should not proceed with land-banking programmes for at least the next 10 years.

Energy and Food

Energy and food are virtually synonymous. Plants and trees convert solar energy, carbon dioxide, water, and nitrogen into food and fuel by the process of photosynthesis. This is the basic process upon which all life on earth depends. Usable solar energy, and the solar energy trapped in plants and trees, is the most important energy source available to man. Indeed, for hundreds and probably thousands of years, because artificial fertilizers were not available, a natural state of balance existed and, in effect, determined the world's level of food supply.

On the average, traditional agriculture was capable of producing only 1,000 kg of food grain per hectare. But the introduction of agricultural technology, using mainly non-renewable energy (especially in the form of fertilizers, mechanical traction processes, and electric power) to amplify solar energy and consequently to increase crop yields, has increased yields more than tenfold.

In effect, non-renewable energy in various forms has increased the productivity of agricultural land. Counterbalancing this trend has been the increase in population and improvements in the standard of living.

To put the energy- and land-use-related aspects of food production into perspective, Table 8.2 shows how much food energy (e.g., protein) in the form of a variety of crops can be produced per hectare of agricultural land, and how much input energy in the form of fossil energy and man-hours is required.[6]

Table 8.2 Vegetable Protein Production per Hectare for Various Crops in the United States and Elsewhere

Crop	Crop yield in food energy (10^6kcal)	Fossil energy input for production (10^6kcal)	Crop yield/fossil energy input	Labour man-hours
Potatoes	20.2	8.9	2.27	60
Corn	17.9	6.64	2.7	22
Wheat	7.5	3.77	2.0	7
Oats	7.4	3.0	2.47	6
Dry beans	5.0	4.5	1.1	15

Source: Data based on "Energy and Land Constraints in Food Protein Production", D. Pimental et al. "Science", November 21, 1978, vol. 190, no. 4216, pp. 754-61.

On the other hand, Table 8.3 shows how much animal protein, for a variety of foods, can be produced per hectare, and how much input energy in the form of protein, fossil energy, and man-hours is required.

Table 8.3 Animal Protein Produced per Hectare in the United States

Animal product	Animal protein yield (kg)	Feed protein input (kg)	Protein yield/ protein input (%)	Fossil energy input (feed and animal) (10^3kcal)	Labour man-hours
Milk	59	188	31.3	8,561	23
Eggs	182	672	27	9,560	174
Broilers	116	651	18	10,233	38
Pork	65	689	9	9,212	28
Beef (feed lot)	51	786	6.5	15,845	31

Source: Data based on "Energy and Land Constraints in Food Protein Production", D. Pimental et al. "Science", November 21, 1978, vol. 190, no. 4216, pp. 754-61.

From the standpoint of optimizing land utilization for the production of food, certainly on a long-term basis, Tables 8.2 and 8.3 suggest that the food habits of Ontarians, and indeed of the industrialized western world as a whole, may have to undergo marked changes if increasing population is to be supported. The time scale involved may be in the order of 50 years; nevertheless, the data provide food for thought. The limiting factor could be, not land use *per se* but the availability of specific non-renewable energy resources. In this respect, the increasing utilization of renewable energy resources in the agricultural industry could provide an important hedge.

Another aspect of the food-energy relationship is the need for nitrogen-fixation in the growing of crops. Nitrogen fertilizers are energy-intensive commodities. The basic feedstock at present, in their manufacture, is natural gas, and as the cost of natural gas increases there will be increasing interest in nitrogen-fixation by biological processes. We refer to the use of legume crops, in this regard, but a number of nitrogen-fixation processes are being investigated by biologists.[7] The argument for increasing our dependence on biological nitrogen-fixation, as a means of reducing the energy input to food production, is gaining strength. Although we are not in a position to assess the land-use implications of increasing dependence on biological nitrogen-fixation, we urge that future agricultural policies should not ignore developments in this area.

Land Stewardship

Agricultural Lands

Urban encroachment on prime agricultural land, mentioned previously, is another manifestation of the complementarity of energy and food. It must be considered primarily as a socio-economic rather than an environmental issue. But it has environmental overtones, especially as far as electric power planning is concerned.

Assuming continuing population growth and continuing urban encroachment, what steps will society take to ensure an adequate supply of food in the future? This question is inextricably linked with the concept of land stewardship (i.e., the preservation of farmland in perpetuity). It is exemplified by the fact that in southern Ontario 13.7 per cent of the land used for farming in 1966 was not in production in 1971.[8] This corresponds to a transfer of about 10 hectares per hour of food-producing land to non-productive land during the same period. However, these were years of low farm income, and some of the abandoned farmland has since been brought back into production.

Ontario occupies a critical position in regard to the preservation of farmland, because almost one-half of Canada's four million hectares of Class 1 land and about one-sixth of Canada's 16 million hectares of Class 2 land are located in the province.[9]

Assuming an increase in the population of Ontario of about 25 per cent (i.e., to 10.6 million people) by the year 2000, and assuming an average population density of about 5,000/km², then to provide habitation for this increase in the population would require about 1.0 per cent of Ontario's agricultural land. Hence the need for land stewardship. But we note that the loss of agricultural land need not result in its total loss for the growing of food. Consider, for example, the vast amount of garden produce that is grown annually in Europe by intensive cultivation of small "allotments". With a comparatively minor change in life-style, and certainly one that enhances the health of the urban dweller, the issue of adequate food production could resolve itself. Perhaps the incentive will be provided by the ever-increasing prices of food. The individual home garden (or communal land allotments for domestic gardens) is a true manifestation of land stewardship; if home gardens proliferate, a considerable amount of energy could be saved by the substitution of gardening for weekend travel to the lake. It is worth thinking about.

If decisions are taken to preserve Ontario's food lands, should they apply only, for example, to Class 1 and Class 2 land? The decisions should be based on the associated social benefits. However, these are difficult to quantify and value judgements of a complex kind will be required. Furthermore, any decision by the Ontario government to preserve high-quality farmland would have to be backed up by regulation, or subsidization of farming, or the maintenance of land banks to offset encroachments, or subsidization to encourage the development of non-agricultural land as more town sites, or by a combination of these.

It is clear that, if a serious attempt is to be made to preserve prime agricultural land, steps will have to be taken within the next few years. The land-stewardship challenge is one that confronts the province, and indeed Canada as a whole, and perhaps the first logical step will be for voluntary organizations of concerned citizens, public interest groups, and appropriate institutions to study the problem in depth. Ontario Hydro would certainly play a key role in such organizations.

Forest Lands

More land, in Ontario, is used to sustain the growth of trees than is used for any other purpose.

We recommend that:

8.4   Ontario Hydro's planning concepts should reflect the primary objective of conserving Ontario's food lands, particularly in southwestern Ontario. Fortunately, most of this land is better suited to growing trees than to any other use. Moreover, the forests are essential for the protection of watersheds and for the sustenance of wildlife and fish. Their contribution to outdoor recreational activities is particularly significant. Furthermore, the impact of Ontario Hydro's bulk power facilities (even the transmission corridors) on the vast forest lands is negligible.

The province's 445,000 km² of forest lands, as well as providing the base for the large pulp and paper and wood products industries, should increasingly be regarded as a major energy resource. Noteworthy is the fact that pulp and paper products constitute Canada's largest single export. If, in addition to supplying the pulp and paper industry, Ontario's forests were to become an important source of solid and liquid fuels (e.g., wood methanol and ethanol), the need for modern forestry management techniques would burgeon.

We draw special attention to the potential of abandoned farmland in eastern Ontario (in the order of 400,000 hectares), for sustained-yield forestry plantations. The resulting forest products might be used for pulp and paper manufacture, for ethanol production, and as fuel for thermal and electric power plants. The Ministry of Natural Resources is undertaking a research programme to assess the potential of hybrid poplar and other fast-growing species of trees. Clearly, land stewardship applies just as much to forest lands as to agricultural lands.

Accordingly, we endorse strongly the concept of the stewardship of agricultural and forest lands and recognize the irreversibility of urban encroachment. We regard the conservation of food and forest lands as being just as important as the conservation of non-renewable energy resources.

Northern Ontario

Although the Commission conducted some meetings in northern Ontario during late August and early September 1976[10], essentially to obtain input relating to electric power planning from native communities, we have assumed that the Royal Commission on the Northern Environment is now responsible for all environmental, energy, and land-use issues in the region. However, in Volume 7 of this Report we have addressed electric power planning aspects of land use in northern Ontario.

Of the 8 million hectares designated as agricultural land in northern Ontario virtually all is classified as Class 4 or lower on the productivity scale. However, several areas, notably in the vicinity of Sudbury and Blind River, on Manitoulin Island, and near New Liskeard and Cochrane in the northeast and Rainy River and Dryden in the northwest, are exceptions, and effective agriculture is practised. Indeed, thriving dairy and market-garden activities exist in the vicinity of several population centres, and profitable farms have been maintained. But the impact of Ontario Hydro's northern bulk power facilities on the agricultural lands is virtually negligible.

The forests of northern Ontario range from deciduous in the southern areas to boreal conifer forests in the north; they cover in the order of 400,000 km². The forest industry provides about 30,000 jobs directly, and another 50,000 jobs are created in related industries. In comparison, the mining industry for the province as a whole employs about 25,000 people. Because of the severe winters and the short growing season, trees may take from 60 to 70 years to regenerate in the southern sections and as long as 120 years in the north, and because the cut-over forest has not regenerated naturally, some uncertainty exists with respect to the future of the industry. Will it be possible to expand, or even to maintain, northern forest output? This question is under intensive study by the government of Ontario. As mentioned in the previous section, there is clearly a need for a forestry policy based on modern forest-management methodologies. In this connection, these questions deserve consideration:

- Is it feasible to develop northern "forest farms" as a basis for a large-scale methanol or ethanol industry?
- Can small-scale, forest-based industries be established to produce a wide range of wood-based products (e.g., homes and furniture) to replace more energy-intensive products such as plastics?
- To what extent are the more valuable tree species renewable?

The forest-farm concept (if viable) upon which a methanol or ethanol industry might be based would be attractive because it would provide a much broader industrial base for northern Ontario. Although not competitive with petroleum fuels, at present prices, northern Ontario alcohols might be competitive at the higher prices for petroleum fuels which now seem likely in the near future. A useful brief on this topic was submitted to the Commission by Energy Probe.[11]

Wood as a fuel for thermal and/or electric power plants presents a different type of problem. Largely as a result of the slow rate of forest regeneration, more than 2,500 km^2 of boreal forest would be needed to fuel a 150 MW plant. Consequently, this is not considered to be a realistic alternative source of energy in northern Ontario. On the other hand, as the Hearst studies have shown[12], there is some potential for the use of forest-industry wastes to supply the forest industry and local communities with thermal and electric power. Apart from northern Ontario's forests and their potential as an energy resource, this great region has large potential hydraulic energy resources. During the early phases of the Commission's inquiry, the undeveloped hydroelectric potential located in northern Ontario was regarded as a feasible source of base-load generation. It has been estimated, for example, that more than 3,000 MW of hydroelectric power could be generated if all potential sites on the Albany and Severn river systems were to be developed. This power is roughly equivalent to that generated at the Bruce A station. However, major studies undertaken by Environment Canada have concluded that the feasibility of the project on economic, social, and environmental grounds is problematical.[13] Furthermore, during our hearings, Ontario Hydro intimated that at present the development of the Albany and Severn river systems for hydroelectric purposes would be uneconomic compared to available alternatives, especially when additional lengthy transmission lines are included. We fully concur with these conclusions, especially in view of the possibility, mentioned earlier, that Manitoba Hydro could be in a position to export hydroelectric power to Ontario on a comparatively large scale during the 1990s.

However, apart from the hypothetical large-scale hydraulic power developments considered above, there are potential sites for small-scale hydroelectric generation, in the range of 1 MW to 10 MW. These are conveniently located close to some communities and, if the power is required locally and the ecological impact is acceptable, they should be developed. While no detailed cost estimates have been made, Ontario Hydro has stated: "There are a number of locations with a capacity greater than 10 MW which may be economic in comparison to fossil generation."[14]

The Onakawana lignite deposit, located about 100 km south of Moosonee and referred to in Chapter 7, constitutes Ontario's major fossil resource. An estimated 190 million tons of lignite (i.e., brown coal) has been discovered under a light overburden of muskeg. We understand that this resource could generate 1,000 MW of electric power for a period of up to 30 years.[15] The generating station would be located at the mine, and power not required in the area would be fed into the Ontario Hydro bulk power transmission system. After examining the Onakawana feasibility study[16], we have concluded that, for the purpose of enhancing power system flexibility (i.e., adaptability to changing load patterns), on grounds of fuel security, and bearing in mind also the employment possibilities, the project should be undertaken. However, it will be interesting to see whether the Royal Commission on the Northern Environment, stressing social and environmental considerations, will come to the same conclusion.

The land-use implications and associated environmental impact of the construction of bulk power transmission lines have not been brought forward specifically as problem areas that adversely affect the people of northern Ontario. On the other hand, we have noted in the past that transmission corridors have been constructed without much account being taken of the social and economic impact on native peoples, wildlife, and the local ecology.

In addition to the above potential energy resources of northern Ontario, we recall that the province's major resources of uranium are situated in that region. The environmental implications of the working of these deposits have been considered in Chapter 5 and in the *Interim Report*.

On the basis of the foregoing, we recommend that:

8.5　The potential of Ontario's forest lands, especially in northern and eastern Ontario, as sources of energy should be the subject of an in-depth feasibility study; and, if the social, environmental, and economic indications are favourable for methanol or ethanol production, a demonstration plant should be built and tested as soon as possible.

8.6　The existing research and development programmes relating to energy plantations, especially the potential of the hybrid poplar in eastern Ontario, with emphasis on abandoned low-quality farmlands, should be expedited.

8.7　On strictly power-systems-planning and economic grounds, the Onakawana lignite deposits should be developed; and an electric power station of 800 MW-1,000 MW capacity should be built at the mine site. However, we recognize that the Royal Commission on the Northern Environment, on social and environmental grounds with respect to both the power station and the associated transmission corridor, may not support this recommendation, and we believe that their views should have precedence.

CHAPTER NINE
Environmental Concerns – Conflict with Nature

It was essentially the potential environmental and social impacts of Ontario Hydro's long-range bulk power facilities programme, published in 1974, as perceived by groups of Ontario farmers with land-use concerns, and by environmentally oriented public interest groups, that prompted the setting-up of this Commission in the first place. "Safeguarding, as much as possible, the environment" is mentioned specifically in the utility's mandate; it is a basic criterion for electric power planning.

By now, we are all familiar with the spaceship Earth concept that springs from the fact that the entire earth-atmosphere-space environment is an ecological unity. Man as an inhabitant of earth interacts with, and is dependent upon, a virtual infinity of micro-organisms, and, of course, untold billions of insects and animals. This incredibly complex network of interdependencies, including man, has been referred to as the earth's ecosystem. We note, in particular, the way the whole earth-space environment is shared. Consequently, man's technologies may not only threaten directly the survival of man himself but also, indirectly, the survival of all earth's co-habitants.

This holistic concept of the environment, central to our inquiry, is nowhere better expressed than in the preamble to the Declaration of the 1972 United Nations Conference on the Human Environment, held in Stockholm:

> Man is both creature and moulder of his environment, which gives him physical sustenance and affords him the opportunity for intellectual, moral, social, and spiritual growth. In the long and tortuous evolution of the human race on this planet a stage has been reached when, through the rapid acceleration of science and technology, man has acquired the power to transform his environment in countless ways and on an unprecented scale. Both aspects of man's environment, the natural and the man-made, are essential to his well-being and to the enjoyment of basic human rights – even the right to life itself.

We need say no more. The message is clearly embodied in Part I of the Environmental Assessment Act, 1975, in which "environment" is interpreted as:
- air, land, or water
- plant and animal life, including man
- the social, economic, and cultural conditions that influence the life of man or a community
- any building, structure, machine, or other device or thing made by man
- any solid, liquid, gas, odour, heat, sound, vibration, or radiation resulting directly or indirectly from the activities of man
- any part or combination of the foregoing and the interrelationships between any two or more of them, in or of Ontario

This act, painted with both a broad and a fine brush, epitomizes the spirit of the UN Declaration. It is probably the most comprehensive piece of environmental legislation to be enacted in Canada and perhaps even in the world.

It is a truism that energy conversion processes, inherent in all of which are varying levels of lost energy,[1] inevitably give rise to environmental impacts. Indeed, this is why so much time during the Commission's public meetings and hearings was devoted to the interrelationship between energy and environment; and why, increasingly, it is being recognized that the planning of an electric power system can be undertaken only in the context of its environmental and social consequences.

But it should not be assumed, *a priori*, that there is inevitably a conflict between the utilization of energy, on the one hand, and the environment, on the other. We note, for example, that the heating of a home, whether by electricity, oil, or natural gas, enhances the environment for those who dwell there. Likewise, sewage works that are energy-intensive improve the environment of a community or city. Nevertheless, even in such apparently clear-cut examples, there are associated adverse consequences. These are obvious when we examine the totality of energy conversion processes involved, beginning with the primary source of energy. Whenever the quality of energy is degraded, through conversion processes, we as individuals may benefit, but at what cost to the environment? Herein lies a major dilemma that faces society as a whole. The balancing of costs and benefits cannot be approached from the standpoint of economic theory and practice alone, although these obviously play an important part, or by the objective assessments undertaken by physical and biological scientists. The reason is that there are no absolute standards for evaluating the quality of the environment. Indeed, in the application of the Ontario Environmental Assessment Act, as well as the enforcement of environmental impact legislation, the cost-benefit balance can only be predicated on the exercise of judgement, i.e., value judgements based on all available information.

The definition of "environment" embodied in the Ontario Environmental Assessment Act, 1975, is broad essentially because there are so many conflicting claims on the environment. In particular, from the standpoint of electric power planning, these claims must be viewed in the light of the province's future need for electricity, the generation technologies that will be employed (e.g., the safety and environmental problems associated with nuclear power and coal-fired generation), and the issues relating to the waste disposal problems associated with these technologies. Consequently, it is increasingly essential to integrate the development of energy policies with the social and environmental policies associated with energy. Furthermore, as we will emphasize in Chapter 10, the conservation of energy, and the means of achieving it, will feature centrally in the resolution of these problems.

As an example of the major financial implications of minimizing environmental impact, the issue of waste heat discharged from thermal generating stations is of special interest. In the United States – this does not apply specifically in Ontario at present – this heat has been defined by law as a pollutant. As a result, many utilities have resorted increasingly to cooling towers rather than using the much simpler and more economical once-through cooling system that involves a large body of water. Thus, to comply with the law restricting heat releases into lakes, etc., an alternative, much more costly, technology has been introduced with different but very real environmental problems. Cooling towers[2] can cause local adverse weather conditions, such as mists and icing conditions, as well as being very unattractive. Not only is the cost of power generation increased because of the many millions of dollars required to build the towers but the efficiency of conversion of primary fuel to electric power is less than with the once-through cooling method.

Clearly, environmental concerns, inherent in electric power planning, are of considerable significance. In the *Interim Report* and in earlier chapters of this volume we have dealt with environmental issues that relate to nuclear power and to transmission lines. This chapter is devoted to the philosophical basis for environmental concerns, and to the issues of air and water pollution in connection with fossil-fuelled electric power generation.

An Environmental Ethic

One of the main difficulties in the majority of environmental assessments is that there are inherent conflicts – the values adhered to by one group are at variance with the values supported by another. These give rise to the adversary nature of environmental hearings and the accompanying atmosphere of confrontation.

The decision-making process, which will be discussed in Chapter 12, is predicated necessarily on choosing one action or process from a set of alternatives. How should relative values be ranked in making this choice? This is not an easy task, especially when some values are not as susceptible as others to quantification. Moreover, such values as human health and happiness, the preservation of beautiful landscapes, the preservation of democracy and human rights, and the value of clean air and water are essentially interconnected and interactive. To rank them is virtually impossible.

Especially noteworthy is the fact that these processes are by no means fully understood, and may not be for many years. And yet decisions involving them must be made by society on an ongoing basis. Obviously, therefore, decisions relating to environmental impacts involve not only qualitative values, as noted above, but also potentially quantifiable factors about which there is a paucity of data (e.g., health effects and climatic effects), and these may carry substantial social penalties. This is one important reason why public input in decision-making processes of an environmental nature is vital. Indeed, when various alternatives are feasible, when environmental values that may change with time are involved, and when decisions once taken are virtually irreversible, the good sense of seeking public input is undeniable.

Although there is consensus that certain environmental characteristics are good (e.g., clean air and water) and others are bad (e.g., air and water pollution), there is no consensus concerning what environmental risks, if any, should be accepted in order to ensure, for example, the continuation of present living standards in Ontario. Clearly, the environmental implications of electric power systems exemplify this point. In other words, if the aggregated effects of excess carbon dioxide in the upper atmosphere are not known, or only partially understood, should we stop burning coal in power plants until the implications are more fully understood, or should we continue to burn coal until it is absolutely certain that serious climatic changes, potentially giving rise to the melting of pack ice in the arctic regions and a reduction in the size of the polar ice caps, are inevitable?

Recognizing that many of the hard decisions ahead, especially in the field of energy policy, will depend on environmental assessments, and recognizing that the high rates of consumption of energy-intensive goods and services continue apace on an individual as well as a societal basis, the Canadian Environmental Advisory Council, a few years ago, formulated an "environmental ethic". Its purpose was to provide decision-makers, and indeed all concerned people, with a basic guideline upon which environmental policy might be developed. The environmental ethic reads:

> Every person shall strive to protect and enhance the beautiful everywhere his or her environmental impact is felt, and to maintain or increase the functional diversity of the environment in general.[3]

It will be noted that the ethic is based, not surprisingly, on the need to preserve the ecological diversity upon which the earth's ecosystem, including man, depends. Without such functional diversity, man's future is threatened. And diversity is always compromised by environmental pollution. Why beautiful? Donald Chant puts it very elegantly as follows:

> The *beautiful* is not only that which pleases our senses in the ordinary meaning but also that which pleases our minds, that which is functional. The beautiful, for example, is not simply a fantastic sunset or a majestic mountain. It also is to be found in a new complex natural system whose multitude of balanced, inter-related parts appeals to our sense of order and the *rightness* of things. Disordered, disturbed ecosystems are ugly: they offend us in the same way that a car buff is offended by a malfunctioning engine. On the other hand, ecosystems in which each intricate part, the species and the roles they play, is functioning smoothly and contributing to the integrity of the whole system are elegant – they have beauty.[4]

Especially from an educational standpoint, the concept of the environmental ethic has profound significance. It facilitates the presentation to the public at large of the concepts of life quality and style and how they are completely dependent upon the way man treats the environment. Like the Universal Declaration of Human Rights, the environmental ethic relates to every human value.

Aesthetic Concerns

Quite apart from the potential toxicity and/or radiation hazards that characterize the effluents and solid wastes from some electric power facilities and mining and milling operations, and the resulting impact on the air, water, and land environments, these wastes usually have undesirable aesthetic and indeed spiritual impacts. We refer, in particular, to the impact on man's psyche and sensibilities of visual, auditory, or odorific environmental insults such as:

- Coal-fired power stations, their fuel dumps and waste disposal facilities – these may cover an area as large as 100 hectares. As well, they are characterized by chimneys several hundred feet high. Note that, in contrast, nuclear power stations and hydroelectric generating stations can be fitted much more effectively into the environment and may not be aesthetically displeasing.
- The corridors required for bulk power transmission lines and lattice towers can sometimes be seen at distances of one or two miles. In scenic and rural areas, it is virtually impossible to blend them into the environment.
- The mine and mill tailings areas associated with uranium mining present a depressing intrusion on an otherwise pleasing landscape.
- Electric power developments in remote areas have major indirect effects on the landscape. For instance, the transportation requirements may be such that new highways and/or railroads will be needed.

It is a truism that Ontario's rural heritage, manifest in its villages, streams, lakes, and countryside, is a highly significant provincial resource that has benefitted many generations of Ontarians and continues to do so. Furthermore, the provincial parks and conservation areas have particular relevance in this regard. The unquantifiable values of solitude and man in communion with nature have special significance. Even the most skilful design and landscaping may fail to mitigate the deleterious visual impact of a bulk power transmission line.

Air Pollution

The term "air pollution" refers to the presence in the atmosphere of contaminants that are harmful to the health of humans, animals, and plants, which may affect the climate (in the sense of giving rise to aberrations in weather conditions and to potential changes in climatic patterns), and which have a generally negative impact on the enjoyment of life. Many contaminants of the atmosphere are produced naturally in vast quantities. For example, a major volcanic eruption, during a single day, can produce an enormous amount of air pollution (e.g., toxic gas and particulates). Furthermore, on average, the natural decomposition of vegetation gives rise, ultimately, through various chemical reactions, to almost 80 per cent of the sulphur dioxide in the atmosphere. The remaining 20 per cent arises from human activities, notably the refining of oil and combustion of fossil fuels. Smoke from forest fires, a wide variety of pollens, and wind-blown dust particles all contribute to what we call air pollution.

In this section we present an overview of our conclusions relating to air pollution, with special reference to the pollutants arising from the coal-fired power generation stations in Ontario. Although the combustion of some coals give rise to traces of radioactive isotopes, and virtually all coals when burned emit low concentrations of such toxic elements as mercury and selenium, we restrict the discussion to the major air pollutants — sulphur dioxide, nitrogen oxides, carbon dioxide, and the particulates. But because the health implications are by no means fully understood it is important to note that the trace elements mentioned above may turn out to be as hazardous as the better-known major pollutants. The annual discharges of pollutants from a typical 2,000 MW coal-fired station operating at 80 per cent capacity factor are given in Table 9.1. Note the marked preponderance of carbon dioxide and sulphur dioxide. The sulphur and ash contents of coals provide good pollution indicators — Ontario Hydro pays special attention to them.

The estimated sulphur content of fossil fuels that have been and are expected to be burned by Ontario Hydro during the period 1970-95 are given in Table 9.2. (In view of the drop in the load forecast, the projected values may be high.) Noteworthy is the decreasing percentage sulphur content. This is largely due to the increasing utilization of low-sulphur United States and western Canadian coal.

Table 9.1 Estimated Air Emissions from a 2,000 MW Coal-Fired Plant at 80 per cent Capacity Factor

		(tonnes)
Carbon dioxide (CO_2)		1,512,000
Sulphur dioxide (SO_2)		200,000
Nitrogen oxides (NO_x)		45,000
Particulates		
Aluminum	1,580.0	
Arsenic	2.0	
Calcium	540.0	
Chlorine	190.0	
Fluorine	13.0	
Iron	910.0	
Potassium	180.0	
Lead	1.0	
Uranium	0.1	
Other	283.9	
Total particulates		3,700
Ash Residue		365,000

Notes: Based on Ontario Hydro's emissions from coal-fired generating stations. Assumed annual coal consumption for a 2,000 MW(e) station with an 80 per cent capacity factor, 4.5 million tonnes. Particulate emissions assumed proportionate to trace analysis of U.S. bituminous coal.
Sources: Donald N. Dewees, "Environmental and Health Issues", "Our Energy Options", Toronto: RCEPP, 1978. Ontario Hydro, "Generation-Technical" memorandum with respect to the Public Information hearings, RCEPP Exhibit 2, pp. 2.2-26.

Health Effects

Although the causal relationship is not fully understood, there is strong circumstantial evidence that exposure to high concentrations of air pollutants can cause acute illness and even death. The most notable single example of the potential lethal implications of air pollution was the December 1952 incident in London, England, when 4,000 fatalities, during a period of four days, were apparently caused by high concentrations of sulphur dioxide and smoke (i.e., particulates). Similar, although not as severe, incidents have occurred in New York City (1953 and 1966). However, in spite of major research programmes into the biological effects of air pollution, there is no clear-cut evidence that sulphur

Table 9.2 Estimated Sulphur Content of Ontario Hydro's Fossil Fuels

Year	Equivalent U.S. coal consumed (million tons)	Sulphur content (million tons)	(%)
1970	8.5	0.213	2.5
1975	10.0	0.200	2.0
1980	21.6	0.342	1.6
1985	22.4	0.345	1.5
1990	26.0	0.394	1.5
1995	32.9	0.428	1.3

Note: Residual oil and natural gas have been converted into equivalent tons of coal using BTU content as a basis.
Source: Ontario Hydro memorandum to the RCEPP with respect to the Public Information Hearings, "Fuel Supply", 1976, Figure 8.7-1.

dioxide is the main hazard because the pure gas is apparently comparatively benign, although the sulphate radical SO_4 may constitute a health hazard.

Our conclusions relating the health implications of exposure to air pollution due to the combustion of coal are:

- There is no scientific evidence to prove conclusively that chronic respiratory illnesses are caused by long-term exposure to low concentrations of air pollution due to fossil-fuelled generating stations. However, the associated epidemiological studies are very difficult to carry out because it is impossible to differentiate between the effects of exposure to air pollution, genetic propensity to respiratory disease, and exposure to pollutants in a work environment. Furthermore, the difficulties are exacerbated by the mobility of the population and its constantly changing character. On the other hand, some studies have concluded that, with normal levels of air pollution in cities, there can be an increase in susceptibility to respiratory infections as well as aggravation of such conditions as chronic bronchitis, emphysema, and asthma. However, as far as we have been able to ascertain, epidemiological studies have failed to distinguish between the individual effects of sulphur dioxide and smoke. But we emphasize that the effects of exposure to a toxic agent in the form of air pollution may have a latency period of as long as 30 years.
- Oxides of nitrogen are a product of all combustion processes. They are the chemical agents that trigger a complex series of photo-chemical reactions that give rise to air pollution conditions characterized by the atmosphere often experienced in Los Angeles. We understand that the key toxic agent is ozone, which is produced when solar energy activates oxides of nitrogen in the presence of certain hydrocarbons. Ozone levels can be many times higher than prescribed levels, especially in the city of Toronto, but these occurrences are comparatively rare and we do not consider average ozone levels in Ontario to be a threat to human health.
- The products of synergistic reactions involving the primary and secondary products of the combustion of coal, especially chemically active hydrocarbons, compounds of heavy metals, and various sulphates and nitrates, may prove to be the most hazardous of all air pollutants. Noteworthy is the fact that some of the compounds, when administered to laboratory animals, are carcinogenic. But it has not been proven that there is a corresponding threat to humans.

Agricultural Effects

It is well known that certain air pollutants are harmful to crops and trees. The two major classes have been identified as acid rain and ozone. Both are attributable, at least in part, to the air pollution caused by the combustion of coal in thermal generating stations.

Above average acidity levels in rainfall (and snowfall) are probably caused by the sulphur oxide emissions of large smelters and coal-burning power stations. However, the formation and transport of acid rain are subjects of intensive research and to date the detailed mechanisms involved are by no means fully understood. But the crop and tree damage attributable to acid rain have been well established and, as we shall explain, technologies for reducing the emission of the sulphur oxides are already available.[5]

The other serious crop pollutant problem, especially in southwestern Ontario, is the damage caused by ozone, which, when present above certain concentrations, seriously affects white bean, tobacco, and potato crops. The ozone, as explained previously, is a derivative of nitrogen oxide emissions from industrial plants, and especially power stations.

It is difficult to estimate the agricultural losses due to air pollution, and especially to identify the sources of some pollutants, because of the complex chemical processes involved. However, we have concluded that air pollution caused in part by Ontario Hydro's coal-fired generating stations is having a negative impact on some commercial crops. In some areas of southwestern Ontario, depending upon weather conditions, this can be serious. The fact that sulphur oxides, nitrogen oxides, and particulates can be transported over hundreds of kilometres and across national boundaries, complicates the problem appreciably.

Methods of Reducing Air Pollution

Although the major sulphur emissions in Ontario are probably caused by the copper-nickel smelters in the Sudbury area, and the nitrogen oxide emissions are due to various industrial processes, Ontario Hydro is contributing significantly. What steps can be taken to minimize these pollutants? An obvious approach is to clean and process the coal before combustion by washing, pulverization, and chemical separation processing. However, the more conventional method is to clean the emissions from the coal-fired boilers.

The methods available for reducing the major air pollutants caused by coal-fired power stations are outlined below:
- The standard method of removing the sulphur oxides (SO_2 and SO_3) is to use limestone as a scrubbing agent. The residual products are calcium sulphite, water, and carbon dioxide. However, a major problem is to dispose of the calcium sulphite. Alternative scrubbing agents and associated cleansing systems that are in various stages of development involve the use of magnesium hydroxide, sodium carbonate, or ammonia.
- All the above "flue gas scrubbers" are expensive and not by any means 100 per cent efficient. Furthermore, because Ontario Hydro is meeting the air quality regulations of the province with a reasonable margin of safety, there appears to be no urgency about introducing the scrubbers. Instead, the utility is minimizing sulphur dioxide emissions by utilizing low sulphur coals (see Table 9.2).
- An outstanding problem is the control of the emission of the oxides of nitrogen. Although conventional scrubbers for the removal of sulphur oxides are reasonably effective, they are not applicable to the removal of nitrogen oxides. Indeed, although several potential techniques show promise, there is no commercially available system for minimizing nitrogen oxide emissions.
- The discharge of particulate matter can be effectively controlled by mechanical filters, by electrostatic precipitators, and to some extent by "wet scrubbers". Ontario Hydro proposes to alleviate the problem by using precipitators.

Because of the complexity of the air pollution problem, not least because of its synergistic aspects, it is extremely difficult to assess the overall impact of Ontario Hydro's fossil-fired stations on the environment.

> *Bearing in mind their very substantial cost, the problems of slag removal they create, and most important the fact that provincial air quality standards are being met at present by Ontario Hydro's policy of using relatively low-sulphur fuels, we recommend that:*

> **9.1 Ontario Hydro should not install sulphur scrubbers at its fossil-fuelled electric power stations as long as the existing policy of utilizing low-sulphur fuels is maintained.**

Although no firm evidence was presented to the Commission in connection with the air pollution effects of burning wood, we understand that the finely divided ash, in the form of particulates, may be difficult to control. Consequently, although sulphur and nitrogen oxide problems would not arise, the particulate problem may, in fact, be exacerbated.

Potential Climatic Implications

During the last few years there has been increasing concern about the potential impact of some categories of air pollution on the future climate of the earth. The topic was raised repeatedly during our hearings. Both particulates and carbon dioxide emissions have been identified as factors that could affect the climate. On the one hand, particulates suspended in the atmosphere absorb and scatter solar energy, thereby reducing the energy reacting on the earth's surface and tending to reduce the earth's average temperature. On the other hand, carbon dioxide molecules in the upper atmosphere trap the low-frequency radiation reflected from the earth's surface and, in effect, give rise to a warming of the surface. This has been referred to as "the greenhouse effect".

At the present stage of climatological research there are conflicting views as to which of these effects will predominate. However, because particulates can be removed from the stack gases (e.g., by precipitators), they are not considered to be as potentially serious as the carbon dioxide problem, which at present can be handled only by reducing the actual amount of fossil fuels consumed – i.e., by energy conservation. Since the carbon dioxide issue is frequently raised during the nuclear versus coal debate, we outline below the nature of the problem.

- Between the years 1860 and 1970, the concentration of carbon dioxide (CO_2) has increased from about 290 parts per million (ppm) in the atmosphere to about 320 ppm. This constitutes a 10 per cent increase. If present trends continue it is considered that "by the year 2020 the amount of CO_2 in the atmosphere could approach twice the current value".[6] A critical factor concerning which there is considerable uncertainty is the degree to which the world's oceans (the ultimate CO_2 sink) can absorb the additional CO_2.

- The CO_2 concentration in the atmosphere could cause significant increases in the surface temperature of the earth within 50-100 years, and give rise to substantial changes in the world climate. An average temperature increase of 2°C could lead to a significant decrease in the size of the polar ice caps with concomitant flooding of millions of square miles of the earth's land surface.[7]

- The CO_2 in the atmosphere plays a critical role in photosynthesis – it provides the carbon which, in effect, consititutes the basis for life. Any major reduction in the earth's forests, especially the tropical forests, which like the oceans constitute a CO_2 sink, could therefore exacerbate the CO_2 problem. It has even been asserted that the net result of deforestation is to create an amount of carbon equivalent to between 80 and 160 per cent of the CO_2 being released by fossil fuels.[8]

- What corrective actions are possible? The most obvious one is to cut down on the combustion of fossil fuels. Another obvious step would be to reduce the level of deforestation, especially in the tropical forests, and to increase reforestation wherever possible.

Land Pollution

Land pollution, associated with fossil-fuelled generating stations, is closely related to air pollution. For example, the products of combustion, especially the sulphur and nitrogen oxides, are the basic constituents of "acid rain". Acid aerosols are created in the atmosphere and fall out, especially during rain showers, on the ground and into lakes. As a consequence, plant growth may be retarded, not least because the acid causes leaching of calcium from the soil. And because calcium is a finely balanced nutrient in many soils, its leaching out inhibits plant growth. As well, as discussed in the previous section, other contaminants that arise directly or indirectly from the combustion of fossil fuels, notably ozone, have highly deleterious effects on certain crops in the farming country of southwestern Ontario.

Increasingly, the problem of municipal solid waste disposal is giving rise to a particularly undesirable form of land pollution. Because large disposal areas are required, and because vermin may constitute an additional health hazard, it is obvious that more permanent solutions to the problem are required. We have noted that technologies are now available for the handling and processing of the combustible component of solid municipal wastes; by combining these with coal and using the mixture to fuel a thermal generating station, a particularly desirable solution is at hand.

Coal-burning generating stations have major inventory problems that do not arise in connection with nuclear power stations of equivalent capacity. Because most of the coal consumed by Ontario Hydro reaches its destination (for example, Nanticoke) via the Great Lakes, a large stockpile of coal must be in place before the Great Lakes shipping season closes. It so happens, of course, that the coal-fired stations are usually in service more in winter than during the summer. Furthermore, the accumulation at the site, in settling ponds, of fly ash and bottom ash and other products of coal combustion occupies sizable areas. This constitutes a form of land pollution as well as being particularly unattractive aesthetically. Fortunately, the building industry and road-building operations (i.e., the major users of furnace wastes) help to keep the waste disposal aspect of the problem under control. However, some of the solid wastes, especially the trace heavy elements, are potentially hazardous and therefore come under environmental legislation. Although present in extremely small quantities, arsenic, mercury, and selenium are toxic elements and there is a possibility that they will be leached into the ground water and inadvertently affect water supplies.

Ontario Hydro is aware of the potential hazards of the solid wastes associated with coal-burning power plants, and steps are being taken to ensure that leaching of the wastes into ground water is minimal. The most comprehensive research programme in North America is being undertaken by the U.S. Electric Power Research Institute (EPRI). The programme has five main objectives:

- definition of the physical and chemical nature of solid wastes
- assessment of the potential impact of utility wastes on human health
- development of resource recovery processes
- development of safe solid-waste disposal systems
- assessment of the economic impact of the management of potentially hazardous wastes

As a corporate member of EPRI, Ontario Hydro is aware of the findings of these research programmes. It is noteworthy that, although power system solid wastes in the form of coal ash have been affecting the environment for many centuries, only during the last few years, as a result of increasingly stringent environmental regulations, has the question of the potential toxicity of these wastes been raised.

Water Pollution

In Ontario, apart from the pollution of some rivers and streams in the Elliot Lake area (e.g., the Serpent River) by radioactive material leached from the uranium mill tailings ponds, the pollution of lakes and waterways that is attributable in part to fossil-fuelled power stations is caused by "acid rain". There is also the important issue of "thermal pollution" of the Great Lakes by thermal power stations.

During the last few years, there has been increasing evidence that Ontario's lakes are being affected by acid rain. As noted previously, this pollutant, which is caused largely by sulphur dioxide emissions from power plants and other industrial sources, causes crop damage and may poison soils and reduce forest yields. It is now believed that acid rain is reducing the pH levels of some Ontario lakes, particularly those in the precambrian shield, and consequently there is a decrease in fish populations.[9] Some fish, notably trout and bass, are very sensitive to pH levels, especially during the spawning phase of their life cycles. However, because sulphur dioxide and associated sulphates can be transported over hundreds of kilometres, it is virtually impossible to ascertain the extent to which the acid rain affecting Ontario's lakes is caused by Ontario Hydro's coal-burning plants, or by the copper-nickel smelters, or by the heavily industralized Ohio River basin and the industralized areas in Michigan.[10] The only obvious solution to this problem is to reduce sulphur dioxide emissions. The problem could be exacerbated by any increase in reliance in the U.S. on coal, as opposed to oil or nuclear energy, for electric power generation.

Thermal Pollution

It is well known that the conversion of energy (e.g., coal to electricity and uranium to electricity) on a large scale at thermal power stations heats the local environment (both air and water). Further, as we have seen, there is a limit to the amount of thermal energy contained in coal or uranium that can be converted to electricity. The second law of thermodynamics indicates the theoretical limits, while the properties of materials and conversion technologies set the practical limits. In the case of fossil-fuelled generating stations, the non-usable energy discharged by cooling water into lakes and rivers accounts for 60 per cent of the potential energy in the fuel. And in the case of nuclear power stations, the proportion of lost energy is 70 per cent. In conventional power generating stations, which use once-through cooling as opposed to cooling towers, the lake or river water used to condense steam from the turbines is heated to about 10°C above the ambient water temperature.[11]

Unfortunately, in conventional thermal stations, although a vast amount of energy is being discharged into the environment, it is low-quality thermal energy and cannot be easily used, for example, for district heating. But in the case of CANDU reactors, the lake water used to cool the heavy-water moderator is at appreciably higher temperatures than the turbine condenser coolant, and it is proposed to use the moderator coolant of a Bruce B reactor to heat several hectares of greenhouses.

The power station warm water discharges are considered to be a pollutant because they speed up bacterial activity, and, indeed, biological activity as a whole. Paradoxically, within limits, the more heat discharged into the lakes, the faster the biological productivity of the ecosystem. But this is an oversimplification of the impact of power system thermal discharges. In particular, we note that, although some mixing takes place, the warm water, being less dense than the cold water, is to be found in the surface layers.[12] Concomitantly, especially in the small bays and inlets in the vicinity of, for example, the Pickering power station, the warm water tends to congregate and may have a harmful effect on the spawning areas near the shoreline. The Commission did not receive definitive quantitative evidence relating to this potential threat, but we believe that research into the subject should be speeded up so that an adequate basis for decision-making is available. The Ministry of the Environment has advocated that the outlet ducts of the condenser cooling system of large thermal power stations should be extended for at least a kilometre beyond the lakeshore. This would ensure much more effective mixing of warm and cool water and would probably eliminate the potential threats to spawning areas. However, not least because of the considerable cost of such extensions, we believe that final decisions should await the results of research programmes being undertaken by Ontario Hydro and by utilities of other countries, aimed at assessing the biological impact of thermal discharges.

Accordingly, we recommend that:

9.2   Ontario Hydro and the Ministry of the Environment should strengthen existing air and water pollution monitoring systems, especially, although not exclusively, in the vicinity of thermal power stations, and environmental impact maps should be prepared for the benefit of the public.

The Nuclear/Coal Choice — Summary of Environmental Concerns

In this and other chapters, and in the *Interim Report*, we have outlined from various points of view (health, environmental, social, and economic) the respective costs and benefits of burning uranium, on the one hand, and coal, on the other. We have stressed also that, to the end of the century and perhaps beyond, and for base-load and intermediate-load generation, Ontario Hydro must rely heavily on nuclear and coal-fired generation, supported by a gradually diminishing proportion of hydroelectric generation. In presenting a summary of the environmental concerns that relate to the nuclear/coal choice, we are aware that, from a conventional environmental point of view, comparisons of this kind are not particularly meaningful (cf., the "apples/oranges syndrome"). But if "environment" is interpreted in the broad sense described earlier in this chapter, we can at least provide a basis for discussion.

The environmental risks associated with nuclear power may be summarized as follows:
- Although very remote, there is nevertheless a finite probability of a catastrophic accident that could eventually cause thousands of fatalities, induce a large number of cancers and cause large areas to be uninhabitable.
- In spite of noteworthy improvement in the short-term containment of uranium mill tailings, the very long-term containment of these tailings remains an unsolved problem. Some of the radionuclides involved have long half-lives and could constitute a threat to the ecosystem. The potential leaching of these radioactive materials into streams, and hence into the Great Lakes and other pathways into the biosphere, could have serious long-term consequences, especially if the worldwide nuclear power programme escalates.[13]
- The problem of disposing of spent nuclear fuel safely has not been resolved. It is a problem with strong social and political as well as environmental aspects. The solution being sought by AECL of depositing such high-level wastes in plutonic rock formations raises the key question of public acceptability of a specific site.
- The environmental risks of decommissioning a nuclear power station are unclear and need to be examined.

The major environmental risks associated with coal- fired stations may be summarized as follows:
- Coal-fired power stations in Ontario emit about 500,000 tonnes of air pollutants annually, which probably already affects the health of some citizens of the province, especially those in the vicinity of the stations who suffer from respiratory disorders.[14] Furthermore, these pollutants, together with air pollutants originating in the United States, have a deleterious impact on the agriculture and the forestry industries of the province. The associated acid rain is proving a threat to many of Ontario's lakes. Also, nitrogen oxides resulting from a variety of combustion processes including coal can interact with sunlight to produce exceptionally high concentrations of ozone in the atmosphere, which can amount to 30 times normal levels during temperature inversion conditions. Such concentrations can be deleterious to some commercial crops.
- The stations produce large quantities of solid wastes that contain, albeit at very low levels, toxic elements such as arsenic, mercury, and selenium, as well as traces of radioactive materials. The safe disposal of these wastes, although they are not considered to be a major hazard, nevertheless poses environmental problems.
- Substantial environmental insults arise in coal-mining areas. The underground coal-miner is subject to a hazardous environment that can result in serious health problems. Moreover, the incidence of fatalities due to accidents in the mines is high.[15]
- Coal-fired stations release large quantities of carbon dioxide. Until comparatively recently, carbon dioxide was regarded as an innocuous trace gas that exists in the atmosphere. But today there is a growing concern that increasing concentrations of carbon dioxide in the atmosphere may give

rise to profound changes in the earth's climate. Few conceivable environmental impacts, excluding those associated with nuclear war, pose a more profound threat to mankind.

As far as health and environmental factors are concerned, the choice between nuclear and coal is obviously not clear-cut. Air pollution, waste disposal, and the hazards associated with mining may be more controllable in the case of the coal option, but we believe that major improvements in the control of the nuclear fuel cycle, especially in the safe operation of the CANDU reactor, are also possible. Consequently, the key environmental concerns in the case of nuclear power relate specifically to nuclear waste disposal at both ends of the fuel cycle. In the case of coal-fired stations, the key concerns are the hazards of toxic effluents; and, although the aggregated releases of carbon dioxide in Ontario are virtually negligible, they nevertheless contribute to the risk of a major alteration of global climate.

Environmental Assessment

The prime purpose of the environmental legislation that is in force in Ontario as well as in many other jurisdictions is, in the words of the United States Declaration of National Environmental Policy, "to maintain conditions under which man and nature can exist in productive harmony, and fulfil the social, economic and other requirements of present and future generations".

We have made frequent reference to the Ontario Environmental Assessment Act, 1975, especially as it relates to the impact of electric power facilities on the environment. We strongly endorse the work of the Environmental Assessment Board which has the key responsibility of "putting the act into practice". But the Board is charged not only with assessing the potential threats to the environment of specific projects, and with advancing ideas for the amelioration of these threats, but also, indirectly, with encouraging environmental studies. The assessment of qualitative factors relating, for example, to choices between alternative sites of generating stations and alternative routes of transmission lines is particularly difficult.[16]

Accordingly, we recommend that:

9.3   Interdisciplinary institutes for environmental research in Ontario universities should be involved more actively in the environmental assessment process.

The Environmental Dilemma

The environmental dilemma associated with energy policy has several dimensions. First, the potential climatic impact of burning coal could lead to a reduction in the land area available for agriculture. Second, replacement of coal-based power generation by nuclear power might lead eventually to widespread use of nuclear fission breeder technology with its concomitant health and environmental threats. Third, heavy reliance on alternative sources of energy involves risks, because, if these sources do not live up to expectations, "crash" energy programmes with minimal concern for the environment would probably be necessary. Fourth, the large-scale development of biomass, utilizing the forests of the earth, would reduce still more the earth's carbon dioxide sink and hence expedite the concentration of atmospheric carbon dioxide.[17]

The implications for future generations will be clear. Is it more ethical to risk insecure storage of radioactive wastes or to risk escalating levels of carbon dioxide in the atmosphere? These are political decisions. Only energy conservation is a certain winner.

CHAPTER TEN

Energy Conservation

The Concept of Conservation

The central idea of energy conservation is that energy resources can be used more efficiently by applying measures which are technically feasible, economically justified, and environmentally and socially acceptable – that is, causing the minimum of change to existing desired life-styles. – Conservation Commission of the World Energy Conference

After many hours of hearings devoted to energy conservation, we have concluded that conservation, in the broadest sense, provides a means for enhancing the welfare of mankind – it is not an end in itself. Its primary purpose is to optimize the world's rapidly diminishing resources of fossil and nuclear fuels so as to ensure that future generations are not deprived of some of the earth's greatest riches or forced to pay unbearably high social and economic costs for them.

In introducing the concept of energy conservation we discussed the laws of thermodynamics in Chapter 4.[1] Because these laws, and the associated concepts of quantity and quality of energy, are basic to an understanding of energy conservation, we will review the salient points here. While quantity of energy can be measured in terms of barrels of oil, tonnes of coal, kilowatt hours of electricity, millions of cubic metres of natural gas, etc., the quality of energy, as we showed in Chapter 4, is a more subtle concept.

Whenever a source of energy, say gasoline, is used to perform a useful task, e.g., propel an automobile, the energy is degraded in the sense that its quality is reduced. The low-quality energy, in the form of conduction and friction losses, cannot be used to perform useful work. This is an extremely important concept. Quite simply, it means that when energy is converted from one form into another some of it is lost and the amount available to perform work is reduced. Energy lost during conversion will have an inevitable impact on the environment.

Available Energy

The need to differentiate between "energy" in the sense of the first law of thermodynamics and "available energy" in the sense of the second law is illustrated by the following example.

Suppose that a car battery is charged with a certain quantity of electric energy and suppose that we thermally insulate the battery to prevent any heat loss. If the battery is then placed on a shelf, the energy of the battery will not change. But, because there is inevitable internal leakage of electricity in the battery, a small amount of thermal energy will be generated, and although this energy cannot escape into the environment, the potential of the battery to do useful work will be slightly reduced. Hence, although no energy has been lost (because of the insulation), the available energy in the battery, and hence the battery's ability to do external work, is somewhat less than it was at the completion of the charging process. Many other examples of how available energy is lost could be cited. At the completion of the process of obtaining electricity from a piece of coal, through combustion, heat exchangers, steam turbines, steam condensers, and electric generators, the available energy represented by the generated electricity is only about 40 per cent of the energy that was in the raw coal.

Any reference to a source of energy that does not take into account quality as well as quantity is meaningless. We illustrate this with another example. It is often stated that the transport of, say, one megajoule (MJ)[2] of energy through a pipeline in the form of natural gas costs less than the transport (i.e., transmission) of 1 MJ of electric energy. This is erroneous when it is considered that the quality of electric energy is greater than the quality of natural gas energy; essentially, all of the electric energy can be used to perform useful work, whereas in the case of natural gas, the process of combustion may reduce the energy available in the gas by as much as 25 per cent. However, we recognize, from a thermodynamic point of view, that it is wasteful to use high-quality electric energy through resistance heating processes, to heat space and water at comparatively low temperatures.

Efficiency of Utilization of Energy

What is the yardstick of energy conservation? It is essentially the efficiency of utilization of a source of energy, and in particular of the "available work" that characterizes the source. Until comparatively recently, the only definition of efficiency of energy conversion was the ratio of the output energy delivered by a process or machine (i.e., an oil or natural gas furnace or a motor car) to the energy input into the process or machine. In a large number of processes, the so-called first-law efficiency turns out to be gratifyingly high – in the case of residential space heating using oil or natural gas, for example, it is about 60 per cent. However, this measure of efficiency does not take into account the quality of the energy source, for a high-quality source of energy may be utilized in a highly inefficient way to carry out low-quality work.[3] Matching a source of energy, bearing in mind both its quantity and quality, to an end use is a very important part of the practice of energy conservation. It puts into focus "efficiency of utilization", which is expressible in terms of so-called "second-law efficiency". This quantifies efficiency as the ratio of the absolute minimum amount of energy required to carry out a specific task (e.g., heat a home, produce a tonne of steel, or propel a motor car) to the available energy actually consumed in performing the task. In other words, with the "available energy" concept, we can measure the efficiency of various processes in terms of the task to be performed rather than in terms of the device used to perform the task. Instead of undertaking tasks in the conventional way by selecting for example, in the case of a space-heating task, an oil or gas or electric furnace, we ask the question, is there a more efficient way? And this question brings us face to face with the technologies of renewable energy resources, heat pumps, optimizing the design of residences, etc.

It turns out that the second-law efficiencies, characteristic of virtually all common tasks, are much lower than the corresponding first-law efficiencies. Indeed, instead of achieving an efficiency of 60 per cent, as mentioned previously, residential and commercial space-heating systems have second-law efficiencies of about 6 per cent; and residential and commercial water-heating systems have second-law efficiencies of 3 per cent.[4]

It is inconceivable that the efficiency of energy utilization, expressed in second-law terms, will ever average even 40 per cent, but the potential for improving the efficiency of most processes and hence for achieving enormous energy savings is nevertheless a realistic goal. We have concluded, therefore, that the concept of energy conservation must increasingly be viewed, not only in terms of minimizing the waste of energy by simple modifications of life-styles (e.g., by adapting during the winter to indoor temperatures of 20°C rather than the normally acceptable 22°C) but, even more important, from the standpoint of choosing the most efficient ways of utilizing energy to perform specific tasks.

However, there is a fundamental constraint that is inherent in the process of matching an energy source to an energy use. It relates to the question of the time needed to perform a specific task. An appropriate high-quality source of energy can undertake a specific task more quickly than a corresponding low-quality source of energy, albeit with less efficiency. The second law of thermodynamics does not involve time. Consequently, in comparing the relative second-law efficiencies of alternative ways of undertaking a specific task, it is most important that an allotted time for completion of the task be stated.[5] Furthermore, the time dimension relates not only to the time required to complete a specific task (in general, the shorter the time, the more energy is required) but also to the economics of energy conservation. The pay-back period required to offset the capital and interest costs of investing in conservation technology is frequently the most important factor in the decision as to whether or not to invest in conservation.

Historically, and indeed prehistorically as well, man has succeeded in harnessing increasingly sophisticated forms of energy for the sole purpose, apparently, of "saving time". However, it is clear that the "least time" criterion of performance will have to give way increasingly to a performance criterion predicated on "least non-renewable energy".

It is important to realize that, although such indices as second-law efficiency can be quantified, other factors relating to energy conservation are essentially normative issues that do not lend themselves to quantification. These include changing mores and life-styles; the protection of the environment; the ubiquity of waste of all kinds, on the one hand, and the need to minimize it, on the other; the ethical implications of high levels of energy consumption in the industrialized nations contrasted with the low levels of energy consumption in the Third World countries; and the social-equity implications of ever higher prices of energy. While most of the arguments relating to the intrinsic value of energy conservation are essentially economic, we hope that in the long run the ethical and moral issues underpinning the concept of energy conservation will be the determining factors in ensuring much more efficient management of the earth's dwindling non-renewable energy resources.

Conservation in Practice

There are two fundamental approaches to conservation: first, we can reduce the energy needed to perform a specific service or task; second, we can gradually transform our society from one relying on energy-intensive industries (e.g., pulp and paper, iron and steel, chemical, petroleum, cement) to a society based on less energy-intensive industry. Clearly, as far as Ontario is concerned, the second possibility is not realistic, at least not on a time-scale of less than, say, 30-50 years. But as the service sector burgeons the per capita utilization of energy will probably decrease. Accordingly, we shall restrict the discussion to the first approach.

In all sectors of society, the most promising energy conservation strategies appear to be:
- investing in higher (than conventional) levels of insulation
- reducing the electric energy consumption by using smaller and more efficient lamps and placing them in optimal locations
- lowering thermostat settings, especially during the night; retrofitting shutters, especially on the north side of houses;
- improving the maintenance of all energy-consuming equipment
- improving the efficiency of electrical appliances, especially electric motors

It is encouraging to note that, by diligent application of such energy conservation practices, energy consumption in the education and health (e.g., hospitals) sectors in Ontario, has decreased markedly during the last few years. This is desirable not only on account of the savings in primary energy consumption, but also as a practical demonstration to student bodies that energy conservation is very worth while.

Programmes of the Ontario Ministry of Energy

The *Ontario Energy Review*, published in June 1979, in addition to providing an overview of Ontario's energy supplies and how energy is consumed, includes an important section on conservation. It begins with the statement:

> The Ministry of Energy would like to see the provincial rate of growth of demand for energy reduced to 2 per cent a year by 1985. This compares with a growth rate of 3.9 per cent from 1966 to 1976.[6]

During the last four years the Ministry of Energy has been conducting a campaign to reduce energy consumption in the province. According to the Canadian Gas Association, the annual consumption of natural gas per residential space-heating customer in Ontario between 1972 and 1977, adjusted for temperature variations, decreased by about 8 per cent. Furthermore, and of special interest to the Commission, during 1978 electricity demand in the City of Toronto increased by only 0.5 per cent (during previous years the average annual growth rate was 5 per cent). According to Toronto Hydro there was also an actual reduction in the peak load for 1978 – the first significant reduction since 1911. Although this may not be a trend, and is very weather-dependent, it is nevertheless encouraging.

Specific steps taken by the government of Ontario to facilitate energy conservation include:
- The removal of the provincial sales tax on energy conservation materials and equipment resulted in exemptions valued at $30 million for the fiscal year 1978-9.
- In co-operation with the Canadian Gas Research Institute, the Ministry of Energy has been undertaking a research programme to improve the design of gas-fired furnaces and hot-water appliances. It is claimed that the commercial development of these improved designs will increase the seasonal efficiency (based on the first law) of gas-fired furnaces (forced-air circulation systems) to above 75 per cent – a marked improvement over the present average of about 65 per cent. Furthermore, the Ministry is actively participating in a long-range research programme aimed at improving efficiency still more.
- Infrared scanners mounted on an aircraft are being used to assess the level of heat losses from buildings and residences in the cities of Lindsay, Stratford, and Peterborough. This information, in the form of thermographs, is being used to demonstrate the positive effects of insulation. To encourage conservation practices in these cities, "thermography clinics" have beeen established.
- A recently completed study commissioned by the Ministry of Energy, "Subdivisions and Sun", sets out various designs of residential subdivisions that capitalize on passive solar energy. This imaginative study shows how energy can be conserved through simple planning measures.
- A study was completed in April 1979 to determine how energy conservation techniques can be applied to skating rinks and arenas in Ontario.[7] There are hundreds of skating rinks and arenas

in Ontario and, if the three case studies presented in this report provide any indication, the potential savings in energy costs would amount to many millions of dollars each year. It cannot be overemphasized that energy-intensive activities such as the operation of ice rinks and arenas are prime examples for in-depth energy conservation studies. The McIntosh and Moeller report is clearly a step in the right direction.

- Several educational and training programmes aimed at enhancing the conservation ethic by providing information and demonstrating conservation techniques are being provided by Ontario community colleges. A major purpose is to train technicians in emerging conservation technologies.
- In the public sector, the Ontario government has established thermal performance guidelines for all new government buildings and, for existing buildings, a target of a 15 per cent reduction in energy consumption over the next five years has been set. In one large government of Ontario office building, for example, energy use was cut in half over a two-year period as a result of "changes to the heating system and the efforts of an enthusiastic operating staff".[8]
- The Ministry of Energy is assisting the Royal Architectural Institute of Canada in the preparation of an Energy Conservation Handbook. This will facilitate the application of energy conservation ideas in the design and construction of new buildings and other developments. In particular, the handbook includes codes, standards, and design criteria for new construction.
- Ontario Hydro is conducting an extensive review of building and street lighting. The programme also covers energy conservation practices relating specifically to major food chains, hotels, restaurants, and arenas.
- During the period 1976-8, Ontario Hydro reduced its annual energy use by 19 per cent.
- In the field of solar energy, the government of Ontario is sponsoring various projects on solar space heating and hot-water heating. These programmes are in a comparatively early stage of development, with the exception of the Aylmer Senior Citizens' Residence – a two-storey, 30-unit building.

There are also a number of government-sponsored energy conservation programmes to help industry:
- To encourage greater awareness in industry of the potential of energy conservation techniques for reducing the energy costs of production, Ontario introduced in 1975 the first energy bus programme in Canada. Buses were equipped to carry out on-the-spot analyses of energy consumption and to identify potential energy savings. During the first four years of operation some 900 companies have participated in the programme and it is estimated that the annual cost savings directly attributable to the programme are in the order of $40 million.
- Complementing the energy bus programme, which is operated by the Ministry of Energy, Ontario Hydro is training more than 200 "energy conservation surveyors" (from Ontario Hydro and the public utility commissions) to determine the extent to which energy consumption in small-to-medium industrial plants in the province can be reduced. Seminars on industrial energy conservation are being conducted in support of this programme.
- A seminar on the economics of industrial co-generation of electricity, co-sponsored by the Ontario Ministry of Energy and Ontario Hydro, was held in December 1978. In view of the potential in Ontario of co-generation as an energy-conservation technology, this seminar was particularly timely.[9]

The Residential and Commercial Sectors

An indication that residential and commercial use of energy will probably grow more slowly during the next 20 years is provided by the projected efficiency improvements expected as a result of changes in the design of equipment, appliances, and structures. For example, the American Society of Heating, Refrigerating, and Air-Conditioning Engineers (ASHRAE) – the professional society responsible for setting thermal energy standards – has set new goals for improvements in efficiency that relate to various key needs. These are given in Table 10.1. The equipment efficiency and thermal performance standards have been shown to be cost-effective. But note that the capital cost of the equipment and appliances is likely to increase as a result of the design changes.

Two notable Canadian examples of the present state of the art of energy conservation in the residential and commercial sector (one actually built and under test, and the other still in the planning stage) deserve special mention. They are the Saskatchewan Conservation House in Regina and the South March Energy Conserving Community in March Township, Ontario. Because both developments are characterized by imaginative social as well as technical innovations, we review them briefly below.

Table 10.1 Efficiency Improvements Expected in New Equipment and Appliances 1975-2000

Item	1975	1980	1990	2000
Space-heating equipment				
Electric	1.0	0.95	0.90	0.85
Gas	1.0	0.80	0.70	0.65
Oil	1.0	0.80	0.70	0.65
Water-heating equipment				
Electric	1.0	0.89	0.80	0.75
Gas	1.0	0.74	0.66	0.60
Oil	1.0	0.74	0.66	0.60
Refrigerators	1.0	0.68	0.60	0.50
Cooking equipment				
Electric	1.0	0.83	0.75	0.70
Gas	1.0	0.67	0.60	0.50
Air-conditioning equipment	1.0	0.80	0.70	0.65
Other equipment	1.0	0.90	0.80	0.75
Single-family units				
Space heating	1.0	0.89	0.89	0.89
Air conditioning	1.0	0.70	0.70	0.70
Apartments				
Space heating	1.0	0.54	0.54	0.54
Air conditioning	1.0	0.45	0.45	0.45

Note: Energy consumption in year given as ratio to consumption in 1975.

Sources: American Society of Heating, Refrigerating, and Air-Conditioning Engineers, "Energy Conservation in New Building Design, ASHRAE 90-75". ASHRAE, New York, 1975. Eric Hirst, "Residential Energy Use Alternatives: 1976 to 2000, "Science", vol. 194, no. 4271, December 17, 1976. p. 1251.

1. The Saskatchewan Conservation House was completed in December 1977, and has been continuously monitored since January 1978.[10] The two-storey house has a total inside floor area of 170 m^2 (1,855 square feet). Its main features include:

- A sealed vapour barrier that minimizes uncontrolled air filtration, and a novel air-to-air heat exchanger.
- Window-shutters that enhance the insulation during night-time hours. Experiments have shown that the operation of the shutters reduces heat loss by approximately 30 per cent.
- Insulation levels in ceiling, walls, and floor that exceed by substantial margins current Canadian standards.
- The majority of the windows have a southern exposure. The passive solar input through these windows accounts for 44 per cent of the heat requirement of the building during the heating season.
- Heat is recovered from laundry and bath water through a waste-water heat exchanger that is expected to reduce the energy requirements for hot water by between 20 and 40 per cent.
- An active solar collector system incorporating 17.8 m^2 of vacuum-tube solar collectors and a 12,700 L water-storage tank designed to provide 100 per cent solar space heating.

It is estimated, with the present configuration, that the space-heating requirement has been reduced to about 5 GJ per year – this compares favourably with approximately 100 GJ per year for conventional houses of the same size. However, this level of energy conservation has been achieved essentially through the improved vapour-barrier technique, greater insulation, insulating shutters, and the air-to-air heat exchanger. The contribution of the active solar system to space heating has been minor.

Although much public attention is focused on the vacuum-tube solar-collection system, it is the energy conservation features which are primarily responsible for the low energy consumption of the building.[11]

The role of the active solar system is to supply energy for both space and water heating. But this has not proved economically viable.[12] The additional cost of the energy conservation features (excluding the active solar equipment) in a $50,000 home – excluding land costs – is in the range of $3,000-$4,000.

The conclusion of Besant, Dumont, and Schoenau is that the economics of the active solar energy component (the solar panels plus associated equipment) compare unfavourably with those of the insulation and passive solar energy components. Of seven alternative energy conservation schemes proposed in the design of a 50,000 square-foot building, an in-depth computer analysis showed that active solar systems (flat plate and evacuated tube collectors) were at the bottom of the list. However, in spite of the fact that active solar space-heating systems did not appear to be economically practical, these systems are appreciably more practicable for providing the building with hot water in summer and warm water, combined with some space heating, in winter.[13]

Although the Regina house was designed specifically to conserve energy during the heating season, it is important to note that, during the summer months, the system provides an excellent level of air-conditioning; the indoor temperature rarely exceeds 25°C even when the outdoor temperature is 34°C.[14]

We understand that approximately 100 passive solar, energy-conserving single-family dwellings were constructed in Saskatoon during 1978 and the early part of 1979. A recent study of the attitudes of home-owners to energy conservation was conducted in the Saskatoon area, and it confirms increasing interest in energy-conserving houses. We quote:

> Energy conservation is of prime importance to this sample of home-owners. That it is a concern should not be surprising, but what is of particular import is the overwhelming desire of home-owners to have energy-conserving houses at present or in the very immediate future.[15]

2. The second example of creativity in the application of energy conservation principles relates to a plan proposed for the South March Energy Conserving Community. The project, which is being sponsored by a private developer, relates to a major subdivision in the Ottawa region. It is planned to build 2,200 housing units to accommodate 6,000-7,000 people. The proposal is to build the community over a 6 to 10-year period. A brief summary of the proposed energy conserving features follows:

- The total community will be designed to minimize the use of non-renewable energy resources.
- The space heating and domestic hot-water requirements for the total community will be supplied by a central district heating system. The boilers are designed to burn wood waste, municipal solid waste, coal, oil, and natural gas.
- The district heating unit will be dual-purpose and will be used for the generation of electricity as well as thermal energy.
- The housing units will be designed to capture as much solar energy as possible and will be insulated to appropriate levels. It is expected that, not including the district heating component, the energy-conserving design features (community compactness and conserving architecture) will result in a 48 per cent reduction in the average energy bill for each unit.
- The community has been designed in a way that keeps all feasible energy-conserving options open.

Although communities of this kind exist in Sweden, in conceptual design the South March total energy-conserving suburban community appears to be unique in North America.

The Industrial Sector

Industry accounts for about 37 per cent of the total primary energy requirements in Ontario and for about 40 per cent of the electric power load in the province. Consequently, even small proportional savings resulting from energy conservation techniques will give rise to large absolute savings in energy. We introduce below some of the techniques that have been brought to our attention.[16]

The energy conservation potential in industry, especially in such industries as iron and steel, pulp and paper, chemicals, and petroleum refining, is considerable. For example, the concept of the waste heat recuperator has been applied in several industries for many years. With recent increases in the price of natural gas, the pay-back period for most recuperator systems is about three or four years. The cement industry is an excellent example – waste heat captured from the exhaust of cement kilns provides potential for recycling energy (in the form of thermal or electric energy) with a pay-back period of about three years. However, even such short periods appear to be only marginally attractive to industry.

Major energy uses such as the mining of raw materials, the manufacture of paper, and the manufacture of plastics suggest potential energy savings that may result from material recycling processes. In Sweden, for example, the pulp and paper industry is virtually self-supporting in its use of energy. We recognize that in some respects the potential of recycling may be marginal because of the costs of shipping recyclable materials, and the cost and convenience factor in the separation of recyclable from non-recyclable materials. Nevertheless, from the conceptual standpoint, the energy conservation advantages of re-using materials is clear, as shown in Table 10.2. The economic viability of recycling is, of course, well known, especially in the steel industry; the recycling of scrap steel is an important component of the industry.

Table 10.2 Energy Savings through Recycling

| Material | Available Energy Cost (kW·h/tonne) | | Savings (%) |
	Original material	Recycled material	
Steel (molten)	14,900	7,300	52
Aluminium (molten)	72,500	2,650	96
Plastics (molten polymer)	14,600	650	96
Paperboard (pulp)	2,120	1,070	50

Note: The available energy required to separate, transport, and process the above materials is not included.
Source: Extracted from R.S. Berry and H. Makino, "Energy Thrift in Packaging and Marketing". "Technology Review", February 1974, p. 41.

The use of industrial waste consisting of crates, boxes, cartons, paper, etc., as fuel for plant heating and cooling also has been reported. Instead of paying the cost of disposing of waste, one company has built a solid waste recycling system that has the following characteristics:

- A saving of 4 million kW·h of electric energy and 15,000 million cubic feet of natural gas annually.
- The change-over from heating to cooling is simple.
- The ash is used as a fertilizer additive – sprayed on a 25 hectare cornfield.

It is notable that, within the last two years as a result of collaboration between industry and the federal government, 15 energy conservation task forces, each identified with a major Canadian industry (e.g., pulp and paper and chemicals), have been established. The purpose of the task forces is to further the voluntary practice of conserving energy (which is preferred by the industries to the alternative approach – government legislation). Each task force sets an energy conservation goal for its industry, and to date energy efficiency increases have been encouraging. Also, tuning boilers, repairing broken windows, and improving the efficiency of lighting can improve an industry's electrical bill significantly.

Agriculture and Food

Ontario's farmers are well aware of the urgent need to conserve energy. During the public hearings, various techniques were mentioned, such as drying corn by solar energy, the use of solar space and water heating in farms and farm buildings, the capture of waste heat (e.g., from the processing of milk) and its recycling, and the generation of methane gas from organic wastes for cooling and space heating. Further, as the price of fertilizers escalates as a result of increasing natural gas prices, the fixation of nitrogen by natural means (e.g., the growing of legumes) becomes increasingly attractive.[17]

The energy involved in the production of food may be regarded as a prime target for energy conservation practices. Energy is required to run tractors and other vehicles, to produce electricity needed in farm operations and in food processing, including refrigeration, to produce fertilizers, and to manufacture farm and food processing machinery, vehicles, and other capital equipment. No in-depth analysis of the energy requirements of food production and processing in Ontario (perhaps in the order of between 10 and 15 per cent of the total energy use in Ontario) has to our knowledge been undertaken. However, some studies in the United States[18] have produced results that are, we believe, significant. In particular, they provide estimates of the aggregated energy inputs (food production and processing) required per gram of protein. For example, the energy input (kilocalories per gram of protein) of feedlot beef is 820; the corresponding input for chicken (broiler) is 220, while for halibut the corresponding energy input (kilocalories per gram of fish protein) is in the order of 120-150 – depending upon the method of home preparation. Clearly the energy needed to produce a gram of protein varies considerably from one source to another; not surprisingly, however, meat is one of the most energy-consumptive foods. On the basis of computations carried out in the United States we have concluded that certain food substitutions and alternative methods of preparing food could lead to appreciable energy savings. For Ontario as a whole, the following examples indicate the level of savings that would be achieved if everyone in the province made certain substitutions:

- Substitution of one pound of halibut (or tuna) for one pound of beef once a month for one year would save approximately one million barrels of oil.
- Substitution of one pound of bread for one pound of beef, on the same basis, would also save about one million barrels of oil.
- Elimination of one hour's use of an electric oven at 200°C once a month would lead to a saving of about 200,000 barrels of oil in a year.
- Substitution of wheat flour for half the nutritional kilocalories at present being derived from sucrose would lead to an annual saving in energy equivalent to about 400,000 barrels of oil.

These conclusions are presented to put energy conservation, with special reference to food consumption and processing, into a different perspective. The conservation ethic has many manifestations in addition to lowered home temperatures and speed limits.

Energy and Communication

Clearly, energy conservation calls for imaginative approaches. The replacement of energy-intensive processes by labour-intensive processes is an obvious approach to conserving energy. Not quite so obvious is the replacement of energy by communication. During the inquiry we have been increasingly impressed with its potential.[19] Society will use highly sophisticated communication and computer technologies in order to stretch out the earth's dwindling fossil-fuel resources. Far less energy is needed to transport "trillions of bits" of information than to move a person from point A to point B.

The costs of physical transport will increase with rising fuel costs, while communication and computer technologies are becoming cheaper (based on the declining cost of logic operations using silicon chip electronics), appreciably more adaptable (as witness the small electronic calculator already widely used by people in all walks of life), and even more relevant. Consequently, there are already strong economic arguments for replacing much of the physical transportation of people and materials by communication processes. For instance:
- With the advent of two-way cable television, coupled with small but nevertheless powerful silicon chip technology-based computers and displays, much more business could be transacted from the branch office and the home rather than at the "central office" thereby reducing the need for travel, "saving time" that could be more profitably used in recreational pursuits
- The concept of the "wired city" is burgeoning and its ramifications and implications for the conservation of energy are clearly impressive.
- Associated with the wired city is the "open university". The success of the Open University in the U.K. has surpassed most expectations. The Ryerson Polytechnical Institute Open College programmes are showing the way in Ontario. A combination of "media instruction" and "live instruction in lectures and seminars" would, we believe, be of great benefit to student and teacher alike and would be energy-conserving.
- The physical manufacture of components can be separated from a central design facility by the use of computer and communication facilities. This will lead increasingly to decentralization of large industrial complexes. The potential is immense, but far beyond the scope of this Report.

For many years, Ontario has been a world leader in computer and communications technologies — indeed, in the early 1950s, a Toronto-based company was ahead of the world in computer logic, modular construction utilizing printed-wiring techniques, and a multiple-address capability.[20] Subsequently, this province has maintained leadership in all aspects of computer technology and as well has pioneered cable TV techniques. Consequently, we believe Ontario is in a strong position to capitalize on the communication-energy trade-off, both domestically and internationally.

The Technology of Conservation

Previous sections of this chapter have emphasized that Ontario is in the early stages of a transition from an energy-wasteful to an energy-conserving society. Not only are we witnessing the emergence of significant changes in human sensibilities and life-styles, but as well the evolution of energy conservation technologies, whose major purpose is to facilitate more efficient use of primary energy sources. Although some, in principle, have been known for upwards of 50 years (e.g., co-generation, solar energy heating, and thermal insulation), their increasing significance during an era of projected non-renewable energy shortage has not been fully recognized until comparatively recently. In this section we reflect briefly on the role of several major energy conservation technologies, and re-emphasize the need for innovative planning, as witness the Saskatchewan Conservation House and the South March Energy Conserving Community.

The technologies we have selected are: insulation and passive solar technology; solar energy water heating; co-generation; electric motors; monitoring and control; heat pumps; and load management. Each has been mentioned previously and each, we believe, will have a burgeoning role to play in the future – see Volume 4. Because it is not the Commission's purpose to discuss technical details we shall restrict the following notes to basic concepts.

Insulation and Passive Solar Technology

The dramatic technical success of the Saskatchewan Conservation House is largely due to the effective use of insulating material, the vapour and infiltration seal, and windows on the south elevation of the house to ensure a high passive solar gain during the daytime in winter (and minimization of thermal losses during the night by use of shutters). A particular virtue of such systems is that the building itself acts as both solar energy collector and storage medium.

The use of cost-effective insulation (in walls, roof, ceilings, floors, and basement) is strongly advocated as a means of reducing space-heating needs. Indeed, through application of the insulation standards suggested below the pay-back could be in about five years. However, to maintain this cost-effectiveness, as fuel and electricity prices increase, the insulation code requirements should be upgraded.[21] It is particularly important to note, however, as far as space heating (of all kinds) is concerned, that the full benefits of high insulation levels and passive solar heating will only be achieved if the building as a whole is designed as an energy-conserving system.

It should optimize the capture of solar energy in winter, minimize heat losses through conduction, convection and radiation in winter (e.g., by minimizing the exterior surface to volume ratio of the building, and optimizing the geometry), and minimize heat gains in summer.

Active Solar Energy Systems

The immediate application of solar thermal energy as an energy conservation technology is generally assumed to be to space heating. However, pending the completion of the National Research Council solar energy studies, and the proposed Science Council demonstration project, there is no conclusive evidence to the effect that active solar energy systems are cost-competitive in Ontario compared with passive solar and insulation technologies. Indeed, the evidence presented in Chapter 4 suggests that the opposite is true. On the other hand, the application of active solar energy configurations to water heating appears to be an increasingly viable technology, with a pay-back period of five to 10 years.[22]

Solar domestic hot-water systems designed to provide water at, say, 60°C must necessarily be active systems. They incorporate solar collectors (either flat-plate or tubular), circulating pumps, and a heat exchanger. The latter isolates the solar-heated water (combined with antifreeze) primary circuit from the domestic hot-water supply secondary circuit. The hot-water tank itself (probably provided with natural gas, preferably, or electric power back-up) provides the storage medium. Since conventional domestic hot-water systems are major users of primary energy resources, it is clear that solar systems have considerable potential as an energy-conserving technology. Similar techniques are applicable to the heating of swimming pools.

The Promise of Co-generation

Conventional thermal generating stations have efficiencies (first law) ranging from 28 per cent to 38 per cent. If, however, the concept of co-generation, in which electric energy and thermal energy are co-generated, is put into practice the efficiency can be in the order of 80 per cent – more than double the efficiency of a single-purpose power plant. In Chapter 4 we introduced co-generation as an alternative method of generating electric power, and either process heat for industry (the preferable mode) or thermal energy for space heating. Co-generation is, not surprisingly, in the forefront of industrial energy conservation technologies.

Ontario Hydro has expressed considerable interest in the technology, even to the extent of organizing a seminar and conducting surveys of the industrial market. Industries that use significant amounts of both electricity and process heat, notably the paper, petroleum, chemical, steel, food, and rubber industries, are prime candidates for co-generation plants. According to Ontario Hydro,[23] it has been estimated that these industries have a potential for industrial co-generation in the range of 700 to 2,000 MW of electric power generation over the next 20 years.

One important attribute of co-generation plants is the flexibility in their primary fuel base. Fluidized-bed combustion processes will burn combustible solid municipal wastes and lignite coals with high sulphur content, and they compare very favourably from an environmental standpoint with most low-sulphur, coal-fired generating stations. Alternatively, more conventional boilers might be fuelled with wood wastes, combinations of wood and coal, and, of course, natural gas.[24]

The range of topics discussed at the seminar on industrial co-generation, co-sponsored by Ontario Hydro and the Ministry of Energy in December 1978, together with the optimistic viewpoints expressed, suggest that this conservation technology will have an increasingly important role in the province. The following points, raised during the seminar, are considered to be particularly significant:
- Ontario Hydro's chief economist, on the basis of co-generation cost studies undertaken in the United States, concluded that "co-generation can generally be an economically viable alternative to purchasing power from Hydro. For such plants operational in 1985, all but the smallest have returns exceeding 15 per cent."[25]
- Co-generation, using waste products, would increase industrial efficiency, cut costs, and even increase the province's competitive position.
- The return on investment in a co-generation system might be as high as 18-24 per cent.

In concluding the seminar, the President of Ontario Hydro, D.J. Gordon, outlined

A programme to be initiated immediately as a result of this seminar:
- The Ministry of Energy can set to work with industry and Ontario Hydro to identify potential co-generation demonstrations and can assess potential incentives that would encourage co-generation.
- Ontario Hydro meanwhile can undertake to visit all potential users of industrial co-generation to discuss in depth the issues raised over the last two days and to assess what assistance we might provide in specific cases.
- And, industry, consultants, and manufacturers can give serious consideration to the formation of a joint venture development for co-generation in Ontario.

The Commission endorses these steps and suggests that they be adopted as guidelines for formulating a co-generation policy for the province. In particular, we believe that the future potential of co-generation will only be realized by the initiation of co-operative programmes between the Government of Ontario, Ontario Hydro, and private industry.

Electric Motors

Electric motors are ubiquitous in home, school, hospital, commerce, industry, etc. Furthermore, they consume a comparatively large percentage of the electric energy generated in the province. Both the Electrical and Electronics Manufacturers Association of Canada (EEMAC) and the National Electrical Manufacturers Association (NEMA) have suggested, and we endorse their proposals, that efficiency indices should be used to characterize motor efficiency. The consumer could then assess the potential savings achievable; it is probable that the demand for higher-efficiency motors would then increase.

Because of its centrality in electric/mechanical conversion processes, the alternating current motor has been subject to considerable research aimed at increasing its efficiency. Although not commercially available at present, but expected in the 1980s, a promising development is the so-called transistorized AC synthesizer (ACS), for use when variable-speed drives are required. The objective is to achieve appreciable reduction in the size and weight of motors and, most important, to increase their efficiency. The authors of a recent article claim: "With ACS technology, it will ultimately be possible to replace large industrial motor installations with comparatively small but equally effective high-frequency units."[26] We urge that Ontario Hydro and the Ministry of Energy should keep in touch with these developments.

Additional factors that relate specifically to energy conservation are:
- The power factor, and hence the efficiency of AC motors, can be improved by installing capacitors, thereby reducing energy losses and costs to the consumer. As the price of electricity increases there will be increasing incentive for this step.

- The conventional induction motor has high core, rotor, and stator losses. Because these motors are "off-the-shelf" items, there is a tendency on the part of the manufacturer to reduce costs at the expense of efficiency.[27]

Monitoring, Metering, and Controlling Energy

One of the fundamental problems relating to the impact of energy conservation on load growth is the monitoring of performance. In particular, the monitoring of consumption trends in the major end uses of electricity is not yet being done. On the other hand, "macro-measures" of the level of energy conservation being achieved by an industry are obtainable in terms of the energy-intensity of the industry.[28]

The concept of energy management and control based on the modern computer is rapidly gaining acceptance. Indeed, some of the energy conservation advances reported in several sectors have undoubtedly been achieved through these sophisticated conservation technologies. Energy management is based on the effective monitoring of energy flows and is especially concerned with the capture of waste energy. Furthermore, just as in the case of Ontario Hydro's Richview Central Control facility, though on a much smaller scale, many large industrial organizations have set up energy control stations with the express purpose of minimizing energy consumption. Based on current research and development programmes, in connection with the metering, display, and communication of information relating to electricity consumption, new techniques with far-reaching implications will be available within the next 10-20 years. We refer, in particular, to the digital chip watt hour meter, combined with digital data storage and a local control facility.[29]

The new meter, using "silicon chip technology", will probably duplicate the accuracy, reliability, and cost of the conventional watt hour meter (which has been in continuous use for more than 50 years). According to a current specification, the single-phase electronic watt hour meter (together with storage, programmable logic, and two-way communication — all based on micro-electronic chip technology), will provide a customer with the capability of monitoring and controlling his load automatically, and will also provide the utility with the capability of shedding specific classes of load when necessary. The new system will incorporate many additional facilities. It will:
- measure and record electricity consumption
- enable meter readings to be taken remotely (thereby enabling accounts to be rendered more frequently than at present)
- retain stored data in case of power failure
- present information to the consumer on demand — e.g., level of consumption and outstanding charges
- in an emergency, enable load-shedding to be effected on a controlled basis
- provide consumers with the facility to record consumption for specific circuits, take advantage of "interruptible load" tariffs, and make payments to the utility through a local control panel.

We do not minimize the magnitude of the task of introducing the new meter technology. It will be at least as massive an undertaking as the conversion from 25 Hz to 60 Hz carried out for most customers in the early 1950s. But the technology is extant, and it will probably enhance energy conservation appreciably as well as the ability of Ontario Hydro to monitor and, if necessary, to control load patterns.

The Heat Pump

In theory, the heat pump (see Volume 4, Chapter 5) is an air-conditioning system run in reverse. An air-conditioning system extracts thermal energy from a room or building and injects the heat into the surrounding environment (the atmosphere), whereas a heat pump extracts heat from a comparatively low-temperature reservoir (e.g., the atmosphere) and uses an external source of available energy (usually electricity) to inject this heat into a building or hot-water cistern. For ambient temperatures in the order of 0°C to 10°C, a coefficient of performance (i.e., the energy delivered by the heat pump divided by the input energy to the heat pump) is in the order of 1.5 to 3. However, as the temperature drops below this, the coefficient of performance drops until it approximates that of the conventional electrical resistance heater, around -10°C. Conventional electrical resistance heaters have a coefficient of performance of slightly less than 1.0. A desirable feature of the heat pump is that, in the summer, it can be used in an air-conditioning mode with coefficients of performance of around 3.0.

Load Management

The importance of load management in total-system planning has been discussed in Chapter 7. However, its role as a conservation technology may not be obvious. The load-management process, in effect, transfers a proportion of the peak load occurring at a specific time to a time when off-peak conditions apply. Because hydraulic and nuclear power, the major base-load generation technologies, provide cheap energy and minimize the depletion of fossil resources, it is incumbent upon the utility to maximize their use. On the other hand, during peak-load situations, a small proportion of the power is generated using gas turbines and diesel generators that burn expensive fossil fuels (notably oil and natural gas). Clearly, therefore, load management *per se* can be regarded as an energy-conserving technology.

Conservation – Socio-Economic Factors

The basic concepts underpinning the relationship between energy conservation and economics, and consequently the social implications (e.g., pricing policies), are treated in detail in the next chapter. However, in view of the centrality of energy conservation throughout the Commission's inquiry, and its socio-economic implications, we introduce briefly in this section some aspects that are not readily separable.

During the last two decades, there has been a trend to reduce labour costs, across the industrial and commercial spectrum of activity, by introducing energy-intensive machines and processes. As recently as 1973, this substitution was expedited by the comparatively low cost of all forms of energy. However, with escalating energy prices, there are now compelling reasons why conservation technologies should be introduced to facilitate a more economical use of energy in the production of goods and services, as well as in the residential, commercial, and agricultural sectors. Until comparatively recently, gross national product (and gross provincial product) has been related linearly to energy consumption. But there is increasing recognition that energy consumption and GNP[30] are gradually being de-coupled in the sense that the energy-intensity of the production of goods and services is tending to decrease. This is a major manifestation of energy conservation as it is interpreted in this volume. Nevertheless, assuming that GNP provides a measure of "standard of living", a certain level of energy consumption is necessary to maintain this standard, and a more efficient use of energy will mean that the standard of living can be maintained in spite of a declining energy/GNP ratio.

While energy conservation will tend to reduce the energy/GNP ratio, we stress that such a change will be gradual. Indeed, this is desirable because a comparatively sudden transition to a "conserver society" could have serious destabilizing effects on the economy.

Thermodynamics and Economics

Optimally, certainly from a thermodynamic standpoint, the price of fuel should reflect the available energy contained in a fuel. It should also reflect the cost of ensuring minimum environmental impact when the fuel is burned, as well as the costs associated with mining, processing, transportation, and handling. Furthermore, the price of the fuel should reflect the resource base and the extent to which it is being depleted.

Awareness of the fact that energy is being used continuously to develop new energy resources is increasingly important because, at the margin, the quantity of available energy used for exploration and development of a new energy source may be almost as great as the available energy inherent in the new source. As non-renewable fuels become more and more depleted and difficult to extract from the ground, the cost in energy, and hence in dollars, per unit of energy increases, and will continue to do so. This is another reason why the concept of available energy and the pressing need to conserve it as much as possible is so relevant today.

Another important economic dimension, with conservation overtones, is "time" – we mentioned its thermodynamic significance earlier. The economic implications of "time to manufacture a product" or "time to perform a service" are well known. Normally reduction of this time constitutes an economic advantage by providing a "competitive edge". However, the more rapidly we undertake some tasks, the less efficiently we use available energy, so balances must be struck.[31] For example, in view of increasing energy prices, there is clearly a limit to production rates on both thermodynamic and economic grounds.

Investing in the Future

Energy conservation practices, such as lowering thermostats at night during winter months, do not require capital investment, while saving non-renewable fuels. Other energy conservation options, such as the improvement of insulation levels and the replacement of low-efficiency equipment, require capital. But precious available energy is saved. The customer must balance the cost of the investment against the resulting saving in energy costs. Taking into account both capital and interest payments, it is relatively easy to determine the pay-back period, subsequent to which the customer's energy bill will be less than it would have been without his investment in energy-conserving technology. Further, and most important, the greater the increase in the real cost of energy, the shorter the pay-back period, and, during the life cycle of the appliance or process, the greater the saving in available energy and consequently in dollars. A critical question is: will capital investments in energy-saving technology yield higher dividends than comparable investments in new supply facilities (e.g., thermal generating stations), especially if social and environmental concerns are taken into account? We believe they will.

By reducing the rate of depletion of the earth's non-renewable resources by energy conservation, we are not only investing in the comparatively short-term future but, far more important, we are investing in the long-term future – in the lives of unborn generations whose uses of the earth's remaining fossil resources could be more effective and desirable than those of their forebears. As Nicholas Georgescu-Roegen rightly points out, "Nature does not have a check-out line for us to pay for the resources we take out; money royalties are set up by people, not by Nature."

Conservation and Employment

While there appears to be universal recognition of the fact that energy conservation can play a significant and positive role in the long-term energy future of the province, the view has also been expressed that energy conservation could slow down the economy, and, concomitantly, the creation of new jobs. The argument is that decreased growth in energy requirements will lead inevitably to decreased growth in the economy as a whole, giving rise to unemployment. In consequence, less income is available for the purchase of consumption goods, and there is a correspondingly reduced demand for these goods, which in turn exacerbates the unemployment situation. The counter argument is that energy conservation measures, coupled with alternative generation technologies, could, in fact, boost the economy and create many new jobs.

Clearly, economic risks are inherent in too precipitate a transition to a conserver society, certainly as far as employment in concerned. And this is a major reason why we urge an energy policy predicated on keeping all realistic options open. In particular, we believe that energy conservation programmes unsupported by evidence demonstrating that their social benefits exceed their social costs should be treated with caution.

Changes in industrial structure will almost certainly accompany changes in energy production and end-use patterns. For example, accelerated construction of co-generation plants, increasing reliance on renewable energy resources, especially wood wastes, municipal wastes, and eventually methanol, will inevitably have an impact on the geographical distribution of industry. Opportunities for small, comparatively labour-intensive business should increase as energy conservation practices proliferate. Further, as intimated previously, the electronics and communications industries, which are relatively labour- as opposed to energy-intensive, should burgeon during the next 20 years. Consequently, both the geographical structure and the mix of industries are likely to be modified appreciably.

Nor should it be forgotten that a lowering of the rate of increase in real salaries and wages is in itself a potent energy-conserving policy. Manifest in the conservation ethic is a change in life-style towards improvement in the quality of work and in education and leisure (i.e., low energy consuming) activities; these need not involve increases in real wages, but they might give rise to enhanced conservation insights and practices.

Conservation through Pricing

There is clearly a close relationship between the price of energy (e.g., electricity), the consumption of energy, and the degree of energy conservation. For example, pricing policies predicated on marginal cost pricing[32] may be regarded as "conservation-inducing mechanisms". The topic will be discussed more fully in Chapter 11.

Three basic electric energy pricing strategies that were brought up during our public hearings may have important implications for energy conservation. Two of these, marginal cost pricing and the phasing out of declining rates (in the sense of the existing inverse rate structure — the more electricity used above a prescribed level, the lower the rate) would probably stimulate conservation. The third, time-of-day pricing, may have an indirect impact since it is expected to change the daily load profile and would therefore facilitate load management. Because its primary purpose would be to reduce peak power levels, and to optimize the use of base-load generation facilities, we can infer that peak power (and hence capital) would be conserved and so would fossil fuels.

Conservation and Government

Conservation, we suggest, and the measures being taken to encourage it, is a central criterion of good government. A pertinent question in any discussion of energy conservation relates to its social and political acceptability. In practice, especially if higher prices of essential commodities result from measures to cut consumption, and if adaptation to different, less wasteful life-styles is involved, the enthusiasm may begin to wane. All levels of government, especially senior levels, have leadership roles to play, through example and demonstration, in ensuring that a conserver society can become a reality.

In a previous section we outlined the conservation programmes that have been initiated by the Ministry of Energy. We believe the Government of Ontario is, in effect, acting as a catalyst. In particular, demonstration projects and provincial government literature relating to conservation are comprehensive. Noteworthy are the special pamphlets and booklets distributed in the school system. Furthermore, the financial support of special interest groups imbued with the conservation ethic has been significant. We cannot do better than to quote from the "Advisory Brief to the Government of Canada by the Canadian Electrical Association":

> There is, however, a limit to what can be achieved by the individual efforts of energy companies or through national associations such as the Canadian Electrical Association. The Federal Government should take a more active and visible lead in the nation-wide promotion of energy conservation through a positive programme which would:
>
> 1) provide funding for rapid development of national specifications governing the design, manufacture and testing of energy-efficient appliances and equipment
>
> 2) establish the requirement that all major electrical goods sold to the consumer provide information on energy consumption and be labelled accordingly
>
> 3) establish tax incentives and depreciation allowances to encourage industrial and commercial enterprises to retrofit existing building structures so as to incorporate more energy-efficient heating/ventilating/air-conditioning systems
>
> 4) carry out a continuing programme of communicating energy conservation practices to the consumer through media- and industry-directed activities
>
> 5) further promote adequate building insulation as a means of saving all forms of energy
>
> 6) continue to set an example to the nation in the manner and form in which it uses energy within its own operations[33]

We strongly endorse these proposals.

We also agree with the Canadian Electrical Association that Ontario Hydro "has been in the forefront in promoting energy conservation as perhaps the single most effective means of immediately coming to grips with our energy problems".

We have concluded that federal and provincial governments should promote laws, not only to provide financial and fiscal incentives, including grants, loans, and subsidies to encourage the development of energy-saving equipment, but also to complement the Canadian Electrical Association proposals through the introduction of efficiency standards and energy audits.

Accordingly, we recommend that:

10.1   Over a period of 10-20 years, efficiency goals for all energy-intensive industrial processing equipment, machines, and systems should be established by the Ministry of Energy. In setting these goals, efficiency standards already being achieved in several foreign countries, notably Sweden and West Germany, should be taken into account. Efficiency goals should be applied in the first place to the pulp and paper industry, the iron and steel industry, the chemicals industry, the petroleum refining industry, and all heat-treating operations.

10.2   Mandatory heating, insulation, and lighting standards should be enacted for new residential

and commercial construction, and these standards should take into account the optimum utilization of passive solar energy measures.

10.3 Progressively stricter efficiency standards for all major energy-consuming appliances, such as water heaters, refrigerators, home furnaces, and air-conditioners should be put into effect through legislation.

10.4 Direct government loans and other economic incentives should be made available to finance the retrofitting of houses, multi-unit residences, and some commercial buildings with conservation equipment, including insulation and, where appropriate, storm windows and shutters.

We are particularily impressed with the measures being undertaken by educational, hospital, commercial, and industrial organizations based on computerized energy-conservation systems. Computer software programmes designed to carry out a multiplicity of energy-management functions, including heating, ventilation, and air-conditioning, have already proved that energy savings in the order of 10 per cent can be achieved. The pay-back period for a system at present being installed by a school board in Ontario is about three and one-half years.[34]

While many imaginative steps have been taken by the government to encourage the practice of energy conservation there are, in addition to those proposed previously, other steps that merit consideration. One of these relates to extending the period of daylight saving time. This we believe would be desirable from energy conservation and life-style points of view. The concept, of course, is not new.[35]

At present, the daylight saving time period extends for six months (from the last Sunday in April to the last Sunday in October). We have concluded that it would be appropriate, as a gesture on the part of a government that it is extremely serious about energy conservation, if the daylight saving time period were to be increased by two months. It would stretch from the first Sunday in March to the last Sunday in October – and eight-month rather than a six-month period.

Conservation and Education

We have concluded, as the major thrusts of this volume make abundantly clear, that energy conservation programmes should be assigned the highest priority and urgency by the government and people of Ontario. Of these we suggest that the most important are the educational programmes in schools, colleges, universities, and the media.

Since its inception, the Commission has been centrally concerned with information, communication, and education. In support of this philosophy, we have initiated a multiplicity of activities – seminars, workshops, public hearings, the educational volume *Our Energy Options*, the educationally oriented Outreach Programme, as well as a plethora of reports and issue papers.

The Basics

Educational programmes, especially in the universities, must be supported by an adequate information base, and should, we believe, be predicated on the following basic needs:

• A recognition of the interdisciplinary nature of conservation and environmental studies is essential. A melding of the traditional disciplines is the first step. The knowledge of specialists in the social sciences, in architectural, mechanical, and electrical design, in materials, in landscaping, and in planning must be expanded so that effective communication across these disciplines is possible.

• In-service training of potential teachers with special emphasis on the development of new theoretical and practical skills and crafts relating to conservation is desirable.

• The development of communication skills so that communication is strengthened between politicians, government ministries, Ontario Hydro, conservation-oriented professionals, and the public with respect to the environmental, economic, and technical dimensions of energy conservation is urgently needed. One of the complaints that was often repeated during many public hearings related to the inadequacy of understandable information, especially concerning such topics as nuclear power and energy conservation, and the inability of most specialists in these fields to communicate with the public.

• Recognition by government that the existence of a high level of conservation and environmental awareness in the general population is a major national and provincial asset. Indeed, without it most efforts on the part of governments and utilities to spawn a conserver and environmentally conscious society will fail. This is why educational forums are so important.

The Students and Teachers

The student body involved in conservation education consists of the entire population of the province from the age of perhaps five years and upwards. This constitutes a special challenge to the press and media. In the formal educational sector — the universities, colleges, and high schools — we suggest the encouragement of:

- student initiatives, experience, and participation as key ingredients
- student participation in community projects
- conservation and environmental educational programmes for qualified students of all ages (The growth of a cadre of professional and non-professional teachers with knowledge and experience in the art and technology of conservation is a prerequisite for a viable educational programme.)

Learning Environments and Processes

Education is neither limited by hours and years nor confined to classrooms, laboratories, and libraries — it is a life-long process. Indeed, as far as conservation and environmental education is concerned, the province as a whole can be regarded as a learning environment in which probably tens of thousands of citizens are actively participating. But they need encouragement, not least in the form of information and access to demonstration projects. Government-sponsored initiatives such as the programmes relating to the free-cutting of designated trees in conservation areas and the innovative programmes associated with the Woodlands Improvement Act are good examples.

We draw attention below to some steps that would probably facilitate energy conservation education on a province-wide basis:

- In fields such as energy conservation and environmental protection we have concluded that the most stimulating learning environments are provided through workshops and seminars rather than formal classes. The "do-it-yourself" philosophy should be encouraged through practical demonstrations of well-established conservation techniques, e.g., insulation needs and techniques and the cost/benefit of various alternative energies. On a more sophisticated level, seminars relating to the management and audit of energy in commercial and industrial organizations, the role of co-generation and how programmes can be implemented, the recycling of materials, and so on, are required.
- The potential of apprenticeships and work-experience schemes in energy conservation technology should be explored with the community colleges and industry. Conservation technology should be recognized as a discipline, albeit an interdisciplinary activity, in its own right. Just as environmental engineering is achieving a measure of acceptability, so will conservation technology.
- In the colleges and universities special learning modules should be developed to enhance student initiatives and all subjects currently being taught, especially engineering, geology, ecology, and geography, should be oriented, at least to some extent, towards conservation issues.

It is fitting that this chapter should be concluded on an optimistic note. A climate of uncertainty such as we are enduring (and which mankind has successfully handled on numerous occasions throughout his history), far from inhibiting learning, stimulates it. Indeed, the more uncertain the environment, the more imaginative our probes. And the probes are the basis of research and education.

CHAPTER ELEVEN

Economic and Financial Factors

According to one dictionary definition, "economics" is the "practical and theoretical science of the production and distribution of wealth". Clearly, without loss of meaning or generality, we can substitute energy for wealth in this definition or, indeed, and preferably, combine them. Not unexpectedly, economic factors have been implicit in many of our previous conclusions.

Although we have restricted the discussion in this chapter to the economic (and financial planning) concepts that relate specifically to electric power planning, we emphasize that electricity must be viewed as a component, obviously an important and essential component, of Ontario's total energy needs. A piecemeal approach to energy policy formulation is neither viable nor meaningful.

In previous chapters, reference has been made to the key role of mathematical models, supplemented by large-scale computers, in the design and operation of a large electric power system. It is perhaps not so well known that mathematical models of the economy, so-called econometric models, are already playing central roles in the development of national and provincial economic policies. For example, because economic policy measures, not least in the field of energy-pricing, influence the dynamics of the economy, it is important to quantify the economic factors and the relationships between them (dynamic as well as time-invariant) as adequately as possible. But herein lies a difficulty. Knowledge of the factors, and even of an adequate structure for the model, is usually insufficient. Furthermore, many influences cannot be explicitly formulated.

An industrialized economy requires continuous growth in order to maintain or increase per capita incomes. This is to be expected because population, with few aberrations throughout the recorded history of man, has increased inexorably and continues to do so. But the key question is – how much energy is really needed if economic growth is to continue? Increasingly, therefore, our interest is being focused on the rate of growth. In Chapter 3, for example, we introduced this concept with special reference to the future demand for electric power, and we developed scenarios that depicted plausible electric power futures, while in Chapter 7 we presented alternative supply scenarios that suggested how these potential demands might be met.

The load forecast[1] is essentially the starting point for the economic as well as the social, environmental, and technological planning of the province's electric power system. The load forecast is the first step in the system planning process. The rate of load growth is partly dependent on the level and structure of electricity rates, which are strongly influenced by the capacity-expansion plan that is developed. The latter, in turn, depends partly on capital requirements relative to capital availability, and partly on the financial policies adopted by Ontario Hydro. We draw attention particularly to the following key points:

- If the load forecast, over a period of a few years, is high, unneeded generation may be constructed. This may be undesirable because idle generating facilities, particularly on a short-term basis, may be costly to the consumer unless some of the energy can be exported.
- If the forecast is lower, over a period of a few years, than that which turns out to be really needed, the utility may have difficulty in meeting the demand, especially during peak periods.[2] Furthermore, if system planning, as is the case with Ontario Hydro, is predicated strongly on nuclear power stations, the long lead times involved in planning, design, construction, and commissioning militate against the provision, in time, of additional generating facilities.
- In spite of the cushioning effect of the reserve margin, and the obviously increasing impact of energy conservation, a lower than necessary forecast could lead to very undesirable circumstances – for example, load shedding, especially during peak periods.
- The major factors that are taken into account in Ontario Hydro's load-forecasting model are economic growth, expressed as growth in employment and productivity, relative energy prices[3] and the short-term load forecasts of the public utility commissions.

Ontario and the Energy Crisis

As long as Canada remains a net importer of oil, and as long as the main source of this oil is OPEC, with the attendant political sensitivities of those countries as well as their ability to raise energy prices, it is justifiable to assume that this country, and more specifically the province of Ontario, will remain vulnerable to disruption of oil supplies. This condition is likely to persist in Canada at least until the mid 1990s, when domestic oil supplies combined with the substitution of natural gas and alternative energy sources may bring about a reasonable level of energy self-sufficiency. But the energy future of the majority of western industrial nations is by no means as promising as that of Canada.[4]

Because Ontario is almost entirely dependent for its fossil-fuel needs on sources outside the province, and because payments for energy (in constant dollars) are virtually certain to increase in the future, it seems inevitable that the province will suffer a transfer of income to fossil-fuel producers. How can this be minimized? There are two basic methods: first, by minimizing the utilization of imported fuels, and, second, by increasing the province's net exports of goods and services.

There are two efficient, reliable, and potentially cost-effective ways of reducing the province's energy deficit.
- Energy conservation, already cost-effective, and likely to become increasingly so, is not only the most reliable and efficient "source" of new energy, but is desirable as well from social and environmental viewpoints.
- Assuming that we wish to minimize oil consumption, there should be stress on inter-fuel substitution – in the short run, natural gas, coal, and uranium for oil, and, in the longer term, the increasing use of renewable energy sources; the cost-effectiveness of some renewable technologies remains to be established but, as the price of oil escalates, there is little doubt that this will be achieved.

With respect to increasing the province's net exports, there is clearly an element of risk; the penetration of new markets, especially during a highly competitive period, is difficult. On the other hand, with imaginative social and technological innovation, especially in areas in which Ontario excels, coupled with an impressive cadre of professional and skilled manpower, particularly in high-technology fields (e.g., electronics and communications), the province could be in a strong position.

A critical factor in the projected fuel-shortage situation is the potential of inter-fuel substitution. We introduced this important topic in Chapter 3 and it is considered in detail in Volumes 3 and 4. It is of particular interest to Ontario because, on the one hand, the province's indigenous energy resources (hydraulic energy, uranium, lignite, and potentially wood and peat) are especially oriented towards the generation of electricity, and, on the other hand, Ontario's major energy imports are oil, natural gas, and coal. The substitution of electric power for oil is clearly of special interest; it is not surprising that many of the public utility commissions are advocating that electricity replace oil in residential space heating.[5] On grounds of thermodynamic efficiency, however, electric space heating is not competitive with fuel oil (or with natural gas) unless heat pumps are incorporated. However, it is now Ontario government policy[6] that, for energy security reasons, oil be replaced wherever possible by natural gas, electricity, and renewable energy forms. Coal and nuclear power will remain the main options for Ontario for many years. Although it is not at present, and will not be in the foreseeable future, a suitable substitute in some major end uses, especially the transport sector[7], electricity will probably prove to be increasingly cost-effective in space- and water-heating applications.

Because the decade 1980-90 will be a period of great uncertainty, it is probable that major conservation measures will be put into effect during this period, alternative sources of energy will be explored in depth, and inter-fuel substitution of an imaginative kind will begin to be implemented. These steps will be necessary because energy demand will have to adjust to whatever supplies of primary energy resources are available. This is the basic reason why all energy options should be kept open. For Ontario Hydro, essentially because the electric power system has surplus generating capacity and in due course will have adequate bulk power transmission capacity to complement the generation capacity, the period 1980-90 will provide a breathing space during which, we believe, certain crucial issues will either be resolved or will be well on the way to being resolved.

Looking ahead over the next decade, it is probable that:
- A comprehensive data base predicated on energy end uses with special reference to electricity will be available and will provide better insights into future energy-consumption patterns.
- The real impact of conservation practices and technology will have emerged, and changing lifestyles will be discernible. We should be able to identify new social indicators relating to both the conservation and the environmental ethics.
- The viability of alternative sources of energy, especially co-generation (including fluidized-bed combustion), solar energy, municipal wastes, and wood wastes, will be established.
- The environmental issues relating to the combustion of coal, and especially those that give rise to

acid rain (resulting from emissions of sulphur and nitrogen oxides) and ozone will be evaluated, and the necessary steps to minimize these impacts will be developed though not yet in place on a large scale in North America. Furthermore, the carbon dioxide issue, manifest in the potentially deleterious impact on climate, should be better understood through the intensive research programmes at present in hand.

• Critical issues relating to the nuclear power programme will be resolved to a much greater extent than at present. For example, if no serious nuclear-power accident (more serious than the Three Mile Island accident) occurs anywhere in the world during the decade 1980-90, we believe that the confidence of the general public in nuclear power will be appreciably enhanced, and nuclear safety as an issue may disappear. This might mean that nuclear power programmes in Canada would not be slowed down on safety grounds.

• The social as well as scientific and technological implications of the management of nuclear spent fuel, including its ultimate disposal, will be much more fully understood. This is a world-wide problem and we expect that considerable progress will be made during the next 10 years. Indeed, as we stressed in Chapter 5, this is a critical issue with respect to the nuclear-power programme as a whole, and its resolution is urgent. We believe, further, that if nuclear power station safety levels, especially in Ontario, are maintained this will enhance public confidence in other nuclear-related issues such as high-level radioactive waste disposal.

• The same remarks relate to the long-term containment of uranium mill tailings. By 1990, definitive conclusions will probably be reached in this problem area.

Ontario Hydro's Impact on the Provincial Economy

Comparative operating, customer, and financial statistics that give an indication of Ontario Hydro's central role in the economy of the province are given in Table 11.1. With the exception of the major banks, Ontario Hydro is the largest corporation in Canada in terms of assets. During the last five years, moreover, about 10 per cent of Ontario's total capital expenditures and more than 70 per cent of the capital spending by all levels of government in Ontario has been undertaken by the utility (see Table 4.1 reproduced from Volume 5). Noteworthy, too, is the benefit derived by the province because of the high Ontario content of Hydro's capital expenditures. Because of the magnitude of these expenditures, and Hydro's status as a Crown corporation, it is clear that the utility plays a major role in the economic life of the province. Indeed, it has become one of the most powerful instruments of economic policy open to the Ontario government.

Table 11.1 Ontario Hydro Comparative Statistics 1976-1978

	1978	1977	1976
Dependable peak capacity (MW)	22,845	21,347	19,677
Primary energy made available (GW·h)	95,373	92,855	90,853
Total Ontario customers (thousands)	2,830	2,775	2,710
Assets (in $ million)	13,163	11,386	9,924
Average number of staff for year	27,850	25,118	24,123

Source: "Annual Report 1978". Ontario Hydro.

Table 4.1 Relative Size of Ontario Hydro's Capital Spending

Capital Expenditures in Ontario ($ million)	1965	1970	1975	1978
Total public and private	4,378.3	6,927.5	12,920.3	15,570
Utilities[a]	689.9	1,334.1	2,930.6	2,972
Government and institutions	884.2	1,299.7	2,003.4	2,094
Ontario Hydro	150.0	511.0	1,442.0	1,537
Percentage of total	3.4	7.4	11.2	9.9
Percentage of utilities	21.7	38.3	49.2	51.7
Percentage of government and institutions	17.0	39.3	72.0	73.4

Note a) Including outlays on heavy-water plants ($250 million in 1975, $254 million in 1978).
Sources: Statistics Canada and Ontario Hydro.

An increase in Ontario Hydro's capital expenditures will eventually induce a greater increase in the gross provincial product (GPP), because the additional expenditures are amplified by complex economic linkages between Ontario industries, and as a result of the circulation of earned income. On the basis of several Canadian econometric models, the so-called "output multipliers" for expenditures in the electricity sector are in the order of 1.5. In other words, an initial $1 billion invested by Ontario Hydro will eventually induce an increase in the GPP of about $1.5 billion.[8] Furthermore, the associated direct and induced employment in Ontario resulting from this investment provides about 48,000 man-years of employment.[9]

To what extent is industry attracted to Ontario on account of the readily available and comparatively low-cost electric power? During the period 1920-40, as well as in the post-war period, many electricity-intensive industries, especially firms involved in processing raw materials, located in Ontario to take advantage of the ready access to markets in central Canada and the United States, abundant raw materials, growing skilled labour force, favourable tax and tariff structures, and low-cost electricity. However, it is now acknowledged by government analysts that electricity prices are of diminishing importance in attracting industry to the province. Because of the province's strength in professional and skilled manpower, it is probably more advantageous to the province to attract industries based on high technology rather than electricity-intensive industries. On the other hand, we do not consider that major out-migration of electricity-intensive industries from Ontario to other provinces or to the United States will take place on account of shifts in the relative price of electricity between utilities.[10] It is noteworthy that employment in the electricity-intensive subsector of Ontario industry has held relatively constant over the last 10-15 years, while growth in value of production of this subsector has been below that of the manufacturing sector average (see Chapter 4, Volume 5).

We have concluded therefore that the price of electricity is unlikely to be a significant factor in the choice of location of a production facility, except perhaps in the case of a very few highly electricity-intensive industries. On the other hand, reliable service is a prerequisite for a broad range of companies — a characteristic that Ontario shares with many other jurisdictions in Canada and the United States. Although a reliable and comparatively inexpensive supply of electricity is a necessary precondition to the establishment of a balanced industrial development programme, this factor alone does little to foster the development.

We therefore recommend that:

11.1 In formulating its industrial policy, Ontario should recognize the need for an adequate and competitively priced supply of electricity, but Ontario should not attempt to compete aggressively for power-intensive industry with provinces with large remaining hydraulic resources.

The Export of Electricity

In Chapter 7 we discussed interconnections between Ontario Hydro and neighbouring utilities from the standpoint of the enhanced reliability and security of individual power systems, especially during emergency conditions. As well, as demonstrated during the last few years, strong interconnecting links provide the potential for profitable export sales, especially sales to the United States (see Table 4.11, Volume 5). In this regard, it is fortuitous that New York and Michigan electric utilities are summer-peaking, while Ontario Hydro is a winter-peaking system. This complementarity facilitates both imports and exports between Michigan and New York and Ontario. At present, Hydro's exports to the neighbouring provinces of Quebec and Manitoba are minimal although imports from them are significant.

The economic and political desirability of Ontario Hydro's export potential is exemplified by the electricity sales to Michigan during 1977-8. These sales involved both capacity and economy power.[11] During the period of maximum sales, Michigan utilities were suffering from a strike in the coal industry and unusual equipment malfunctions, and their system-expansion programme was behind schedule. However, it is considered that the factors that gave rise to the emergency were temporary.

As for the New York Power Pool, the Commission has held discussions with several NYPP members and we have ascertained that the outlook, assuming that the nuclear-power programme is on schedule, is for healthy reserve margins in their system. But since almost half of the generating capacity in the State of New York is oil-fired, we believe there is a comparatively large market for economy sales. In this regard, Hydro-Québec, because of its almost 100 per cent reliance on hydraulic power, and the resulting stable, low, marginal cost, is in a strong position to capitalize on the New York market.

Because the likelihood of spare hydraulic or nuclear power for export is low, export sales in the 1980s will be predominantly from coal-fired stations. With the increasing component of high-cost Alberta coal in Ontario Hydro's coal stockpile, the profitability of export sales may be reduced as the decade progresses. Nevertheless, we have concluded that the export market in the first half of the 1980s looks promising. During the late 1980s, it is more problematical not least because of limitations in the utility's ability to sell competitive surplus, transmission limitations, and the improved system characteristics of potential buyers. The reasons for optimism are:

Fig. 11.1: p. 163

- Ontario Hydro will have excess generating capacity, in the East System, during the next 8-10 years. Figure 11.1 shows Ontario Hydro's estimates of the excess generating capacity in the East System and the utility's expectations for firm and interruptible sales. Note that the West System is interconnected with Manitoba Hydro which, in turn, exports electric power to Minnesota and Nebraska. It is highly improbable that the West System will become a net exporter of electricity.[12]
- Uncertainties associated with the United States nuclear-power programmes, especially as an aftermath of Three Mile Island, as well as the difficulties associated with the rapid replacement of oil-fired generation with coal-fired generation, suggest that during the 1980s our expectations for sales to Michigan, New York, and Pennsylvania may not be unfounded.
- Ontario Hydro announced in early December 1979 that it is negotiating with General Public Utilities (GPU), the parent company of the utility that owns the Three-Mile-Island nuclear plant, for the sale of 1,000 MW of firm power over the period 1985-90. This unexpected development illustrates well the unpredictability of the export market for Hydro power. It also reinforces our view that the costs of maintaining temporary surpluses of generating capacity are less than they may at first appear. Ontario Hydro is studying the economics of a direct tie to the GPU system by a direct-current cable under Lake Erie between Nanticoke, Ontario, and Erie, Pennsylvania.[13]

The outlook for significant export sales to the United States in the 1990s is less predictable. While export sales of nuclear power would obviously be desirable for the Ontario economy, it does not seem a justifiable risk for Hydro to build a nuclear power station (say, of the Darlington type) essentially for the export market. The reason is that, because of the long lead time involved in the building of a nuclear power station, there is a risk that the power will eventually be unexportable (e.g., as a result of major conservation measures in the United States). Furthermore, it may not be feasible to arrange a long-term 30-year firm power contract 10-12 years in advance, considering that Hydro-Québec and Manitoba Hydro may have significant surplus hydraulic capacity and may be in a strong position to handle increasing export commitments. On the other hand, advancing the in-service dates of Ontario Hydro power stations relative to their required in-service date to meet Ontario loads alone, in order to maintain an export potential, would present considerably less financial risk. For example, during the early 1980s, if Hydro's projected load growth does not materialize, then, instead of stretching out or deferring capacity already under construction, the utility could enter into firm sales agreements with United States utilities to export that capacity, albeit on a short-term basis, pending the need for the additional power in Ontario.

We believe that Ontario Hydro has a special responsibility to ensure that firm power commitments do not jeopardize either the generation reserves or the transmission system needed to meet possible Ontario commitments. But our analysis suggests that, even if the Ontario load grows at somewhat greater than the 4 per cent per annum rate we consider to be the high end of the range, Ontario Hydro would, for example, be able to supply a 1,000 MW export load without a decrease in its planned 25 per cent reserve margin (see Figure 11.1, which is based on an average annual load growth of 4.5 per cent).

In its 1979 annual review, the Economic Council of Canada expressed concern that, if current trends continue, Canada will move by the mid 1980s from a position as an energy exporter to a major energy importer and thus contribute significantly to the deficit on current balance-of-payments account. It noted, however, that increased electricity and natural gas exports in the medium term "could reduce that import balance – perhaps even eliminate or reverse it".

We recommend that:

11.2 **The Ontario government should continue to support Ontario Hydro's efforts to utilize its surplus generating capacity by undertaking interruptible or firm sales to neighbouring utilities that are both profitable and in the best interests of the people of Ontario. No firm sale commitments should be made that might jeopardize the generation reserves required to meet Ontario requirements or tie up needed transmission capacity.**

The Cost of Electricity

In this and the next section, we introduce the concepts that underpin the costing and pricing of electricity. During 1977-9 these important topics were the subject of public hearings by the Ontario Energy Board (OEB). The purpose of the hearings was to evaluate a major study undertaken by Hydro on the costing and pricing of electricity. The subject is technically complex, as witness the length of the OEB hearings, the more so because, as we demonstrated in Chapter 9, the external social costs [14] associated with the generation and distribution of electricity are increasingly difficult to quantify. Below, we identify the major components of existing costs, and subsequently we consider the incremental costs involved in expanding the total electric power system.

- Overhead costs are the costs incurred by the utility whether or not any electric energy is used by a specific customer. They include, for example, the cost of administration, accounting, and billing.
- Capacity costs relate essentially to the depreciation and interest charges associated with the provision of facilities to meet the peak load and to provide an adequate margin of excess capacity to ensure reliable service.
- Energy costs relate to the cost of the electric energy actually generated, transmitted, and distributed. Fuel is the major component of these costs and hence the energy costs vary with the fuel mix — from water rentals for hydro stations, at one extreme, to oil and gas fuels, at the other.
- Operational and maintenance costs relate to the cost of operating and maintaining the physical plant as a whole, i.e., generating stations, transmission lines, and distribution networks.
- Social costs are much less determinate than the above costs. Ideally, the optimum operational level of any societal productive activity can be regarded as that level of production whereby incremental social costs are exactly balanced by the incremental social benefits. We recognize that, while theoretically achievable in a highly homogeneous society, the perceptions and values that democratic society places on, for example, a plentiful and reliable supply of electric power are in fact indeterminate.

To put bulk power costs into perspective, we note that over the six-year period 1975-80 there is expected to be almost a threefold increase in bulk power cost. The cost of providing the necessary physical plant (capital and interest) and the cost of fuel are clearly the main components of the total cost of electricity. Illustrative of the impact of the generation mix on the cost of electricity is a recent Ontario Ministry of Energy report that notes that residential electricity bills in Toronto in July 1978 were the second lowest of 15 cities listed, while industrial electricity rates in Ontario were the third lowest out of 18 jurisdictions shown. [15] The even lower electricity rates for Quebec and Manitoba reflect the fact that the utilities of these provinces are based almost exclusively on hydroelectric power, whereas the high electricity rates in New York, Charlottetown, Boston, and Hamburg (West Germany) arise because of a high reliance on oil-fired, and to a lesser extent coal-fired, generating stations. Ontario's electric power system, being based almost equally on hydroelectric, coal, and nuclear power, thus results in power rates that compare very favourably indeed with those of the rest of the world, a fact that may not be fully appreciated by the majority of consumers in the province.

After 1990, at which point Ontario Hydro's base-load capacity will consist largely of hydraulic power and nuclear power, the cost of base-load energy is likely to remain comparatively invariant in terms of constant dollars. On the other hand, because of the continuing escalation in the cost of coal and oil, the cost of energy at peak times on the system will increase markedly. Our analysis suggests that the cost of supplying additional resistance space-heating loads electrically at a 30 per cent load factor would only become competitive with oil or natural gas in equally insulated homes if the prices of the latter fuels were to rise to the equivalent of about $40 per barrel (1978 dollars) compared with $15 (January 1980). This divergence between incremental peak- and base-load costs, essentially between the fossil and non-fossil components of the system, is a departure from historical trends that should in future be reflected in Hydro's rate design.

Of special significance in the planning of an electric power system is the incremental (or marginal) cost of the additional facilities required to service the projected demand for electricity and to replace obsolete facilities. We have been particularly struck by the high levels of uncertainty surrounding these estimates that arise because of the long lead times and the uncertain economic future. In spite of these uncertainties, alternative strategies and plans must be weighed in relation to the cost of additional physical plant and the projected social costs.

Tables A.1, A.2, and A.3, reproduced from Volume 5, show the system costs (in 1987 dollars) corresponding to the following specific demands:

- incremental electrical space-heating demand assuming a 30 per cent annual load factor
- incremental industrial power demand with a 65 per cent annual load factor
- incremental industrial power demand with a 100 per cent annual load factor

In addition to providing the comparative costs of the alternative generation scenarios, the tables also include transmission and distribution capital costs (and losses) on an annual basis as well as the system total power costs and energy costs. These comparisons, although hypothetical if only because the associated social costs in each case are not taken into account, provide important insights into Hydro's cost structure. They were designed to give an indication of trends in the utility's cost structure as well as to serve as a basis for the comparison of electric with non-electric energy (e.g., solar energy) investments when both are evaluated on the basis of public-sector financing conditions. The method we use in developing the comparisons is described in detail in Appendix A, Volume 5. It is based on the determination of the total annual unit energy charges associated with the life-cycle energy costs of the incremental capacity that is required to meet specific end uses. We believe this approach to the evaluation of comparative investments in alternative energy sources to be superior to the so-called "front-end cost" methodology.[16]

Table A.1 System Costs Incurred in Meeting an Increment to a Space-Heating Electricity Load with a 10 per cent Annual Load Factor[a]

	New nuclear			New coal			Existing coal		
	$/kW	$/MW·h	$/million BTU	$/kW	$/MW·h	$/million BTU	$/kW	$/MW·h	$/million BTU
Base station									
Capital[b]	85.5			48.9			0		
Operation and maintenance									
(including heavy-water upkeep)	23.0			14.1			0		
Fuel	13.2			53.8			59.2		
Subtotal	121.7			116.8			59.2		
Reserve station									
Capital[b]	0			0			0		
Operation and maintenance	0			0			0		
Fuel	21.7			23.3			23.3		
Total	143.4	54.6		140.1	53.5		82.5	31.4	
Transmission and distribution losses		4.7			4.6			2.7	
Transmission capital costs[b]		10.9			10.9			0	
Distribution capital costs[b]		4.1			4.1			4.1	
System Total – 1987$		74.3	21.8		73.1	21.4		38.2	11.2
System Total – 1978$		43.7	12.8		42.9	12.6		22.4	6.6

Notes:

a) Generation costs in 1987, expressed in 1987 dollars unless otherwise noted.

b) Annual capital charge calculated using a capital recovery factor. See "Assumptions Underlying Tables A.1, A.2, and A.3".

Sources: Ontario Hydro System Planning Division, Report 584SP, January 1979; Economics Division, Economic Forecasting Series, October 1978; and Memorandum to RCEPP, Generation Planning Processes and Reliability, 1976.

Table A.2 System Costs Incurred in Meeting an Increment to an Industrial Electricity Load with a 65 per cent Annual Load Factor[a]

	New nuclear			New coal			Existing coal		
	$/kW	$/MW·h	$/million BTU	$/kW	$/MW·h	$/million BTU	$/kW	$/MW·h	$/million BTU
Base station									
Capital[b]	85.5			48.9			0		
Operation and maintenance									
(including heavy-water upkeep)	23.0			14.5			0		
Fuel	26.6			109.8			120.7		
Subtotal	135.1			173.2			120.7		
Reserve station									
Capital[b]	24.5			24.5			0		
Operation and maintenance	7.0			7.0			0		
Fuel	27.4			27.4			30.2		
Total	194.0	34.1		232.1	40.8		150.9	26.5	
Transmission and distribution losses		3.0			3.5			2.3	
Transmission capital costs[b]		5.0			5.0			0	
Distribution capital costs[b]		1.9			1.9			1.9	
System total – 1987$		44.0	12.9		51.2	15.0		30.7	9.0
System total – 1978$		25.8	7.6		30.1	8.8		18.0	5.3

Notes:
a) Generation costs in 1987. expressed in 1987 dollars unless otherwise noted.
b) Annual capital charge calculated using a capital recovery factor. See "Assumptions Underlying Tables A.1, A.2, and A.3".
Sources: Ontario Hydro System Planning Division, Report 584SP, January 1979; Economics Division, Economic Forecasting Series, October 1978; and Memorandum to RCEPP, Generation Planning Processes and Reliability, 1976.

Table A.3 System Costs Incurred in Meeting an Increment to an Industrial Electricity Load with a 100 per cent Annual Load Factor[a]

	New nuclear			New coal			Existing coal		
	$/kW	$/MW·h	$/million BTU	$/kW	$/MW·h	$/million BTU	$/kW	$/MW·h	$/million BTU
Base station									
Capital[b]	85.5			48.9			0		
Operation and maintenance (including heavy-water upkeep)	23.0			14.8			0		
Fuel	37.4			160.4			176.4		
Subtotal	145.9			224.1			176.4		
Reserve station									
Capital[b]	24.5			24.5			0		
Operation and maintenance	7.0			7.0			0		
Fuel	48.6			50.7			55.7		
Total	226.0	25.8		306.5	35.0		232.1	26.5	
Transmission and distribution losses		2.2			3.0			2.3	
Transmission capital costs[b]		3.3			3.3			0	
Distribution capital costs[b]		1.2			1.2			1.2	
System Total – 1987$		32.5	9.5		42.5	12.5		30.0	8.8
System Total – 1978$		19.1	5.6		25.0	7.3		17.6	5.2

Notes:
a) Generation costs in 1987. expressed in 1987 dollars unless otherwise noted.
b) Annual capital charge calculated using a capital recovery factor. See "Assumptions Underlying Tables A.1, A.2, and A.3".
Sources: Ontario Hydro System Planning Division. Report 584SP, January 1979; Economics Division, Economic Forecasting Series, October 1978; and Memorandum to RCEPP, Generation Planning Processes and Reliability, 1976.

In summary, if the incremental load has a 30 per cent annual load factor (e.g., electric space heating), existing coal-fired generating stations are much more economical in terms of system cost than either new coal stations or new nuclear stations. New coal and new nuclear stations have very similar unit energy costs in meeting incremental space-heating loads, given that the escalation factors for coal fuel costs and nuclear construction costs are as assumed. In the case of incremental industrial power demands having a 65 per cent annual load factor, existing coal-fired stations (in respect of both energy and power production) are more economical than new nuclear and new coal-fired stations. But a "new nuclear" station is to be preferred on economic grounds to a "new coal" station in 1987 and increasingly thereafter. To meet incremental industrial electricity demands with a 100 per cent annual load factor, new nuclear power stations and existing coal stations appear to be reasonably competitive. However, it should be emphasized that our estimates are very approximate. The purpose of the comparisons is to demonstrate a principle and a methodology rather than to develop actual numbers.

In connection with the space-heating example, it is important to note that the seasonal nature of additional space-heating load gives rise to a decrease in the annual system load factor. This could be offset to some extent by the exporting of surplus power to meet the air-conditioning loads of the United States utilities during the summer months. While additional expenditures on load management could improve the daily load factor, only exceptional measures would have any appreciable impact on the annual load factor.

In view of the results presented above, we believe that the methodology has broader applications in the energy policy and decision-making field.

We recommend that:

11.3   Ontario Hydro should perform system simulations to estimate more accurately the incremental costs of encouraging the substitution of electricity for fossil fuels, especially oil.

The Ministry of Energy should develop comparable cost estimates of alternative means to supply, or save, the same energy at point-of-end-use.

Rate Structures — Concepts and Strategies

Having established how the cost of electricity is determined, we consider now the equally complex subject of pricing electricity, noting that Ontario Hydro is required to supply the people, institutions, and industries of the province with electricity at the lowest feasible cost. As mentioned previously, the Ontario Energy Board has just concluded a major public inquiry into the whole question of the costing and pricing of electricity.[17] Their decision having come down so recently, the Commission has not attempted a detailed assessment of it. The studies we have done and the conclusions we have reached have been limited to the broad principles of rate policy rather than the details needed for rate-setting in practice. Since electricity rate structures, and the associated price of electricity, affect every consumer in the province, some understanding of the changing philosophy of electricity pricing is desirable.

We begin with the concept of the marginal (or incremental) cost of electricity.[18] How do we reconcile current and projected prices of other energy forms with the marginal costs of generation and distribution of electricity? Economic logic suggests that, unless energy policy is predicated on informing the consumer as accurately as possible through the price mechanism what the cost of supplying an increment to energy consumption actually is, there will not be any real incentive for the optimum utilization of energy. The consumer should not be given a distorted view of the true scarcity value of energy as a result, for example, of the subsidization of energy. We have concluded that this argument has considerable merit, especially during a period in which there is virtually universal agreement concerning the desirability of accelerating energy conservation.

Electricity Pricing Policy Issues and Strategies

Some factors of special concern in the establishment of equitable electricity price structures are:
- It is important to recognize the inevitable association between the price of electricity and the demand for it. A recent conclusion based on Hydro's econometric model of electricity demand (used in the 1979 load forecast) suggested that, in the long run, a 10 per cent increase in average real electricity prices would lead to a 6 per cent decrease in the annual peak demand for electricity. Price "elasticity" information of this kind for specific end-use markets or times-of-use is essential, not only in load forecasting, but also in the establishment of electric price structures.
- Essentially because of the relative invariance of the base-load cost of electricity (i.e., the cost of nuclear and hydraulic generation) during the next decade or two,[19] base-load electricity costs will become increasingly competitive with oil and natural gas costs.
- The total unit energy cost of meeting daily or seasonal peak loads will continue to escalate because of rising coal costs and also because of the higher unit capital charges associated with each unit of energy, assuming that nuclear stations, operating at comparatively low annual capacity factors, provide at least some intermediate capacity.
- In the late 1980s there will probably be a marked divergence between marginal peak-load costs and base-load costs. This will arise because of the differential between the cost of operating fossil-fuelled generation and that of operating non-fossil generation. We believe there should be recognition of this factor in Ontario's electricity rate structure.
- The implementation of a change in rate structure is difficult from social, economic, and technological points of view. Customers will somehow have to be made aware of rate projections several years into the future so that they will understand, for example, that when demand has caught up with Ontario Hydro's capacity, probably in the early 1990s, the costs of seasonal peak supply will again begin to rise significantly.[20] Adjustment to higher prices or changed rate structures of any energy form takes time because of such factors as the end-use technologies involved and appliance turnover rates. We recommend to Hydro the strategy proposed by the Economic Council of Canada (in relation to oil pricing) in their 1979 annual review of "steadiness on a well-plotted, well-posted course".
- Energy price increases are quite properly assumed to add to inflationary pressures. However, because electric energy constitutes only a small percentage of the GPP (about 2 per cent of the value-added of goods and services), the impact of electricity price increases on inflation is probably less than generally feared.
- We are convinced that much more credible information, meaningful to the public, should be

available, to ensure that there is a general understanding of the reason for electricity price increases.[21] It is clear that the "lack of trust" syndrome applies both to the major oil companies and to the utilities. A first step in attempting to exorcize the distrust is to provide adequate information and to allow opportunity for the public to participate in the debate.

• We have heard many criticisms of traditional rate structures, and of the criteria upon which they are based. In particular, it has been argued that the traditional declining-block rate structure, in which the unit rate falls for each successive increment of electricity consumption, is neither representative of the true supply cost of electricity nor conducive to energy conservation. Also, the desirability of marginal cost-based rates was raised on numerous occasions during our public hearings, usually as part of the argument that electricity rates should foster energy conservation. Social values, or "externalities", it was argued, have also been inadequately recognized and accounted for in existing rate structures.

• Marginal cost pricing may refer to rates based on either long- or short-run costs. Long-run marginal costs refer to the cost of meeting a given increment to the load once the entire power system has adjusted to the new load. Short-run incremental costs reflect only the energy costs of meeting an increment of demand today, with no increase in system capacity. Seasonal and time-of-day rates (collectively, "time-differentiated rates") are usually justified on the basis of either long- or short-run variations in marginal costs in the different seasons or divisions of the day.[22] On either a short- or long-run basis, these criteria suggest higher rates in Ontario in the winter season than in the summer, and during the day than at night. Such rates should lead in time to an improvement in annual and daily load factors and to a reduced rate of growth of peak demand and, hence, to reduced revenue requirements. Time-differentiated rates would tend to increase the ratio of nuclear generation to fossil-fuelled generation, because the impact on consumption patterns would be to shift the load to off-peak periods thereby increasing the need for base-load capacity, which, by the 1990s, will be served incrementally by nuclear power. This shift would help to reduce the environmental impacts associated with the combustion of coal.

• A conservation rate structure has been suggested whereby a positive rate incentive would be available to consumers who meet specific standards of utilization, either through efficiency-in-use or through low levels of utilization. While this idea clearly has merit from the point of view of encouraging conservation, and although power-factor-correction clauses in rate contracts for industrial customers are an example of this, we believe its implementation on a wide scale would be extremely difficult.

• Life-line rates, wherein a minimum block of electricity is provided each month at subsidized rates, have been suggested as a means of assisting low-income families and senior citizens to pay their hydro bills. It has been pointed out, however, and we agree, that the income tax system (especially if it included a negative income-tax feature) would be a more efficient and effective way to achieve this objective.

• "Load management rates" provide a price incentive in exchange for which the utility has the option of interrupting certain classes of service, such as water heaters or storage furnaces, during certain peak hours or when system reliability is at risk. We have concluded elsewhere that greater ability to control certain types of loads is particularly useful as a way of reducing the need for generating capacity, increasing the resiliency of the system, and lowering overall system costs.

We recommend that:

11.4	Time-differentiated electricity rates (seasonal and time-of-day) should be introduced as soon as possible to as many classes of customers as practicable. Seasonal rates should be introduced first, to ensure that the higher long-run costs of supplying low-load-factor space-heating loads are properly recovered. Time-of-day rates should be phased in as day-night electricity supply-cost differentials become significant and obstacles to metering small customers are overcome.

11.5	Means should be sought to ensure that all customers are made aware of the likely future trend in the costs of providing electricity service in each of the rating periods and end uses selected.

11.6	For rate-making purposes, Ontario Hydro should calculate marginal electricity supply costs in each "rating period" on the basis of the current system expansion plan, for comparison with the expected near-future accounting costs proposed by the Ontario Energy Board.

11.7	Ontario Hydro should include, in its tests of time-of-use rates, assessments not only of customer response concerning willingness to change personal energy habits, but also the required technology.

11.8	To encourage the prudent and efficient use of electricity, such features as declining block rates, uncontrolled flat-rate water heaters, and bulk metering of new electrically-heated apartment buildings should be modified or eliminated.

The Technology of Implementation

As is well known, existing metering methods, based on the traditional kilowatt-hour meter, manual meter-reading, and associated accounting procedures, would be inadequate to deal with the requirements of time-of-day rates. This does not apply in the special case of the large industrial and commercial consumers, where the cost of sophisticated digital-demand metering is not an obstacle. On the other hand, the introduction of a new meter (see Chapter 10), based on silicon chip micro-electronic technology, together with programme storage and logic and communications technology, would not only handle the requirements of time-differentiated rates from an accounting point of view but would also facilitate load-management and conservation. We believe that the basic concept of the silicon chip meter and its potential applications is a breakthrough of considerable significance in the field of electric power system operation.[23] We have noted, for example, that Ontario Hydro, in conjunction with the municipal hydroelectric commissions, plans to test time-of-use rates in residential, commercial, and industrial applications across the province. A range of meters are being investigated for use in the experiment.

We recommend that:

11.9 Ontario Hydro should pursue vigorously the potential of the miniaturized solid-state (silicon chip) meter for mass application and include such meters in its current tests of load-management systems and time-of-use rates. A demonstration project involving perhaps 100 residential consumers should be set up during the next few years.

The Nuclear:Coal Choice

Ontario Hydro's policy, as stated in its 1976 submission to the Commission on generation planning processes, was to build additional capacity in the ratio of 2:1, nuclear to coal. With the reduction in its load forecast over the last three years, the Ontario Hydro system incorporates sufficient coal-fired capacity to ensure that the utility should not require another large-scale coal-burning generating station before the end of the century. Consequently, over the period covered by the Commission's mandate, the economic choice between building nuclear as opposed to coal-fired stations is essentially of academic interest only. Nevertheless, because we expect that these options will continue to have central roles to play until well into the next century, a brief review of their comparative economics as summarized in our *Interim Report* is not out of place.[24]

- Comparison of the relative merits of the nuclear and coal generating options should be conducted on the basis of the full fuel cycles – from exploration and mining at the front end of the cycle to the ultimate decommissioning and disposal of waste at the back end. Both coal and nuclear options have major front-end costs, but the back-end costs of nuclear generation, especially those relating to the disposal of spent nuclear fuel and the decommissioning of the reactor, remain uncertain at this date.
- Nuclear and hydraulic generating stations, once constructed, are less susceptible to inflation. For instance, the capital cost of a nuclear power station comprises about 60 per cent of the station's lifetime costs (including fuel and heavy water replenishment), while in the case of a coal-fired station the corresponding percentage is 15.
- Notwithstanding the fact that the international spot price of uranium increased by a factor of more than 10 during the period 1973-8, while the cost of coal increased by a factor of only three, the comparative economic advantage of nuclear over coal has not been greatly affected. Over an assumed 30-year operating lifetime, the cost advantage of nuclear power compared with coal-fired generation is about 65 per cent at a capacity factor of 77 per cent.
- From an economic standpoint, the obvious choice for Hydro's base-load generation is CANDU nuclear power units in the 850 MW range, i.e., the "Darlington-type" reactor.

Because of Hydro's existing and potential excess generating capacity during the 1980s, and the low order level facing the nuclear industry, it is pertinent to consider the question – is the lifetime economic advantage of nuclear so great that it is justifiable, on economic grounds, to build an additional nuclear station to replace existing coal-fired base-load capacity? The results of an analysis undertaken by Commission staff are summarized in Table B.1 of Volume 5, which is reproduced here. It is assumed that the capital cost of a 4 × 850 MW station is $4.7 billion (1987 dollars), that real coal costs increase at

0.5 per cent per annum, and that capital costs are recovered in equal real amounts over the 30-year economic life of the plant.

Table B.1 Comparison of System Economics: First Year of Operation of a New Nuclear Station and an Existing Coal-fired Station (1987 $/MW·h)[a]

	New nuclear		
	65% ACF	80% ACF	Existing U.S. coal
Capital	14.8[b]	12.4[b]	–
Fuel[c]	5.9	5.9	26.5
Operation and maintenance	4.1	3.4	–
Subtotal	24.8	21.7	26.5
Transmission capital costs[d]	5.0	4.1	–
Total	29.8	25.8	26.5

Notes:
a) Omitted cost items are common among alternatives in this case.
b) Capital cost of $4.7 billion for a 4 × 850 MW station, applying a capital recovery factor of 0.0614 or $84.9/kW per year for 30 years. In the first year, the sinking fund capital charge is $144/kW and straight line depreciation gives about $185/kW.
c) Source – Ontario Hydro, System Planning Division, Report 584SP, January 1979: Coal costs increased 10% to reflect lower combustion efficiency of marginal stations.
d) RCEPP estimate, see Assumptions for Tables A.1-A.3.

Based on this analysis, we have concluded that:
- Assuming a 65 per cent, or higher, annual capacity factor, and neglecting incremental transmission costs, the new nuclear station demonstrates a small economic advantage over the existing coal-fired capacity. If, however, additional bulk power transmission expenditures are required, the balance shifts slightly in favour of the existing coal station. Because of the uncertainties inherent in the analysis, and the closeness of the results, we expect that considerations other than economic optimization would determine which of the alternatives would be preferred.
- By the point at which one-quarter of the capital expenditures on a nuclear power plant have been sunk, it becomes a fairly clear-cut economic decision whether to finish the plant on schedule, regardless of load-growth considerations. This is contingent upon the nuclear plant, as it is brought into service, replacing base-load coal-fired generation. One-quarter of the capital costs of a nuclear plant probably would be spent within three or four years of the start of the construction phase. During this initial period, construction of the nuclear station could be deferred without a cost penalty. This also assumes that other constraints such as delays in the completion of associated transmission facilities do not arise.

Ontario Hydro's target generating mix, in which nuclear and hydraulic capacity will supply almost all base loads, will be attained during the commissioning of the Darlington Generating Station, i.e., in the late 1980s or early 1990s. Subsequently, there will not be a coal-fired station on base-load duty that could be displaced. Maintaining the system expansion programme on its present schedule, if load growth is in the range we predict, would create a real risk of having excess nuclear capacity in off-peak periods in the late 1980s and 1990s. This may require "throttling back" some nuclear units at nights and shutting them down on weekends. This is not impossible, but it is a sub-optimal operating mode for CANDU units. The only committed reactors whose timing remains open to question as a result of declining load growth forecasts are the last two units at Bruce B and the four Darlington units. On the basis of the previously stated "sunk-cost criterion", the construction of the Bruce B units could remain on its present schedule in spite of a declining load-growth forecast, without a cost penalty. The first two units of Darlington are perhaps two years away from the point at which their completion on the present schedule may be justified on the basis of the "sunk-cost criterion". Unless the load-following capabilities of the nuclear units can be improved the physical incorporation of Darlington units 3 and 4 could be difficult, if load growth is only 3 per cent and those units are constructed on their current schedule. Without assuming export markets for excess base-load nuclear generation, or greater load-following capability, the ability to advance nuclear stations to help keep the nuclear option open, and the nuclear industry healthy, is severely limited.[25] If, as expected, annual load growth falls below 4 per cent, so that base-load capacity beyond Darlington would not be needed to meet the Ontario requirement until the late 1990s, large cost penalties could arise as a result of advancing the next nuclear station after

Darlington. Furthermore, if better alternative investment opportunities in Ontario were to be foregone because of these additional capital expenditures on nuclear power, such a policy of advancement would not enhance provincial economic performance.

The Economics of Alternative Energy Sources

Recognizing that the economics and financing of alternative and conservation technologies is likely to determine the extent to which they are adopted on a commercial scale during the next 20 years, we have selected and analysed in Volume 5 of this Report energy technologies that, to the year 2000, appear to hold particular promise for Ontario: co-generation, insulation, and solar space heating and water heating. We believe that each of these is technologically viable and that each has a role to play in enhancing Ontario's overall energy situation, especially with respect to energy conservation. Each technology will act to reduce the demand for centrally generated electricity and hence to stretch out the need for additional bulk power facilities.

Industrial Co-generation

In Chapters 4 and 10 we discussed the co-generation of electricity and thermal power as essentially a conservation technology, emphasizing its thermodynamic efficiency. About one-half as much fuel is used to produce the electricity and steam together as would be required to produce the two separately. We also indicated how co-generation might fit into Hydro's future generation mix, pointing out that because it can be added in comparatively small increments with short lead times, co-generation can enhance the responsiveness, or fine tuning, of the total electric power system. It remains to assess, however tentatively, the economics of co-generation from the standpoint of Ontario Hydro's long-range planning programme. We note, for example:

- In the 1980s and early 1990s, natural gas would probably be chosen in preference to coal as the boiler fuel for co-generation on the grounds that the capital costs of the co-generation units will be sufficiently lower than for coal-fired units to offset higher fuel costs. We have concluded that gas-fired co-generation appears to be a thermodynamically and economically efficient addition to the Ontario natural gas market, assuming adequate supplies.
- Because natural gas prices are expected to escalate more than coal prices, the choice, on economic but not necessarily on environmental grounds, will likely be to revert to coal during the 1990s, or earlier if the increase in natural gas prices is sufficiently dramatic. Coal's attractiveness would be enhanced if economical fluidized-bed-based co-generation systems became available. Although not comparable, on environmental grounds, with the excellent characteristics of natural gas, fluidized-bed combustion of coal offers increased efficiencies and reduced sulphur oxide emissions compared with conventional coal-fired boilers.
- Assuming the same financing and coal-purchasing conditions enjoyed by Ontario Hydro, we have concluded that coal-fired co-generation will remain cost-effective in comparison with nuclear power in base-load applications for some time to come.
- The maximum potential for retrofitted and new industrial co-generation projects might realize as much as 2,300 MW of additional electric generating capacity in Ontario by the end of the century under the assumed conditions.[26] We estimate that the capital cost of this upper bound would be in the range of $1.3 to $2.6 billion (1978 dollars), depending on the mix of gas- and coal-fired co-generation.

Insulation

The concept of conservation of home heat by means of insulation, vapour barriers, and the capture of otherwise wasted heat by means of novel heat exchangers has been introduced in Chapter 10, with special reference to the Saskatchewan Conservation House. In view of the obvious success of this experimental home, and of related conservation systems based on insulation and related technology, we have undertaken an analysis of the costs and benefits of the approach. Full details are included in Volume 5, Chapter 5. Our main conclusions are summarized below.

- We have shown that an investment in the order of $1.3 to $1.8 billion (1978 dollars) over the next 20 years to provide higher insulation levels and improved air infiltration controls in new non-apartment dwellings in Ontario would be cost-effective, and appears to be highly desirable. We have concluded as well that an investment of $1 billion on home attic insulation through retrofitting would also be cost-effective. Both electrically and non-electrically heated homes are included in these totals. A general conclusion is that, where such cost-effective conservation measures are available, they should be implemented prior to introducing electric resistance heating to displace fossil fuels for home heating. Although we do not believe that reduction of residential heat losses by the above techniques is likely to be achieved entirely through the action of market forces, we nevertheless suggest that it would be a waste of non-renewable energy resources if generating capacity were installed by Hydro to meet a home-heating load that had not been minimized by conservation measures to the extent justifiable on a life-cycle cost basis.

• The province may be better served by developing thermal energy conservation standards for all new housing stock in Ontario that are consistent with the cost of capital to the government and are based on the same 30-year pay-back period that is applicable to nuclear generating capacity.

• In assessing the insulation standards for new electrically heated homes, and perhaps even for gas-heated homes, we have concluded that it would be appropriate, because of the comparatively low load factor associated with space heating, to base these on the long-run marginal cost of electricity. For example, nuclear power with a 30 per cent annual load factor is competitive with natural gas for home heating only when gas is priced at a 100 per cent energy equivalence with oil at $40 per barrel (1978 dollars – assuming a 60 per cent conversion efficiency in the home).

• Although the insulation measures outlined above would boost the cost of new houses, the increment in cost might be mortgaged separately and, perhaps, guaranteed by the province. Furthermore, the financing arrangements could be such that the incremental costs involved would be less than the differential involved in paying a higher fuel bill.

The comparatively low cost of the insulation and associated measures built into the Saskatchewan Conservation House, together with a correspondingly low pay-back period, suggest that our analysis of this alternative form of "energy generation" is credible and justifies a major development programme in this field.

Solar Space Heating and Water Heating

As we have pointed out, solar energy is both plentiful and free. The problem is how to collect it efficiently and economically. Our analysis in Volume 5 is based on the data provided by three solar-energy studies undertaken on behalf of the Commission.[27]

The third study concerned the cost-effectiveness of solar space heating and water heating. Our main conclusions are as follows.

• Solar space- and water-heating systems are characterized by high front-end costs. They become economically viable only when life-cycle costing is performed using low discount rates and high conventional fuel price scenarios. For example, assuming Hydro's real discount rate of 4.5 per cent, combined solar space- and water-heating systems in new multiple-family dwellings only become cost-effective when home-heating fuel bills (based on 1978 dollars) are doubled in real terms.

• Although combined space-heating and hot-water systems in new or existing single-family dwellings do not appear to be justified on economic grounds at this price level for conventional fuels (i.e., double the 1978 dollar real cost), solar water-heating systems may be viable for most single-family dwellings or existing multiple-family dwellings.[28]

• On the basis of the aforementioned studies, and assuming that our long-run incremental cost estimate of electric space heating is correct, solar space heating and water heating could become competitive with electricity by 1990. But because more than one-half of the new houses projected for the period 1980-2000 will have been built by the time solar systems are cost-competitive, it is improbable that there would be large-scale implementation of solar energy before 2000 (even assuming that the infrastructure could be in place – an unlikely assumption). Consequently, cost-effective investment in solar energy to the turn of the century would probably not exceed $300 million. Furthermore, the studies undertaken by our consultants were completed before new evidence became available concerning the viability of active solar space-heating systems (see Chapter 10). In particular, the studies did not take into account solar energy system requirements (especially as they relate to the need for large thermal storage systems) to cope with "worst credible weather conditions".[29]

Implementation of Alternative Energy Sources

In our view, industrial co-generation, home-heat conservation (i.e., insulation, passive solar, vapour barriers, and heat exchangers), and solar water heating may constitute three of the most important approaches to the improvement of the efficiency of energy use in Ontario. We did not analyse the so-called passive solar systems, which probably rank ahead of any active system. Each could (and, though on a small scale, is doing so at present) reduce the need for centrally generated electric power. Although these alternative energy technologies have reached a comparatively mature stage of development [30], there continues to be a need for large-scale demonstrations at the frontiers of the technology with special emphasis on the social, economic, and environmental implications. We urge collaboration between the federal and the provincial governments in their sponsorship.

We recommend that:

11.10 In analysing the options for increasing the province's capacity for energy self-sufficiency, a systems approach should be adopted in which the incremental costs of conventional electrical generation are compared with the unit costs of conservation or renewable energy technologies, taking into account the load characteristics of each end use.

11.11 Because of institutional and financial obstacles facing decentralized, heavily "front-ended", alternative energy and conservation programmes, and in view of the redeeming social importance of reducing Ontario's oil dependency, provincial loan guarantees, tax and fiscal incentives, and direct financial support should be made available to promote industrial co-generation, heat-loss and building-design standards aimed at optimizing energy-conservation investments, solar water heating, and passive solar systems. A mini-utility, backed by the Ontario Energy Corporation, should be considered, to support industrial co-generation initiatives.

At this time we do not advocate additional provincial government assistance for more than research and demonstration of active solar space-heating systems. Because several fully monitored (by the National Research Council) solar-heated homes are operating at present, and in view of the possibility of a major demonstration programme being initiated, as suggested by the Science Council of Canada, we believe important new information will be available within a few years. The alternative energy and conservation programmes described in Volume 5 involve spending between $4 and $7 billion (1978 dollars) over the next 2 years. Financial assistance from the province would probably be needed for something less than one-half of the total cost. Though significant, such a financial burden would be less than the $6 billion (1978 dollars) reduction in Hydro's capital expenditure programme for 1979-94 that occurred as a result of the downward revision of the load forecast between 1978 and 1979.

Employment and Energy

According to a study commissioned by the Ontario Economic Council[31] during the period 1950-77, the Ontario labour force grew from 1.8 million persons to over 4.0 million persons. This represents an average annual growth rate of 3 per cent. During the past decade the average annual growth rate approached 4 per cent. One reason for this increase has been a marked increase in female participation rates since the mid 1960s. However, because the overall population growth rate has slowed noticeably and is expected to stabilize at about a 1 per cent average over the next 20 years, there will be a corresponding drop in the annual labour force growth rates to about a 1.6 per cent average to the year 2000. Even so, according to the projection, there will be an additional one million persons in Ontario's labour force by the year 1987.

Although an essential component in job creation and the maintenance of employment, the availability of energy is obviously not the only requirement. Nevertheless, the short-term impact on employment of shortfalls in the supply of energy can be severe, as witness the situation in the northeastern United States during the winter of 1977-8. On the other hand, over the long term, many adjustments in lifestyle, especially those relating to energy conservation which will have an impact on the relationship between energy and jobs, will evolve gradually and not disruptively. And, as we have suggested, new employment opportunities will emerge as new technologies become commercially available. We have alluded to some of these in Chapter 10, together with our reflections on the employment implications.

As an example of the complexity of the energy/employment relationship, consider how labour and/or capital can be substituted for energy in many productive operations. But even when cost-effective, given the recent energy price increases, such substitution may give rise to falling labour productivity levels. This trend has been reinforced by the gradual transition in Ontario from manufacturing jobs towards lower-productivity, service-oriented jobs — at present 60 per cent of jobs in the province are in the service sector. Statistics presented in Volume 2, Tables 2.7 and 2.8, show that the same trends are taking place across Canada, especially in the commercial sector.[32]

Direct Employment — Ontario Hydro

High technology, such as that embodied in a modern electric power system, requires cadres of highly skilled technicians, engineers, and managers. As well, and gratifyingly, during the last few years Ontario Hydro has hired an increasing number of social scientists, biologists, and economists to deal with the social, environmental, and economic aspects of the corporation's activities. Because it frequently takes many years to assemble a team of competent professionals and technicians, it is particularly important that a reasonable level of stability of employment be assured. During a period of rapid growth in the demand for electric power, the problem was not so much one of maintaining stability of employment but rather one of acquiring competent new staff. However, with a slowing-down in the rate of load growth the utility anticipates a reduction in total manpower requirements (staff and construction employment) during the next four years, followed by five years of slow expansion and then rapid growth subsequent to 1988. Total regular staff requirements show a steady increase of between 1 and 2 per cent per year to the year 1988 followed by a rate of increase that approaches 6 per cent per annum. These projections are predicated on an average annual load growth to the year 2000 of 4.5 per cent and on the generation plan adopted in 1979 by the Board of Ontario Hydro. Staff requirements will obviously be less under Ontario Hydro's 1980 forecast of 3.4 per cent.

Obviously, Ontario Hydro will have to undertake a searching scrutiny of its manpower resources and requirements over this period. However, as a consequence of the energy-demonstration projects (e.g., co-generation, fluidized-bed combustion, solar water heating, and micro-electronic metering) in which Hydro will be an active participant, some new employment opportunities will be created.

Employment Impacts of Energy Investments

Several submissions to the Commission (that of Energy Probe on August 7, 1979, was a notable example) argued strongly that a programme of renewable energy and conservation investments generates more employment than a similar investment in electricity generation and fossil-fuel extraction. Other advantages claimed for alternative energies and conservation were that employment in these activities will be more dispersed geographically, will have lower skill requirements, will favour small business, and will be somewhat more permanent than employment in the conventional energy sector.

On reviewing the literature on this issue, we have concluded that such unambiguous conclusions can only be supported by analyses that focus excessively on first-round employment effects. More sophisticated analyses suggest that the regional employment impact per dollar invested cannot be distinguished when equal expenditures of roughly comparable regional content are made on conventional energy supply systems or conservation/renewable energy systems.

Any preference for one or the other method of serving the same end use arises chiefly from differences in cost-effectiveness as determined by a uniform evaluation of the alternatives. As a priority, undertaking those energy sector investments that either supply or conserve energy at the lowest unit cost will allow the energy purchaser to spend a larger share of his budget than he otherwise could on other goods and services, which most likely will have a much higher employment intensity than prevails in the energy sector.

When two otherwise equivalent investment opportunities differ in Ontario content (e.g., conservation/renewables vs. imported oil and gas), clearly the investment with the greater Ontario content will have the greatest impact on provincial employment. Should nuclear power and conservation/renewables become competitors in supplying the same end use (e.g., space heating), the Ontario content of each is so high that it is unlikely that this factor will make much difference to the estimate of jobs created.

Equally important is the distribution of jobs over time and the resistance that can be expected to occur as jobs in some energy sectors are lost, even if more employment opportunities are created in other sectors. Such transitional problems will require imaginative manpower retraining programmes and, possibly, relocation assistance.

Earlier in this chapter we concluded that there is a considerable potential for conservation/alternative energy investments to compete effectively with nuclear power. We can now add that from Ontario's perspective the important task is to continue to study such options and to concentrate on implementation strategies.

Our general conclusions are:

- The existing generation of economic models cannot satisfactorily compare the employment impact of energy supply, particularly at the provincial level, and conservation/renewable energy investments, even assuming that expenditures and regional content are equivalent.

- The choice of energy technology to serve a specific end use, on economic but not necessarily on social and environmental grounds, will depend largely upon the relative cost-effectiveness of the alternatives. The encouragement of investment in energy sectors that either supply or conserve energy at low unit cost will result in the energy consumer spending his fuel savings (relative to the available more expensive alternatives) on other goods and services. In turn, these expenditures should result in the creation of more employment than the equivalent expenditures on the energy itself.

- Assuming that nuclear power and conservation/renewable resources become competitors in serving the same end use (e.g., space heating), we have concluded that the Ontario content of each is so high that it is unlikely that this factor alone will make much difference to the estimate of jobs created. Indeed, the distribution of jobs over time, and any transitional problems that arise between existing and new industries, will likely be more critical considerations.

- Although energy, employment, and economic growth are clearly interdependent, we believe that Ontario can, in the long run, grow and prosper, continue to create employment opportunities, improve the quality of the environment, and still conserve energy on a massive scale.

Financial Factors

An overview of Ontario Hydro's financial and operating position for the year ending December 31, 1978, as presented in the Corporation's annual report, revealed that the total assets of the Corporation were $13.162 billion, the annual revenues totalled $2.137 billion, the long-term debt was $10.226 billion, and the equity totalled $1.802 billion. These are impressive figures − Ontario Hydro has the largest assets of all non-financial Canadian corporations.

The Corporation is financed essentially through internally generated revenues (rates) and the issuance of long-term debt. Its current debt/equity ratio is about 84/16, and it hopes to reduce this over the next few years. How this might best be done is a matter for the Ontario Energy Board and beyond the scope of our inquiry. Nonetheless, we recognize that decisions relating to revenue, expenditure, and financing are interdependent and, in particular, essentially because of the uncertainties relating to the load forecast, that financing policies should be as flexible as possible.

Because of the centrality of nuclear power in the utility's short- and long-term power generation programme, a good starting point is the potential problem of a "capital gap", which we introduced in the *Interim Report* (Chapter 7). The following points were made:

- Capital borrowing restraints imposed by the Provincial Treasurer, especially during the period 1976-8, gave rise to significant adjustments to the plant expansion programme and there were substantial cutbacks.

- The view of the Provincial Treasury, then and now, is that "we have entered an era of continuous borrowing constraint in which capital availability must be a key element in the planning process".

- On the basis of the LRF 48A forecast plan (extant at the time of writing the *Interim Report*) a shortfall of capital funds appeared to be likely in the early 1980s. Indeed, it was anticipated that, by 1990, there could be a cumulative shortfall of $7 billion and by the year 2000 a shortfall of $25 billion.

- The projected capital requirements (based on LRF 48A) of the utility were such that all the capital available to the province was required for Ontario Hydro expansion programmes. This would have had a direct impact on the extent to which the provincial government could indulge in non-public borrowing from various pension funds, and, indeed, on the desirability of achieving a balanced provincial budget.

- Ontario Hydro, and concomitantly the province, places substantial reliance on non-Canadian capital markets, with the attendant risk of major foreign exchange fluctuations.

However, the dramatic drop in Hydro's 1979 load forecast (compared with those of 1975 and 1976), and in consequence a marked stretching out of the intermediate- and long-term planning programmes, suggest that capital availability will not be a planning constraint during the 1980s. Nevertheless, the basic concepts outlined above, relating to capital needs and their fulfillment are unchanged.

The escalating capital expenditures in Ontario during the period 1965-78, and the relative magnitude of Hydro's capital spending, are shown in Table 4.1 (reproduced from Volume 5). The concept of capital availability itself is serving a valuable purpose in restraining provincial government and Hydro expenditures during a transition period from a high-growth economy, with an even faster-growing government sector, to a slow-growth economy with a roughly constant government share of provincial income. Noteworthy, too, is the fact that, through the adoption of a prudent long-term borrowing strategy, Ontario has gone a long way towards ensuring that debt service payment will not impose too heavy a burden on future taxpayers.

The Availability of Capital

Within the financial community there appears to be little doubt that, over the long term, the debt financing available to Ontario Hydro will be maximized if Ontario Hydro bonds continue to be guaranteed by the Province and the Province retains its triple A rating. Ontario Hydro's long-term debt of some $12 billion accounts for over 85 per cent of the publicly held debt of the province. Since the two are regarded by the financial community as essentially a single "credit", it is a truism that excessive borrowing by the utility would affect Ontario's credit rating and consequently the province's ability to raise capital at comparatively low interest rates.

Cutbacks in Ontario Hydro's generation expansion programme are continuing in the face of reduced load forecasts. The 1979 programme, for example, cut about 10,000 MW and $15.3 billion over 20 years from the programme that had been approved only a year earlier. The reduction in borrowing requirements is estimated at $9 billion over the period. Nevertheless, Ontario Hydro's outstanding debt is projected to grow from $12 billion currently to about $60 billion by 1999. Figure 2.2 of Volume 5 compares the projected increases in the amount of debt outstanding year-by-year to 1999 with expected increases in capital availability. The analysis assumes that any additional borrowings by the province will not affect amounts available to Ontario Hydro, and that funds used to retire existing debt remain available through roll-overs but that unused funds in any year are not available the following year. No serious financing problems are suggested by the comparison, but Ontario Hydro expects some difficulties between 1988 and 1994 when markets will have to be found to accommodate increases in debt outstanding which is growing at an average rate of 21 per cent per year.

Nevertheless, in spite of the province's excellent credit rating in the public capital markets, increasing restraints on Ontario Hydro's borrowing could arise as a result of the diminishing availability of non-public funds to finance the provincial government's deficit. For example, according to Professor Pesando:

> The dramatic reduction in the future availability of non-public sources of funds is clearly the most important development affecting the outlook for the finances of the Province of Ontario in the next decade. As a result, the input of financing considerations into decisions with regard to tax rates and/or expenditures will presumably assume increasing importance in the years ahead. In view of the interdependence of borrowing requirements of the Province and Ontario Hydro, the potentially constraining impact of financial considerations on the planned capital expenditures of Ontario Hydro must also be noted.[33]

Unless the province succeeds in balancing its budget during the next few years, the decreasing availability of non-public sources of funds may force it to turn to public sources to finance its own deficit, or to make other energy-related capital expenditures.[34] This could impinge on Ontario Hydro's borrowing potential and could act as a restraining influence on the capital expansion programme. This is another reason why a policy of energy conservation (especially in the sense of increased efficiency of utilization) coupled with more effective means of load management and control is so important. It is also the reason why close financial co-ordination between the province and Ontario Hydro on financing questions must continue.

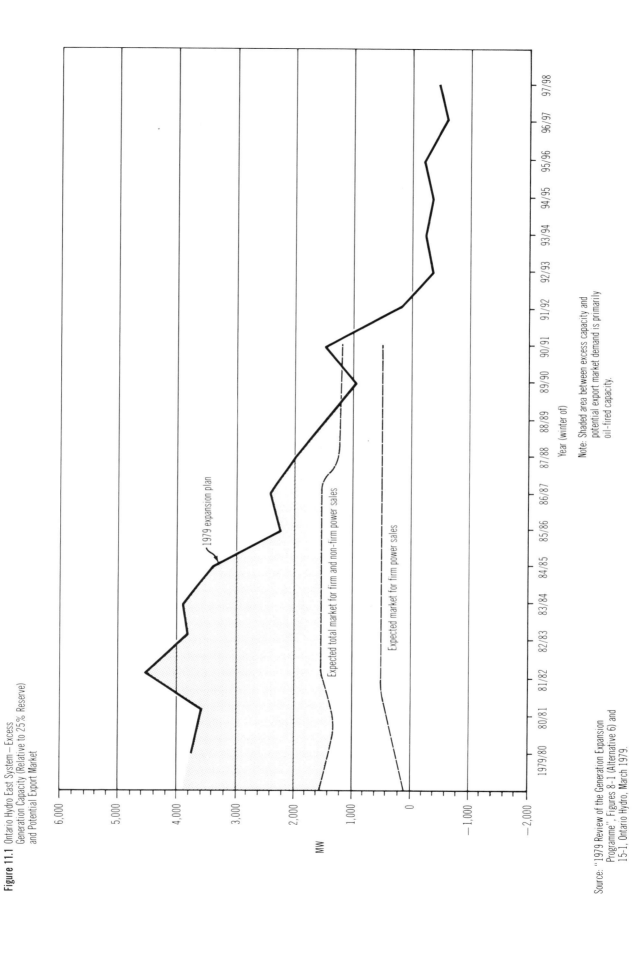

Figure 11.1 Ontario Hydro East System—Excess Generation Capacity (Relative to 25% Reserve) and Potential Export Market

1979 expansion plan

Expected total market for firm and non-firm power sales

Expected market for firm power sales

MW

6,000

5,000

4,000

3,000

2,000

1,000

0

−1,000

−2,000

1979/80 80/81 81/82 82/83 83/84 84/85 85/86 86/87 87/88 88/89 89/90 90/91 91/92 92/93 93/94 94/95 95/96 96/97 97/98

Year (winter of)

Note: Shaded area between excess capacity and potential export market demand is primarily oil-fired capacity.

Source: "1979 Review of the Generation Expansion Programme," Figures 8-1 (Alternative 6) and 15-1, Ontario Hydro, March 1979.

CHAPTER TWELVE

Decision-Making

There is no reason to believe that major physical, social, economic and psychological adjustments to new energy policies, including those related to the wiser and more efficient use of energy, will be easy, especially in view of the increasing complexity and built-in inertia of modern society. The conditions to be expected during the period 1983-93 and beyond will necessitate determined search for innovative and more sustainable sources of energy and, most important, a realization that the energy balance may depend on a remoulding and restructuring of our institutions, organizations, and value systems. Diversity, flexibility, and resilience should characterize both our energy supply systems and the decision processes related to them.

Generically, decision-making is the process of selecting (on the basis of accepted criteria) from a range of viable alternatives to achieve a specific purpose. The process is ubiquitous. It is manifest, for example, in the choice of a scenario (e.g., a demand or supply scenario), and in the choice of a generation technology. In this chapter we review the basic concepts and principles that underpin decision-making, with special reference to energy policy and to the evolution of Ontario's electric power system. Also, as required by our mandate, we propose a new decision-making framework. In particular, we stress the increasing significance of public participation.

The energy problems and the inextricably linked economic implications of the industrialized western world are increasingly problems of decision-making rather than of technology and systems operation. As we have stressed previously, they are exacerbated by supply constraints, and by environmental concerns aroused by pollution and associated health risks. Where the issues can be expressed adequately and exclusively in physical terms, the decision processes are comparatively straightforward and can be handled unambiguously. But where the central issues are the social, environmental, ethical, and economic implications of, for example, electric power planning, then, because of their extreme complexity, they can only be handled by social and political means. Indeed, this was recognized by the Government of Ontario in 1975 when the Royal Commission on Electric Power Planning was established.

The Basic Concepts and Principles

Energy-policy formulation and implementation is necessarily pluralistic, as exemplified by the diversity of issues that relate to electric power planning. In developing an integrated framework for decision-making, we have been guided by certain basic concepts and principles, introduced briefly below.

Risk and Social Responsibility

Especially when a high level of uncertainty exists (because of a lack of an adequate information base), there are inevitably risk assessments associated with decision-making. This was demonstrated in Chapter 5 when we discussed the risks associated with the exposure of humans to low levels of radiation. It is a truism that there are risks associated with the production of all forms of usable energy, and with a failure to produce enough usable energy; these are essentially "involuntary" social risks.

A highly pertinent question is: To what extent should the regulation of social risks — and it is universally recognized that such regulation is essential — be left in the hands of "experts" who understand the physical and biological nature of many risks even though they may be unable to quantify them? The difficulty arises because many decision-making bodies, responsible for the regulation of technological risk, depend almost exclusively on scientists and technologists — even social scientists are usually excluded. This exclusion is paradoxical, especially in view of the fact that, although a statistical assessment of the frequency and severity of a specific social risk may be available, the politician (perhaps against his better judgement) may have to take uninformed public opinion into account.[1] The public's perception of a risk relates much more to its potential severity than to its frequency, and this perception is powerfully influenced by press and media. Bearing in mind that some risks are inevitable, the challenge to the regulatory process is to focus on the important question of the level of risk (i.e., the product of the frequency of occurrence of an event and its severity) that is acceptable to the general public.

One aspect of the regulation of risks in electric power planning deserves special comment. During our hearings, from time to time, the ethics of nuclear engineers, in particular, and professional engineers and scientists in general, were called into question. Can the public trust the judgement of professional engineers responsible, for example, for the design of a nuclear reactor? In respect of the design to meet specified safety standards, we emphasize, *a fortiori*, that the answer must be unequivocally in the affirmative. But in the overall decision-making processes there remains the question — "How safe is safe enough?" This rests in the domain of the regulating body, in this instance the Atomic Energy Control Board (AECB). We note, for example, that the professional engineer's code of ethics,[2] directly or by implication, requires him or her not only to meet adequate safety standards but to do so at reasonable cost. What is adequate and what is reasonable? Here is where value judgements are essential, and, most important, they must be made by politicians, social scientists, the general public, and lawyers, as well as by scientists and engineers. We cannot emphasize too strongly that the key decisions inherent in the planning of Ontario's electric power system would not vanish even if all professional engineers were required to adhere to a much stricter code of ethics than that in place at present.

There is another concern. It is conceivable that an engineer, without contravening the ethical code of the profession, may consciously or inadvertently withhold information. It is clear that the fears of the public relate just as much to the possibility of information about a potential risk being suppressed as to the risk itself. This is why complete openness of information is so essential. Related to this is the fact that, throughout the inquiry, we have been conscious of the centrality of trust; in particular, public trust in the decision-makers of Ontario Hydro. (This is all the more important since most of the major decisions, excluding specific design decisions, have qualitative as well as quantitative dimensions and hence involve value judgements.) Unless this trust (both *de jure* and *de facto*) is assured, major social difficulties in the planning of the electric power system will persist. Our proposals are aimed very strongly at this aspect of decision-making.

Information — Facts and Values

It is indisputable that, in the past and until comparatively recently, the government and Ontario Hydro have had a virtual monopoly on technical information relating to the planning and operation of the province's electric power system. This was particularly apparent during the Commission's public hearings on the role of nuclear power. We believe profoundly that the democratic process, and indeed the evolution of our society, requires that the proponents of policies that differ from those of Ontario Hydro and/or the government should have access to all pertinent information and to the methods and the skills that are available to the government. If this is not the case, it follows inexorably that both sides of a debate will be weakened. Further, the policy-makers will tend to ignore criticisms coming from a public whom they deem to be uninformed, and there will be a tendency for the public to reject policies and decisions that they cannot verify. Accordingly, we have concluded that full disclosure of technical, economic, environmental, health, and all other information in understandable form that relates to future energy policies would enhance trust and broaden the base of public understanding.

Energy decisions inevitably are based on esoteric technological and social factors. For this reason, the problem of communication between, for example, Ontario Hydro and the public has become increasingly difficult. It is exacerbated by the justifiable concern over the fact that science and technology deal essentially with facts, while human values are usually ignored. Additionally, in the eyes of the public, even the interpretation of scientific facts is often suspect when differences of opinion arise between groups of experts. These are not shortcomings of the scientific method *per se*, but often relate to the incompleteness of information. For instance, because the latency period for many radiation-induced cancers is in the order of 30 years, the epidemiological data relating to workers in nuclear power plants is inadequate for an assessment of the risks associated with exposure to low-level radiation. Similarly, not enough information is available at present for an objective and unambiguous assessment of the safety of nuclear reactors. The confidence levels increase, however, as more and more reactor years that are free of major accidents are recorded. Nor is the epidemiological data base relating to exposure of the public to low levels of toxic gases resulting from the combustion of coal any more adequate. And this is fundamentally why the nuclear-coal choice must to some extent involve value judgements.

Because the planning of the electric power system requires recognition of value priorities (and indeed often brings into the open hidden value conflicts), a key consideration is the ranking of values. For example, how can the maintenance of health, the preservation of community, the conservation of soil, and the rights of the individual be adequately weighed in decision-making processes? Suffice it to note that value choices of these kinds, and associated levels of acceptable risk, must ultimately rest in the political domain.

Political and Social Factors

- The ultimate responsibility for energy-policy decisions, in Ontario, rests with the government, and in particular with the Minister of Energy. However, the Minister alone cannot assemble all the research and information needed to make sound decisions. He or she must also rely on the independent advice of commissions and committees charged with examining specific issues or areas in energy policy (the RCEPP is a case in point). This is a sound principle.

The decision process must be adaptive and therefore must have two basic capabilities. First, it must react as promptly as possible to crisis situations (e.g. a sudden shortage of fuel oil or a major nuclear power station accident). Second, and just as important, by capitalizing on "feedforward" principles, it must anticipate predictable, in contrast with unpredictable, events by having "contingency plans" in place. Both the feedback principle (i.e., reacting to a crisis) and the feedforward principle (i.e., anticipating a crisis) are essential attributes.[3] Through its analyses of various future planning options (and its insistence on "hearing the voice of the people"), the Commission has been concerned essentially with the feedforward aspects of planning.

Because of environmental (in the broadest sense) uncertainties, the complementary principles of resilience and flexibility in the decision-making structure must be invoked.[4] A key characteristic of a resilient system is a certain diversity, in the sense that decision-makers should preferably have diverse points of view, diverse backgrounds, and diverse value priorities. Note that the prevailing long lead times involved in putting new bulk power facilities into service are not conducive to resilient behaviour, certainly not in the sense of responsiveness. A lack of responsiveness can often result from redundancy in the decision process. Such overlap inevitably gives rise to hearing bodies with a propensity always to begin at "square one".[5] The long history of Ontario Hydro's attempts to obtain approval of a transmission line from Bradley Junction to Georgetown is a classic Ontario example.

The Rationale for Public Participation

While openness and public involvement are not new concepts for the 1980s, we believe they should continue to be important aspects of future electric power planning processes. We believe there are four basic reasons why the public should have the right to participate in decisions relating to energy planning. These are:

- Because each citizen of the province is affected by energy decisions, the individual should have the right to participate in the making of the decisions.
- Decisions in which the public has actively participated will probably be more widely acceptable to the public at large.
- The increased diversity of opinion, and knowledge, resulting from participation of the public in many cases will probably improve the ultimate decisions.
- Public participation is synonymous with openness and consequently enhances the accountability of government.

However, we are also conscious of the various "costs" of public participation, for example:

- Lengthy public participation processes (perhaps involving a considerable amount of redundant information and argument) can cause undesirable delays in the construction of facilities.
- Although the administrative costs of lengthy public hearings are usually insignificant in comparison with the costs of the facilities under consideration, these are nevertheless not trivial.
- By increasing the time between the making of a specific proposal and its ultimate implementation, the public participation process usually increases uncertainty in the overall system planning process.
- There may be unjustifiable delays as a result of public participation that may prove costly to the taxpayer.

Clearly, the benefits and costs of public participation must be weighed in respect of each specific proposal. However, we have concluded that the principle is sound, and, especially in the field of energy decision-making, public participation should be encouraged. The challenge is to minimize the unnecessary and therefore undesirable costs while at the same time ensuring that the right to participate is protected.

Ontario Hydro and Public Participation

Through the election of representatives to provincial and municipal governments, the "public" already participates, albeit indirectly, in Ontario Hydro's decision-making processes. Furthermore, broad sectors of Ontario society, through representation on the Corporation's Board of Directors, also participate in policy formulation and decision-making. Nevertheless, it is increasingly clear that direct public participation (in decision-making) is an essential complement to the activities of government and its agencies, in assuring that decisions and plans are subjected to adequate public scrutiny. This is not surprising in view of the social and environmental implications of energy planning and the fact that these affect the lives of all people in the province. In recognizing this need, Ontario Hydro has stated: "The function of public participation is to involve directly or indirectly those people who are willing to, or should, participate in the planning of man's work which may alter community fabric or influence change in the lifestyle of its citizens."

Notwithstanding the utility's efforts to encourage public participation[6], however, these have been only partially successful, largely, we believe, because of deeply ingrained suspicions (dating back several decades) in the minds of many prospective participants, especially in the farming communities. Indeed, these problems were anticipated by Task Force Hydro (1972) which, in urging continuing efforts on the part of the utility to involve the public in its affairs, stated: "Procedures [i.e., public participation] used will produce little in the way of positive results in the absence of widespread commitment to the principles involved and a response to the changing social environment by a majority of those responsible for [Ontario] Hydro's operations."[7]

We acknowledge that since 1972 Ontario Hydro has utilized some imaginative approaches to involving the public in its planning process (see Volume 8, Chapter 4).

Accordingly, we recommend that:

12.1   Ontario Hydro should be encouraged to continue and, where necessary, to expand its public participation programme to ensure that the public is fully involved. Ontario Hydro should adopt joint planning processes whereby real decision-making authority is shared with, and in some cases (see recommendation 6.3) left to the initiative of, citizen representatives.

An unresolved problem relates to the question: Participation of whom? Concern with the consequences of decisions or the "ends" is usually represented by the views of politicians and the concerned public, especially the public interest groups. But the ends are unachievable without the "means", and the means are the concern of the experts. Consequently, Ontario Hydro feels fully justified, and we concur, in bringing forward highly qualified expert witnesses when occasion demands. However, there is an inevitable communications gap between the expert and the non-expert, and herein lies the key dilemma of public participation. How can this gap be adequately bridged? We have concluded that Ontario Hydro can, if committed, reduce this communications gap.

Accordingly, we recommend that:

12.2   Ontario Hydro should ensure that the participants in the utility's participation programme have access to independent expertise whether the expertise is supportive of or opposed to Ontario Hydro's planning concepts.

During the public hearings of the Commission, we received several briefs and other representations from the municipalities, and citizens particularly interested in municipal government, relating to the involvement of municipal government in the planning of the province's electric power system. After reviewing the evidence, we have reached the conclusion that, in the future, Ontario Hydro should work much more closely with the municipalities and, in particular, that there should be in-depth consultations relating to long-range planning, for example, for the routing of bulk power transmission lines (or the siting of generating stations) within a municipality's area of jurisdiction. In addition, we believe that Ontario Hydro should work closely with public utility commissions (PUCs) on information dissemination. For example, the PUCs should participate in and even sponsor educational programmes in their areas. This is an obvious step, because the PUC constitutes necessarily the "grass-roots" of the electric power system in Ontario. A similar role should also be undertaken by the Regional Offices of Ontario Hydro in respect of direct customers and especially, and appropriately, in the rural areas where the farming communities provide an important component of the demand for electricity.

Accordingly, we recommend that:

12.3   In order to enhance the optimum utilization of electricity, both PUCs and the Regional Offices

of Ontario Hydro should be adequately financed and encouraged to sponsor, in their areas, educational programmes, seminars, and workshops in energy utilization and conservation.

In the *Interim Report* we concluded that, to have meaningful public involvement, new and imaginative ways must be found to inform and involve the public, and greater and freer public access to information must be provided. During the course of the inquiry we came to the conclusion that Ontario Hydro has been missing opportunities to share its unquestioned technical expertise with the broader community of scientists and informed laymen.

Accordingly, we recommend that:

12.4   Ontario Hydro should find practical means to give effect to its commitment to greater openness by commencing to publish a technical-papers series, containing accounts of technical, scientific, and socio-economic research in language understandable to the layman. These publications should be made widely available to libraries across the province.

The Existing Decision-Making Structure

In earlier chapters, we introduced the key decision areas associated with electric power planning: the future need for electricity in the form of the load forecast; the generation mix; security of fuel supplies; environmental impacts of electric power facilities; the location of generating stations and the siting and routing of bulk power transmission systems; the rate structures; interconnections with neighbouring utilities; and the financing of major facilities. In this chapter we will discuss the institutions involved in electric power decision-making.

In addition to the Cabinet and the Ontario Legislature, which have been involved in electric power decision-making since the creation of Ontario Hydro, and, of course, the Corporation itself, the key actors are the Ministry of Energy, the Ministry of the Environment, the Ontario Energy Board, the Environmental Assessment Board, municipal governments, the Ontario Municipal Board, the Atomic Energy Control Board (AECB), and, most recently, the Select Committee on Ontario Hydro Affairs. The role of the AECB in Ontario Hydro's operational decisions has been considered in Chapter 5.

Ontario Hydro

Fig. 12.1: p. 180

The Ontario Hydro Board of Directors has prime responsibility for the development of the operational policies of the utility. However, as shown in Figure 12.1, there are a multiplicity of other bodies and groups which influence Ontario Hydro's operational decisions. What Figure 12.1 does not display is the complex interdependency that exists between the key components. The multiplicity of players and the complexity of the interdependencies has often led to confusion as to the focus of responsibility for policy formulation about electric power developments.

As was evident throughout the inquiry, this has resulted in a questioning of the powers and responsibilities devolving on Ontario Hydro. It has been suggested, for example, that the Corporation is too large, technocratic, inflexible, insensitive, remote, and, above all, arrogant. In particular, some participants have argued that Ontario Hydro has too much power over important decisions of social policy that should be left to government. We have reviewed these matters and have concluded that the Corporation has not in fact been allocated too much responsibility.

However, we recommend that:

12.5   A clear statement of the objectives and responsibilities of the utility, especially as they relate to the social objectives as endorsed by government, should be issued by the Ministry of Energy.

As far as the accusations of "arrogance" are concerned, these relate, we believe, almost exclusively to specific local incidents. During the conduct of our inquiry, we have noted a desirable trend towards a more humanistic approach by the utility and a correspondingly smaller number of accusations of "arrogance". This we believe is a direct result of the utility's education, information, and public participation programmes.

Government Ministries and Agencies

Ontario Hydro is closely linked with three government ministries, the Ministry of Treasury and Economics (in respect of questions of capital borrowing), the Ministry of Energy (in respect of general energy policy questions), and the Ministry of the Environment (in respect of questions of environmental impact due to electric power facilities). There are also less clearly defined ties with other ministries, notably those of Health, Industry and Tourism, Natural Resources, and Housing.

The responsibility for general policy on energy is clearly in the hands of the government, and in particular the Minister of Energy. Although, as we indicated in earlier chapters, the Ministry of Energy is undertaking impressive and rewarding programmes in the field of energy conservation, its credibility in respect of long-term energy policy formulation is much less evident. This is not surprising in view of the fact that the Ministry has had neither the resources, nor the staff, nor the time to undertake extensive futures studies, not least because of its heavy commitment to day-to-day questions and short-term problems. Furthermore, by its nature, the Ministry's business must be conducted in private and its performance, although subject to scrutiny and criticism in the Legislature, is not subject to direct public scrutiny. The result is that energy policies have been somewhat fragmented, predicated on a short-term basis, and *ad hoc*.

Because the environmental impacts of Ontario Hydro's facilities are so very visible, it is not surprising that the Ministry of the Environment, especially in so far as questions relating to air quality and water quality are concerned, has been very active in reviewing the utility's long-range plans. During our inquiry, the Ministry participated actively in the examination of all major topics. Indeed, it was encouraging to note such openness of approach and such dedication to a holistic approach to electric power planning.

Since 1972, when the Ontario Energy Board Act was amended, the Ontario Energy Board (OEB) has conducted, on an annual basis, a series of public hearings in respect of Hydro's proposed rate increases. More recently, it undertook an in-depth public examination of a major costing and pricing study prepared by the utility. In consequence of the detailed analyses required in connection with rate setting and related questions, the OEB has gained considerable insight into electric power planning processes and has been particularly conscious of the interdependence between the load forecast, the system expansion programme, the financing constraints, the revenue requirements, and the rate structure. Indeed, in the light of its experiences, the OEB has reached certain basic conclusions with respect to the rate-review process. These are:

- The OEB's value is limited on account of the requirement that it must consider changes in the rate structure for a single year at a time. This gives rise to excessive fragmentation of the rate review process. For example, if a period of several years were to be used as the base line, and if the examination were to take into account key system growth parameters such as system capacity and financial expenditures, then the process probably would be more productive.
- The OEB considers that a more positive policy statement by the government on financing constraints and objectives would be desirable.
- The OEB has only limited capability in the more technical aspects of electric power planning and rate review, and feels that this capability should be strengthened.
- The OEB endorses the concept of public participation in the rate review processes.

These conclusions are consistent with our findings, especially in so far as they recognize the need for a more integrated approach, for example, through recognition of the interdependence of rate structures for both electricity and natural gas, and system planning. Although public participation in the hearings of the OEB has been limited (apart from that of special interest groups normally represented by counsel), perhaps because of the comparatively formal hearing format that is essential when money matters are under consideration, we believe that the process would be enhanced through, for instance, the holding of in-depth pre-hearing workshops at which the issues could be identified and clarified.

In Chapter 9, we emphasized the centrality of Ontario's Environmental Assessment Act, passed in 1975, in Ontario Hydro's planning processes (and, in general, its importance in all energy resource planning). This Act is perhaps the most comprehensive of its kind in the world. Of special significance to decision-making is the creation under this Act of the EAB, the basic mandate of which was given in Chapter 9. To date, however, although the EAB has examined some environmental problems related to electric power planning, notably the siting of additional uranium mill tailings ponds in the vicinity of Elliot Lake, it has not undertaken under the Environmental Assessment Act any review of Ontario Hydro proposals for major generating and bulk transmission line facilities.[8]

Furthermore, our experiences suggest that, in respect of major Ontario Hydro proposals, the EAB may have difficulty in fulfilling its mandate. We refer, in particular, to the following points:

- As at present constituted, the EAB lacks the proper composition, staff, and resources to undertake a thorough analysis of large complex undertakings. Moreover, undue reliance on the expertise of the staff of the Ministry of the Environment might compromise the findings of the EAB, and, in particular, might detract from the essential openness of an inquiry.

- The EAB has relied on the traditional adversary hearing processes which, however appropriate for extant environmental issues, may not be appropriate for hearings in which long-range planning considerations are involved.
- As was the case with Darlington Generating Station, the Act permits the government to exempt undertakings from all or part of the environmental assessment process. We emphasize that the use of this escape clause might endanger two very desirable aspects of the assessment process, namely, provision of an open and independent review, and acceptance of sensitivity to environmental considerations in the planning process. Ontario Hydro's major projects are of such significance, in view of their environmental implications, that exemptions should not be granted. For example, if all of the utility's projects that have environmental implications were submitted to open and comprehensive public discussion, the public's trust in the planning processes would, we believe, be appreciably enhanced.
- Under the Act, the EAB is required to inquire into the rationale[9] for a specific undertaking, to consider alternative methods of carrying out the undertaking, and to consider alternatives to the undertaking. In effect, therefore, the EAB would inevitably become involved with major energy policy issues (e.g., the nuclear-coal choice and the future role of alternative generation technologies). These would be difficult undertakings for the EAB as presently constituted, and could potentially lead to considerable overlap with other Ontario government institutions.

Especially in the light of the "need problem" outlined above, and in view of the interdependence between the technological, economic, and environmental aspects of Ontario Hydro's major undertakings, we have concluded that the EAB is not an appropriate agency to consider these issues from a holistic standpoint.

The Select Committee

The concept of the "select committee" is particularly significant; for example, the Select Committee on Ontario Hydro Affairs (an all-party committee) constitutes a central political body for the review of Ontario Hydro's plans and procedures.[10] In a real sense, it is the Legislature's watchdog of Ontario Hydro (the province's largest corporation). But it is much more than this, because it provides a catalyst in the Legislature for energy debates. However, as at present constituted, the committee has basic limitations:
- Its investigations are of an *ad hoc* nature, are restricted to the electricity sector, and are inevitably fragmented.
- The committee is limited as a vehicle for direct public involvement in the debate on electric power planning (e.g., the public cannot cross-examine Ontario Hydro or other witnesses). Public participation, we stress, is essential in energy policy formulation and in the consideration of the nature and siting of facilities that may have an impact on nearby communities.
- The existing staff support, albeit excellent, is nevertheless inadequate, especially in the key areas of science and technology (e.g., nuclear power and areas related thereto).

Other Institutions and Pertinent Legislation

As noted in Chapter 2, Ontario Hydro must conform to the provisions of the Planning Act in the planning and construction of its facilities. Where a municipal restricted area by-law conflicts with an Ontario Hydro proposal for the location of facilities, and an amendment is not acceptable to the municipality, the utility presents its case to the OMB. (An example is the Halton Hills hearings relating to the Bradley – Georgetown 500 kV transmission line.)

The duties of the OMB are specified in the Ontario Municipal Board Act. We note, in particular, that the activities of the OMB relate specifically to municipal planning and to the approval of by-laws.

We have concluded that, while decisions most pertinent to local municipalities relate essentially to the location of facilities (and hence land use), the municipal planning process is not appropriate for the making of these decisions. A basic reason is that Ontario Hydro facilities, especially transmission lines, usually traverse several municipalities, and clearly a single municipality will not be in a position to assess the facility on a sufficiently comprehensive basis. It is essentially for this reason that we have concluded that the municipal planning process should not involve decisions relating to the need for or the location of electric power facilities. However, and most important, regardless of how comprehensive new decision-making processes turn out to be, it is essential that the interests of the municipalities should be addressed and that their input to the decision-making process should not be impaired. In this regard, we have noted a conclusion of the Ontario Government's White Paper on the Planning Act in which it is recommended that "broad provincial interest" including "the efficient supply and use of energy" be protected from impairment by municipal planning.

It is stressed in the same paper that:

> Ontario Hydro is not presently exempt from municipal controls, but its works are subject to the Environmental Assessment Act in the same manner as major provincial public works. In keeping with the proposals to avoid unnecessary duplication between the planning and environmental assessment processes, Ontario Hydro undertakings will in the future be exempt from municipal planning controls, but Ontario Hydro will be required to consult with municipalities prior to taking specific actions.[11]

We fully agree with these general conclusions.

Recently, there has been some evidence of misunderstandings that have arisen with respect to the actual scope of the Expropriation Act and how it relates to Ontario Hydro's planning processes. We understand, for example, that intervenors at a recent expropriation hearing requested that the hearing officer consider the general location of facilities and the routing of transmission lines. Because the hearing of such issues would give rise to still further duplication, and because the hearing officer would normally not possess expertise in the routing of bulk power transmission lines, this kind of hearing would not be desirable, or, indeed, meaningful. On the other hand, hearings under the Expropriation Act would be perfectly appropriate for the consideration of minor modifications to the location of facilities.

The Present Decision Framework

The present institutions and process for electric power decision-making have fundamental shortcomings, not the least of which results from the fact that no general provincial energy policy has been formulated (although the recent Ontario Ministry of Energy Report entitled *Energy Security for the 80s* spelled out some broad objectives). Such a policy would provide guidance for the reviewing of specific projects and an indication as to how each project might fit in. Related to this is our conclusion that, in principle, while each review board cannot be faulted for its specific contributions to decision-making, the existing structure as a whole does not and cannot take into account the interdependencies existing between them (e.g., decisions on the system expansion programme must take into account environmental considerations, land use, the availability of alternative energy sources, and, above all, energy conservation technology). The result has been undue fragmentation of decision-making processes. The activities of the Select Committee are also highly fragmented, because the committee, no doubt justifiably, responds to media pressure; this in turn results in an *ad hoc* approach to the political assessment of Ontario Hydro's plans.

Inevitably, essentially as a result of the fragmented approach to decision-making, there is undesirable duplication and overlapping in the review procedures – a tendency for each review body always to "begin at square one". This gives rise to a lengthening of the lead times involved in putting major facilities into place. As well, the present procedures can lead to decisions being taken in a non-optimum order. An example is the decision a few years ago by the Government of Ontario to delay the approval of a second line out of the Bruce Generating Station, in spite of the fact that Ontario Hydro was proceeding on schedule with the the construction of both Bruce A and the Bruce B generating stations.

For the above and other reasons, the existing decision-making structure has been referred to as the "multiple-window" approach. The analogy, in so far as it pictures several essentially independent boards (and on occasion the Select Committee) sequentially examining specific aspects of Hydro's programme, is appropriate. We have attempted to portray it in Figures 12.2 and 12.3. Figure 12.3, Fig. 12.2: p. 181 which uses a Venn diagram form of representation, is especially enlightening since it shows how the Fig. 12.3: p. 182 windows (decision processes) overlap, as mentioned above, and yet how, at the same time, there are significant gaps in the processes. In the next section, we will show how the multiple-window approach, especially in connection with public participation in decision-making, can be changed into a one-window approach.

The need for additional facilities is a central theme. Although an appropriate mechanism is in place, as mentioned in connection with the work of the EAB, there are major difficulties. This is because "need" is inextricably related to the load forecast and consequently to the potential future role of electricity in meeting the province's total energy requirements. The problem is exacerbated by the fact that no comprehensive data base aggregating electricity end uses is at present available, nor has there been any adequate analysis of the extent to which one form of energy might be substituted for another (e.g., electricity for oil, and, to a lesser extent, electricity for natural gas). It is noteworthy also that, because all aspects of power planning subsume need, there is an inevitable tendency for boards and agencies to consider the need issue, however tentatively. In turn, this can give rise to inconsistencies in decisions involving several agencies, and at the same time lead to still more uncertainty.

Summary of Shortcomings of the Present System

- It does not provide for the development of a general policy on energy.
- Historically, government intervention in Ontario Hydro's planning processes has been essentially after the fact. The government and the public have not been involved in the early consideration of long-range electric power needs (e.g., there is no opportunity to debate in public the basic assumptions that underpin the utility's load forecast) or in the early assessment of the cumulative effects (environmental, economic, etc.) of individual projects.
- Existing ministries, institutions, boards, and committees (e.g., the Ministry of Energy, Ontario Hydro, the EAB, the OEB, the Select Committee) do not provide an open, independent, comprehensive review process involving the public, nor can they act in an advisory capacity to the Minister of Energy on the future development of the electric power system.
- Existing procedures for environmental assessment, municipal planning, zoning, and expropriation, as well as for rate-setting, necessitate some level of needs analysis, not least in the broader context of Ontario's total energy needs. The associated boards do not possess adequate technical expertise to undertake these analyses.
- The multiple-window approach creates undue delays, especially because of the time required for public participation and for dealing with the difficult communications problem between the boards, or agencies, and the public.
- The sequencing of recommendations, and the associated decisions, is not always logical; this exacerbates the uncertainties inherent in the planning process.

How can these shortcomings be circumvented? We deal with this central problem in the next section.

A Decision-Making Structure for the Future

We have been particularly conscious, in proposing a new decision-making framework for energy, that both government and governed must be concerned just as much with human values as with the questions of resources and technologies, and perhaps even more. Reflect on the following:

- the critical importance of energy supplies to the provincial economy
- the diversity of public bodies and private interests involved in ensuring the security of these supplies
- the existing lack of co-ordination between these bodies, agencies, and private interests
- the compelling need to establish policies and plans that will be responsive to the contingencies of the 1980s and beyond
- the crucial importance of imaginative leadership to which, we believe, the people of the province will respond, and in which they will have explicit and implicit trust

We have concluded that these all point in one direction — to the creation of a new institution that integrates decisions related to electric power development. Such an institution would provide a "single window" to decision-making, would provide for comprehensive review, would be open, and would provide a vehicle for information dissemination, technology-assessment, policy advice and regulation.

Since 1975, according to a reasonable cross-section of the citizens of Ontario, the RCEPP has fulfilled a real need, in the provincial context. In particular, it has provided a vehicle for public discussion of key electric power planning issues, and for undertaking important power planning research projects. Furthermore, and most important, by making available through its information centre a great deal of hitherto restricted information, it has facilitated more openness in decision-making. Increased openness is manifest in the changed outlook, with respect to public involvement, that has evolved in Ontario Hydro, and which we believe is being accepted increasingly by the AECB.

Now that the RCEPP has fulfilled its mandate, we believe there is a need to establish a permanent institution with similar aims but with a somewhat broader base. The case for such an institution was argued convincingly by several intervenors during the hearings on decision-making and public participation.[12]

Because of the interdependencies between the primary energy sources (including the potential for inter-fuel substitution), the future need for electric power must take into account the requirements for coal, natural gas, and oil. Consequently, although a major portion of the new institution's work will relate to the electric power sector, it will inevitably be concerned with other energy sectors as well. It is therefore appropriate that the new institution be identified with "energy" rather than specifically with "electric energy". Furthermore, in advocating an energy demand management approach (in contrast to a supply management approach), we recognize the need for comprehensive public input and for the development of extensive data banks that catalogue and quantify the end uses of primary energy resources.

In Volume 8, the feasibility of a range of options with respect to the establishment of the proposed energy agency are examined in detail. We have concluded, bearing in mind the government's programme of budgetary constraint, that several existing agencies could provide a suitable base; of these, the two most obvious are the EAB and the OEB.

We recommend that:

12.6   The status of the existing Ontario Energy Board should be enhanced through expanded membership, representing a broad range of interests and disciplines, and the agency should be renamed the Ontario Energy Commission (OEC). It should be an authoritative and independent body.

12.7   The Chairman of the recommended Ontario Energy Commission should be a person well known to the public and not associated with any of the special interests that should be represented.

12.8   As well as providing a vehicle for the consideration and examination of rate structures for both electricity and natural gas, the Ontario Energy Commission should be responsible for advising the government and people of Ontario on energy policy in general and on electric power planning in particular. The OEC should be strongly future-oriented and just as strongly people-oriented.

12.9   The Ontario Energy Commission should be provided with a modest increment in the staff and consulting budget over and above that of the existing OEB. The designation Commission as compared to Board was selected not only to suggest a break from the past but also to provide a broader umbrella to embrace a policy advisory function as well as a traditional regulatory function. The indications are that the additional staff requirements would be small.[13]

Although, as we suggest later in this chapter, a so-called provincial "energy ombudsman" would not be appropriate in the present context of energy planning and review, we draw attention to the fact that, because of the OEC's potentially broad access to the public and to all levels of government, in some respects it would act in such a role.

The Major Activities

A major objective of energy policy, as we have stressed repeatedly, must be to plan for a resilient system, and one that keeps all energy options open. To facilitate this objective, the OEC should have four primary functions; first, to review the province's prospective demand for, and supply of, energy in all forms; second, to propose general policies relating to the province's electric power system; third, within the context of the general policy, to examine the need for, and, if appropriate, the nature and location of specific electric power facilities; and fourth, to examine the rate levels and structures for natural gas and electricity and to continue the other programmes of the OEB within the framework of the aforementioned general policies. In respect to each of these functions, public hearings would be held, and recommendations would be submitted to the government, except in the case of existing OEB regulatory functions, where the OEC would assume the Board's decision-making powers. To fulfil these functions, the OEC would:

- assemble, review, and disseminate data (e.g., end-use data) and information required for energy-policy formulation, especially as it relates to electric power planning
- develop plausible medium- and long-term energy demand scenarios for the province, taking into account potential saving in the use of energy resulting from conservation technology and the use of alternative energy sources, subject these to public scrutiny, and then submit them to the government. (The policy preferences of the OEC should be clearly enunciated.)
- develop corresponding electric power scenarios that take into account health, environmental, land-use, economic, and financial factors and constraints, subject these to public scrutiny, and submit recommendations to the government, again indicating the OEC's policy preferences
- every three to five years, undertake a major review of Ontario Hydro's general policy and long-term plans, in the light of developments in the total energy sector, and prepare a comprehensive

report for use as a central working document by the Ministry of Energy, Ontario Hydro, and other agencies involved in energy planning

• within the context of general policy and future plans, conduct hearings concerning specific electric power facilities (These would encompass municipal, environmental, socio-economic, and financial aspects of these projects. Similarly, again within the context of general policy and future plans, the OEC would conduct hearings on rate-setting and rate structures.)[14]

Some of the foregoing functions require clarification. We refer, in particular, first, to the comprehensive periodic (every three to five years) review of general energy policy, especially as it relates to electric power planning, and the development of associated demand and supply scenarios; second, the complementarity of the OEC (which would incorporate the existing OEB), the EAB, the OMB, and other agencies involved with electric power decision-making; and, third, the rate-setting processes.

The periodic review would, in effect, be a continuation of the work of the RCEPP, as summarized in this volume and presented in detail in the other volumes of this Report. However, there would be important additions to the RCEPP's mandate. For example, although we have attempted (see Chapter 3) to put electric power planning into a total energy framework by taking into account, albeit summarily, the interdependencies between the potential provincial demand for electricity, on the one hand, and for oil and natural gas (e.g., for space heating and water heating) on the other, much more in-depth study, especially from end-use and demand management points of view will be required in the future. Indeed, we envisage the OEC evolving within the next decade from an agency oriented towards electric power planning to a "total energy" commission. We recapitulate below some of the major objectives, all involving public participation, that will be basic to the periodic review.

• The enhancement of the resilience of the electric power system, and indeed the provincial energy system as a whole. This applies to both the adequacy of the mix of generation technologies and just as important, the achievement of resilience in decision-making (e.g., by increasing the diversification of the process).

• The development of energy end-use patterns with special reference to the potential of inter-fuel substitution (e.g., the balance between the use of electricity, oil, natural gas, and solar energy in space heating and water heating).

• The development of plausible demand scenarios and, in particular, the setting of peak demand and electric energy targets.

• The assessment of alternatives (energy conservation, renewable energy resources, and co-generation) as compared with further development of centralized supply systems, for the principle purpose of determining when in the future these alternatives will be commercially available in the Ontario context.

• The development of criteria for the siting of electric power facilities, including the routing of transmission lines, and for the storage of spent nuclear fuel.

The above components of OEC's activities are all essentially related to general policy. In particular, they would provide important inputs into Ontario Hydro's long-range load forecast. Indeed, the demand scenarios (resulting from the periodic review analyses) may be regarded as "goal-oriented" forecasts, while the utility's econometric modelling procedures, combined with the public utility commissions' forecasts, are essentially "trend-driven" forecasts. The two are complementary and together will provide a powerful tool for planning the electric power system, as well as providing guidance for energy policy as a whole.

We have referred to the existing decision-making structure as a multiple-window approach, and contrasted it with a one-window approach. As we envisage the OEC, it will necessarily provide a one-window approach to the assessment and adjudication processes that are at present undertaken by several agencies. In effect, therefore, the OEC will perform tasks previously carried out by the EAB, the OEB, and the OMB, including that of providing for public participation with respect to electric power developments in the province.

Fig. 12.4: p. 183 The one-window approach is shown diagrammatically in Figure 12.4. As we indicate below, the one-window approach applies to specific projects as well as to general policy considerations. A desirable feature of the proposed OEC, as a one-tribunal (i.e., one-window) approach to electric power decision-making, is that the body would be responsible for developing demand and supply scenarios in an overall energy framework, and at the same time would conduct hearings to consider the need for and the nature of individual facilities. Unlike the existing approach, in which the consideration of a specific bulk power facility is, *de facto*, taken out of context, the OEC would be in a position to provide the appropriate context. Furthermore, the duplication of hearings (by two or more boards) would be avoided and the excessively long lead times associated with the planning and construction of bulk power facilities would therefore be reduced to some extent.

We endorse the principle of cross-appointments. For specific hearings of the OEC, membership of the the appropriate panels might be strengthened by including members of the EAB, the OMB, government ministries, and/or the private sector. Flexibility in both appointments and procedures must be sought as a principle of the OEC. Noteworthy, too, would be the continuous interplay between the ongoing review of general policy, the development of plausible demand and supply scenarios, the detailed consideration of individual facilities, and the review of rate structures; this would foster resilience in the decision-making framework.

The same general principles apply to the rate hearings. As we have noted, the OEB has already pointed out the constraints under which it operates at present; to a large extent, these are a result of its comparative isolation from other key aspects of electric power planning. If the existing OEB were to be expanded, as we have recommended, the rate-setting investigations would be part of a total decision-making framework – indeed, a highly significant part – and one that would reflect future planning philosophies (e.g., demand management concepts) rather than being constrained by the existing fragmented approach.

We turn now to the role of the OEC *vis-à-vis* that of the final arbiter – the government. While the OEC would have a central role in policy development, the final responsibility for policy determination would rest with government. As to decisions relating to specific facilities, we believe the OEC should have powers similar to those vested in the Environmental Assessment Board through the Environmental Assessment Act. Concerning electricity rates, we believe the responsibilities should remain as they are now – that is, the OEB expresses an opinion on the appropriate level of bulk power rates but the final decision rests with the Ontario Hydro Board. Ultimately, the choice among the options is a political prerogative – the prerogative of government.

By publishing its recommendations in the form of reports relating to rate levels and structures or the need for individual facilities, or through interim reports on general policy (including appropriate scenarios) as well as major reports pertaining to the periodic review, and not least by staging energy workshops, seminars, hearings, and conferences, the OEC would ensure that the government, members of the Legislature, and the public are informed of the options open and of the consequences of each option. In this way it would also act as a catalyst in the formulation of viable energy policy. As a result, the energy debate both within and outside the Legislature would be enhanced and the ultimate determination of policy would be based on the most comprehensive information available.

The effectiveness and the usefulness of the OEC will be significantly influenced by the procedures employed to ensure openness and public involvement. During the course of the RCEPP's inquiry, we had a unique opportunity to experiment with a number of formats and with the concept of public funding. Accordingly, for the benefit of the government and for the proposed OEC, we report in the next section on our experience with public participation.

The Royal Commission and Public Participation

You will be given every possible opportunity to express your opinions on the use of electricity, on the rate of growth of the electric power system, on environmental questions which relate to the generation and distribution of electricity, and most importantly, on how electric energy can be conserved.[15]

Through the Preliminary Public Meetings, the Public Information Hearings, and the Debate Stage Hearings (a total of more than 2,000 hours of sittings), numerous seminars, several major symposia, and three workshops, the Royal Commission attempted to fulfil this promise. In most respects, it was a rewarding experience; however, in some respects it was disappointing. We summarize below salient features of our experience.

Attendance at Hearings and Meetings

The average attendance at the preliminary public meetings (1975-6), which were held in eight locations across the province, was higher than that at the public information hearings (1976-7), which in turn was higher than at the debate stage hearings (1977-9). The attendance at the "regional hearings" in southwestern Ontario and in eastern Ontario was consistently excellent.

In spite of the fact that, from a decision-making point of view, the debate stage hearings, not least the many hearings devoted to nuclear power, were far more important than, for example, the preliminary meetings, this was by no means reflected in public or media interest. Indeed, during many crucially important debate stage hearings, in which topics of considerable significance to the future of the province were under examination, the attendance (apart from official representatives of Ontario Hydro, etc.) was rarely more than 10 persons. In view of the not inconsiderable efforts by the RCEPP to encourage the involvement of the public, even to the extent of providing modest financial support, on the whole we were disappointed with the response.[16]

While, as noted above, little public or media interest was aroused by the debates on long-term planning concepts, local interest whenever an apparent threat appeared to be posed by Ontario Hydro was noteworthy. We have described this phenomenon as "the backyard syndrome".[17] For example, an incipient threat to a specific group, or to a specific locality, usually aroused high emotions and resulted in high attendance at our hearings. We do not deplore this tendency; indeed it is a healthy sign. But we are concerned that the time horizons of so many people are very limited.

The Spectrum of Concern

Fig. 12.5: p. 184

Throughout the inquiry, we were conscious that many intervenor groups, both well disposed and antithetical to Ontario Hydro, as well as pro- and anti-nuclear, complained about the apathy of the majority of the population. Each side claimed to represent this majority. We have attempted in Figure 12.5 (as suggested by Lord Ashby during a public seminar) to illustrate that the "apathetic majority" is by far the largest sector of the population. The hard-core proponents and opponents, which are two of the groups portrayed in Figure 12.5, are defined as those people who, regardless of the evidence, will not change their minds – in the words of Lord Ashby, their "prejudice is impenetrable". On the other hand, the "open-minded" proponents and opponents are clearly a very significant, though comparatively small, proportion of the total population. Their contribution to the hearings was particularly important. For example, subsequent to the publication of the *Interim Report on Nuclear Power in Ontario*, several public interest groups that could be categorized, by and large, as anti-nuclear leaned markedly towards the RCEPP's "middle-of-the-road" position. Unfortunately, on the other hand, the prejudices of several strongly pro-nuclear proponents, and several strongly anti-nuclear groups apparently were not influenced by the *Interim Report*.

Formats and Environments

The Royal Commission experimented with a variety of techniques in its attempt to optimize the public's involvement in the inquiry. For example, the early letters of the Chairman, the press releases, the establishment of an information centre, the deposition in public libraries across the province of key RCEPP material, the information and seminar programmes aimed specifically at the schools (e.g., the Outreach Programme) set the stage for (or proceeded in parallel with) the more formal meetings and hearings.

The requirements that relate to public hearings are specified by common law and by legislation (the Public Inquiries Act). The essential rights are the right to know the issues, the right to present evidence, the right to cross-examine, the right to openness, and the right to counsel. However, the law permits considerable diversity and flexibility in the choice of procedures, and the Royal Commission welcomes this freedom. Nevertheless, we have observed that most major public inquiries involve courtroom-like procedures – they adopt an "adversarial approach". In many cases this is necessary; in others (such as this Royal Commission), with a marked futures orientation, a courtroom-like atmosphere, although desirable for some hearings, is not conducive to the creation of innovative learning environments. We believe, in general, that hearings involving speculations concerning the future should adopt a more inquisitorial-type approach.[18] On the basis of our experiences, we have concluded that:
  • Effective public participation necessitates informality – environments should be sought that do not inhibit even the most reticent witness from coming forward and enjoying the experience.
  • Identification and clarification of the issues through informal preliminary public meetings is essential. In retrospect, we believe that, in addition to the preliminary public meetings, some well-planned and strategically located workshops, for the purpose of clarifying the issues and making them comprehensible to a broad range of people, would have been very helpful.
  • Adversary procedures proved to be only partially successful, because the cross-examination process was inhibited, e.g., a witness would hesitate to question an examiner. We note, however, that the right to "cross-examine" enshrined in the Public Inquiries Act creates an adversarial

approach in proceedings under the act. If the act were amended to provide for the right to "question" a witness rather than to "cross-examine", the courtroom-like atmosphere would be dispelled and it would encourage more public participation and informality, without eliminating cross-examination if that becomes necessary during the proceedings. With a mandate as broad as the Commission's, restriction of the hearings to essentially adversary proceedings would have militated against the learning environment we had created. Indeed, the inquisitorial format served us well, especially during the public information hearings. Nevertheless, we believe, much more experimentation is required before optimum hearing processes, and particularly learning environments, can be achieved in the conduct of inquiries such as the RCEPP's. Experience has shown that the most efficient utilization of time is achieved through workshop formats, panel discussions, and informal seminars, especially in the early phases of the debate, as alternatives to the conventional hearing procedures.[19]

Some of the participating public interest groups are nationally based (e.g., the Consumers' Association of Canada the Canadian Coalition for Nuclear Responsibility, and the Canadian Federation of Agriculture). Others are organized on a provincial basis (e.g., the Sierra Club of Ontario, the Ontario Coalition for Nuclear Responsibility, and Pollution Probe), and some have strong local and regional associations (e.g., the Concerned Farmers of the United Townships and the North York University Women's Club). Many of the groups are formed on an *ad hoc* basis in response (usually in opposition) to a specific project. On the other hand, the special interest groups, some nationally based (e.g., the Canadian Manufacturers Association, the United Steelworkers of America, and the Electrical and Electronic Manufacturers Association of Canada) and others provincially based (e.g., the Association of Municipalities of Ontario and the Ontario Municipal Electric Association), are concerned with the health of a particular industry (a trade association or a trade union) or with the well-being of a municipality, co-operative society, or profession. Both public and special interest groups contributed centrally to our inquiry.

The following observations apply for the most part to the public interest groups. As our lengthy series of hearings, meetings, seminars, etc., demonstrated, the majority of participating public interest groups have forged a significant niche for themselves in the energy debate by developing expertise in a broad range of energy-related issues. Because they are essentially voluntary organizations, with limited time and funds, but with impressive potential to participate effectively in decision-making, we have been acutely aware since the inception of the Royal Commission of the need to foster and encourage such groups:

• The first workshop organized by the RCEPP (September 1975) brought together a representative sample of public interest groups, with the object of determining how the groups might contribute most effectively to the inquiry. A major conclusion was that some financial support for the groups would be essential to ensure optimal participation. Moreover, it was agreed that certain groups with common interests and concerns would enhance the inquiry by co-operating through the formation of consortia. These would undertake research, present briefs, and undertake examinations and cross-examinations during public hearings, as well as participating in seminars, symposia, and workshops. Two consortia that were particularly effective throughout the inquiry, were the Food Land Steering Committee and the Ontario Coalition for Nuclear Responsibility.

• The establishment of the Office of the Public Interest Coalition for Energy Planning in June 1976 was intended to encourage the public to participate in a variety of ways in the work of the RCEPP. One specific purpose of the office was to help individuals in the preparation of briefs, a second was to publicize the work of the Commission across the province, and a third was to present key electric power planning issues in understandable form through the publication of a newsletter. Although the office was financed by the Commission, it was completely independent and worked under the general direction of a steering committee ultimately chaired by Dr. Arthur Timms.[20] The office operated on a full-time basis for 18 months. It was closed down in December 1977 as a result of a consensus amongst the steering committee who concluded that, because of the completion of the information hearings (and the early stages of the debate stage hearings), the work of the office was essentially complete. While the experiment demonstrated considerable promise, especially during the early months under the leadership of Ms. Dolores Montgomery, it never fully recovered from her tragic loss in September 1976.[21]

• In the light of our general assessment of decision-making in the energy field, we have concluded that an "energy ombudsman-type" operation, similar to the Public Coalition Office, would only be successful if backed by larger financial resources than were available, and by a larger staff. However, the OEC, as intimated, could fulfil this role.

- As we noted earlier, effective participation in inquiries necessitates not only a high-level of commitment of time for studying and preparing briefs but also a reasonable degree of continuity in attendance at hearings and meetings. Excepting special circumstances, such commitment of time and resources may constitute a serious economic hardship. On the other hand, a balanced assessment of critical issues (e.g., the role of nuclear power) is only possible if the views of a broad spectrum of concerned citizens are available. It was in recognition of this need that the Government of Ontario suggested that it would be appropriate for the Royal Commission to provide funds to support suitably-qualified public interest groups, in order that they could participate effectively in the inquiry. How were the qualifications and potential contributions of these groups assessed? In the awarding of funds, the Royal Commission was guided broadly by the principles established by the National Research Council of Canada in connection with its grants-in-aid programme. Historical information relating to a specific public interest group, its membership, its administrative structure, its general aims, its level of participation in previous public hearings, and its publications provided a measure of the credibility of the group. In addition, we required a detailed proposal relating to research projects, and an estimate of expenses in connection with attendance at public hearings. A committee of the RCEPP, chaired by a Commissioner, scrutinized each proposal and assessed its merit, especially from the standpoint of the potential capability of the group. The Chairman of the Royal Commission was responsible for the final decision.
- In general, the major public interest groups applied for funds to hire consultants (usually academics) to undertake specific studies relating, for example, to the environmental impact of bulk power transmission, the case against nuclear power, etc. Most important, the experts employed by the public interest groups were required to be independent both *de facto* and in the eyes of the public. Certain it is that, had these studies not been undertaken by the groups, they would have been commissioned by the RCEPP. This should be borne in mind in assessing the real cost of the funding programme, as outlined in Appendix C. Moreover, because the specification and supervision of a study was undertaken by the group or consortium concerned, the imprint of the latter was very visible, and this added to the credibility of the studies and indeed to their importance during the debate stage hearings. Two submissions (selected from many excellent contributions), made possible by the funding programme and exemplifying its significance, deserve special mention. They are the two-volume submission "Food Land and Energy Planning", by Professor Norman Pearson, prepared and presented on behalf of the Food Land Steering Committee, and the submission "Half-Life", by Ralph Torrie, prepared and presented on behalf of the Ontario Coalition for Nuclear Responsibility.

On the basis of our experience and the fact that expert assistance by *bona fide* public interest groups, in research and in the preparation of briefs, is required to facilitate the critical scrutiny of proposals and issues, we have concluded that appropriate funding programmes are essential.

Accordingly, we recommend that:

12.10 **The principle of funding of public interest groups from the public purse should be adopted in connection with energy and environmental hearings in the future. Only in this way will it be possible for disparate views to be aired adequately in public hearings.**

The public interest funding programme should be improved in two areas;

- **The requirement of adequate accounting practices should be written into contracts between the groups and the funding body.**
- **Wherever appropriate, an essentially inquisitorial rather than adversarial approach should be adopted in order to reduce the expenses incurred by participating groups.**

Educational Implications

Although, during the last five years, we have left few stones unturned in order to stimulate public interest in energy questions, we have had limited success. We attribute this to deficiencies in the education of the public, both formal and informal, in energy matters.

In Chapter 10, we emphasized the central role of education in energy conservation programmes. The arguments we presented there apply equally to the development of responsible public participation in decision-making. Clearly, without both an intelligible information base and an adequate level of understanding of the basic issues, effective public involvement in decision-making will be very limited. In spite of the comparatively large-scale information dissemination programmes of Ontario Hydro, few members of the public (with the exception of members of public interest groups) appear to be informed concerning the basic concepts of electric power planning or energy questions in general.

We have concluded, moreover, that with notable exceptions [22] the schools and universities do not appear to be particularly interested in establishing energy courses. For example, courses in energy, including conservation, health and environmental impacts are urgent needs. Motivation for our own educational initiatives resulted in part from our review of programmes aimed at improving the public's perception of energy needs in other countries, notably in Sweden. The Swedish programme, launched in the early 1970s, involved trade unions, adult education institutes, and public interest groups, which co-operatively organized study circles; these were financed with government funds. [23]

When the future is so uncertain, educational institutions in general, and universities in particular, should interact more meaningfully with issues that relate inextricably to human survival. Energy policy, including electric power planning, is such an issue. Nor should society be deterred by the fact that there are major dichotomies of opinion concerning electric power planning and energy issues in general. Indeed, this should be regarded as a social advantage that should spur educational projects of the kind envisaged above. Lord Ashby spelled this out succinctly in his book *Reconciling Man with the Environment*:[24]

> But – and this is the heartening thing – people are not indifferent to these conflicts nor do policymakers override them; the conflicts themselves are in my view immensely useful for they provoke a continuing debate about moral choice: choice between hard and soft values, choice between indulgence in the present and consideration for the future.

Figure 12.1 The Key Components in Ontario Hydro's Decision-Making Processes

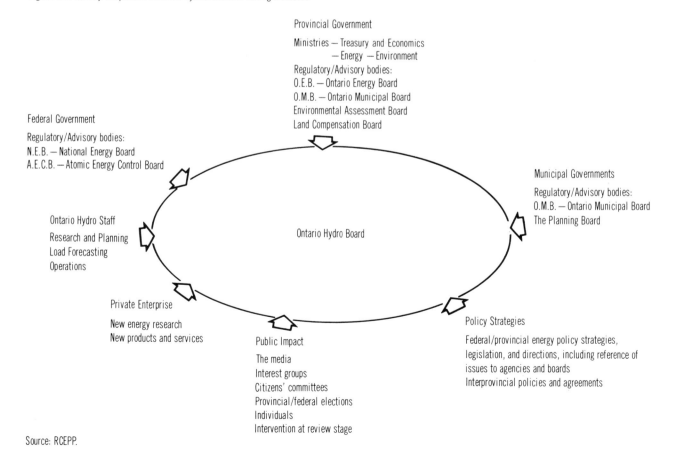

Source: RCEPP.

Figure 12.2 The Current Decision Process (Simplified)

No regular public review of general electric power programme

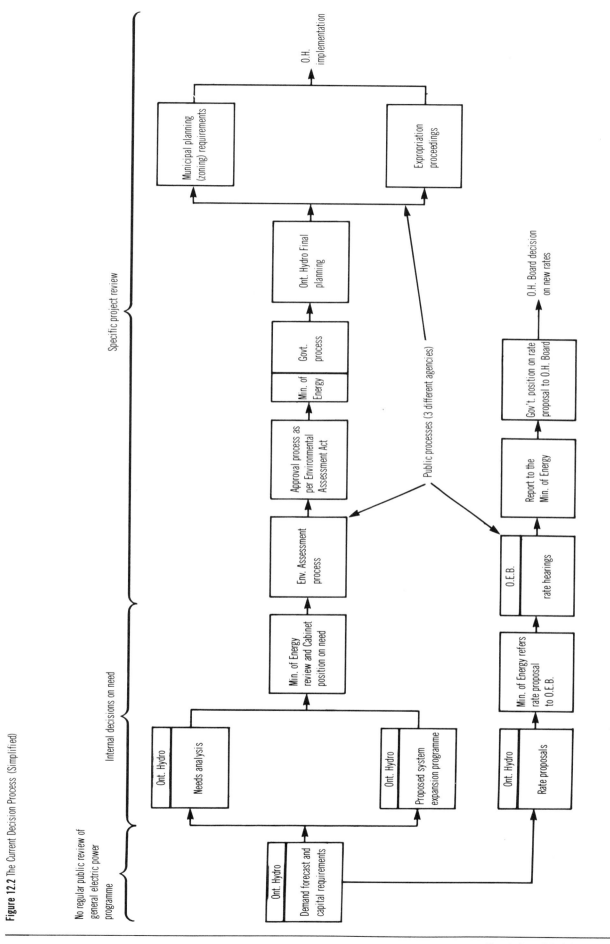

Source: RCEPP.

Figure 12.3 Multiple-Window Approach to the Decision-Making Process

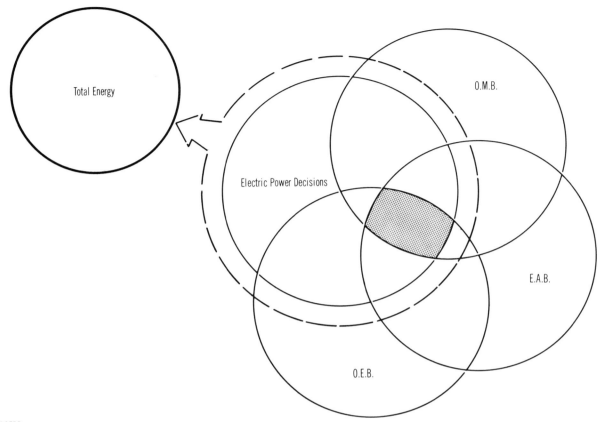

Total Energy

Electric Power Decisions

O.M.B.

E.A.B.

O.E.B.

Source: RCEPP.

Figure 12.4 Ontario Energy Commission — a "One-Window" Concept of the Regulation of Electric Power Developments

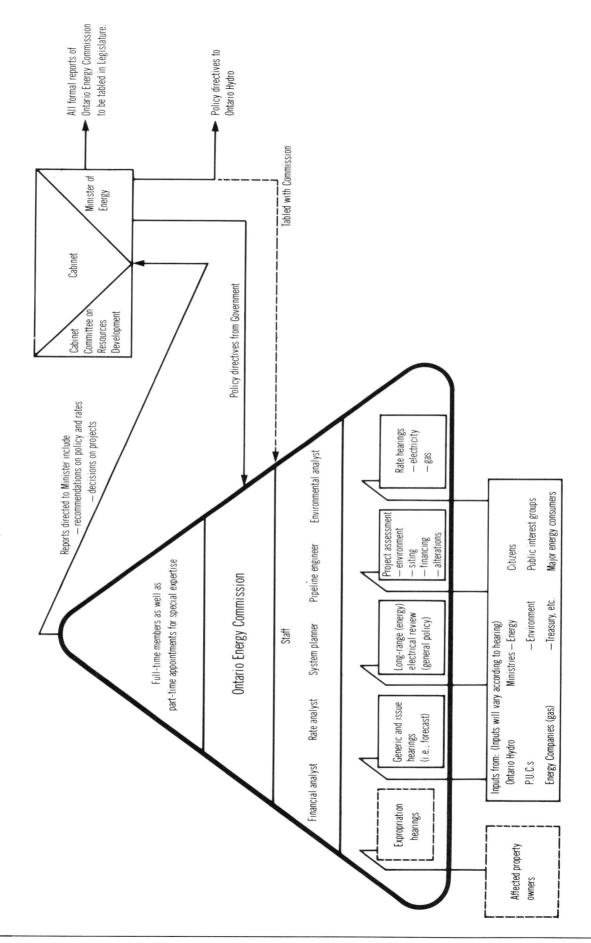

All formal reports of
Ontario Energy Commission
to be tabled in Legislature.

Policy directives to
Ontario Hydro

Tabled with Commission

Minister of
Energy

Cabinet

Cabinet
Committee on
Resources
Development

Policy directives from Government

Reports directed to Minister include
— recommendations on policy and rates
— decisions on projects

Full-time members as well as
part-time appointments for special expertise

Ontario Energy Commission

Staff

Financial analyst   Rate analyst   System planner   Pipeline engineer   Environmental analyst

Expropriation
hearings

Generic and issue
hearings
(i. e., forecast)

Long-range (energy)
electrical review
(general policy)

Project assessment
— environment
— siting
— financing
— alterations

Rate hearings
— electricity
— gas

Affected property
owners

Inputs from: (Inputs will vary according to hearing)
Ontario Hydro   Ministries —Energy   Citizens
P.U.C.s   —Environment   Public interest groups
Energy Companies (gas)   —Treasury, etc.   Major energy consumers

Source: RCEPP.

Figure 12.5 The Spread of Public Opinion

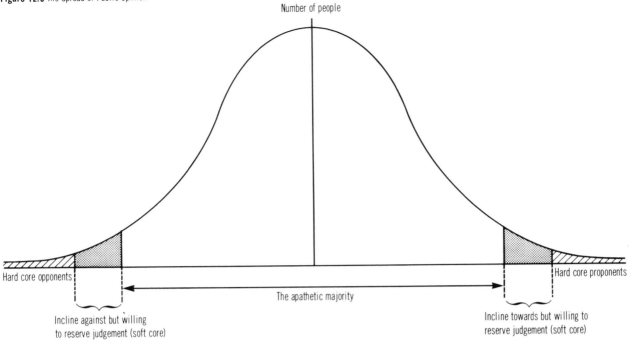

Notes: The curve is a Gaussian Distribution (Bell Curve) — the areas under various portions of the curve correspond to "numbers of people".

The area under the curve as a whole corresponds to the "total population".

Hard-core — minds are made up; cannot be dissuaded by the evidence.

Soft core — "the Key Group".

Source: "Seminar on Public Participation in Decision-Making", Lord Ashby, RCEPP, March 30, 1979.

Postscript

While it can be argued that man's evolution as a biological species has changed little during the period of recorded and indeed pre-recorded history, it can be argued with equal force that man's social and technological evolution has been dramatic, especially during the last few decades. This has given rise to greatly increased uncertainty, but a climate of uncertainty stimulates learning. This has been manifest throughout our inquiry. We have been exposed to the ideas of many imaginative people, people who are concerned, people whose attitudes reflect the values of a new society that has just become aware of the world around it, of the environment in which it lives, and of the value of natural resources.

If we have succeeded in fashioning even a fraction of their creative inputs into a framework that will facilitate the future planning of the province's electric power system, we shall be well content.

Royal Commission on Electric Power Planning

Terms of Reference

The Royal Commission on Electric Power Planning has been empowered and instructed under Order-in-Council number 2005B/75 dated the 17th day of July, A.D. 1975 to:

1) Examine the long-range electric power planning concepts of Ontario Hydro for the period of 1983-93 and beyond and to report its findings and recommendations to the Government, so that an approved framework can be decided upon for Ontario Hydro in planning and implementing the electrical power system in the best interests of the people of Ontario;

2) Inquire comprehensively into Ontario Hydro's long-range planning program in its relation to provincial planning; to domestic, commercial and industrial utilization of electrical energy; to environmental, energy and socio-economic factors, including load growth, systems reliability, management of heat discharge from generating stations, interconnecting and power pooling with neighbouring utilities, export policy, economic investment policy, land use, general principles on the siting of generating utilization of electrical energy and wise management (conservation) of primary energy resources, power generation technology, security of fuel supplies and operational considerations;

3) Deal primarily with the broader issues relating to electric power planning, and thus serve to alleviate the need for re-examination of these issues at subsequent hearings of other hearing bodies on specific details such as siting, rates, etc.;

4) Consider and report on a priority basis on the need for a North Channel Generating Station, a second 500 k.V. line from Bruce, a 500 k.V. supply to Kitchener, a 500 k.V. line from Nanticoke to London, and a 500 k.V. line in the Ottawa-Cornwall area, and other projects as may be directed by the Lieutenant Governor in Council.

Paragraph 4 was amended and supplemented under O.C.3489/77 dated the 14th day of December, A.D.1977 to include that the Royal Commission on Electric Power Planning be instructed and empowered to complete its examination of issues relating to nuclear power, to prepare an interim report of its opinions and conclusions in this area, including the extent of the need for nuclear as a component of Ontario's future energy supply and the proportion of nuclear power in Ontario Hydro's future generating capacity, and to provide such report on or before the 30th day of June, A.D. 1978.

Paragraph 4 was further amended under Order-in-Council number 2065/78 dated the 12th day of July, 1978, as follows:

A) Having concluded its hearings with respect to paragraphs 1, 2, and 3 of its terms of reference;

    i) For the geographic area of Ontario south of Bruce Nuclear power development and west of a line between Essa transformer stations and Nanticoke generating station, consider and report to the Minister of Energy on or before May 31, 1979 on load growth in the area up to the end of 1987 and from 1987 to the year 2000, the capability of existing and committed bulk power generation and transmission facilities to supply this load to the area taking into account Government policy with respect to the use of interconnections with neighbouring utilities, and the resulting date at which additional bulk power facilities, if any, will be needed, but excluding consideration of the specific nature of the additional bulk power facilities which may be required and of their locational and environmental aspects; and

    ii) For the geographic area of Ontario east of Lennox generating station, consider and report to the Minister of Energy on or before June 30, 1979 on load growth in the area up to the end of 1987 and from 1987 to the year 2000, the capability of existing and committed bulk power generation and transmission facilities to supply this load to the area taking into account Government policy with respect to the use of interconnections with neighbouring utilities, and the resulting date at which additional bulk power facilities, if any, will be needed, but excluding consideration of the specific nature of the additional bulk power facilities which may be required and of their locational and environmental aspects;

B) Provide the Government with its report and recommendations on paragraphs 1, 2 and 3 of these terms of reference on or before October 31, 1979.

Order-in-Council 2065/78 was further amended by Order-in-Council 2837/79 dated October 24, 1979 such that the requirement that the report and recommendations due on or before October 31, 1979 be altered to make said report due on or before February 29, 1980.

Ontario Hydro's Vital Statistics

Statement of Operations for the year ended December 31, 1978

	1978 ($ thousands)	1977 ($ thousands)
Revenues		
Primary power and energy		
Municipal utilities	1,275,107	1,108,099
Retail customers	442,224	407,382
Direct customers	261,816	243,560
	1,979,147	1,759,041
Secondary power and energy	288,533	210,046
	2,267,680	1,969,087
Less excess revenues	130,292	122,093
	2,137,388	1,846,994
Costs		
Operation, maintenance, and administration	501,800	414,307
Fuel used for electric generation	487,037	441,902
Power purchased	97,949	75,842
Nuclear agreement — pay-back	46,936	49,643
Commissioning energy	21,866	52,322
Depreciation	265,060	215,601
	1,420,648	1,249,617
Income before financing charges and extraordinary item	716,740	597,377
Interest	519,449	407,552
Foreign exchange losses (gains)	29,346	(3,724)
	548,795	403,828
Income before extraordinary item	167,945	193,549
Extraordinary item	20,500	—
Net income	147,445	193,549
Appropriation for:		
Debt retirement as required by The Power Corporation Act	113,446	98,078
Stabilization of rates and contingencies	33,999	95,471
	147,445	193,549

Source: Ontario Hydro. "Annual Report 1978".

Statement of Financial Position as at December 31, 1978

Assets	1978 ($ thousands)	1977 ($ thousands)
Fixed assets		
Fixed assets in service, at cost	9,549,008	8,423,173
Less accumulated depreciation	1,859,391	1,607,067
	7,689,617	6,816,106
Fixed assets under construction, at cost	3,651,344	3,137,872
	11,340,961	9,953,978
Current assets		
Cash and short-term investments	692,884	447,973
Accounts receivable	254,785	256,035
Fuel for electric generation, at cost	409,781	357,502
Materials and supplies, at cost	112,129	99,271
	1,469,579	1,160,781
Other assets		
Advance payments for fuel supplies	140,703	95,077
Long-term investments	59,555	68,623
Unamortized debt discount and expense	105,635	91,003
Long-term accounts receivable and other assets	46,073	16,173
	351,966	270,876
	13,162,506	11,385,635
Liabilities		
Long-term debt		
Bonds and notes payable	10,129,119	8,640,531
Other long-term debt	269,556	268,232
	10,398,675	8,908,763
Less payable within one year	171,912	212,910
	10,226,763	8,695,853
Current liabilities		
Accounts payable and accrued charges	512,843	428,086
Short-term notes payable	25,415	44,935
Accrued interest	273,579	217,647
Long-term debt payable within one year	171,912	212,910
Excess revenues payable	132,544	122,093
Estimated liability on cancellation of capital construction projects	16,657	7,348
	1,132,950	1,033,019
Equity		
Equities accumulated through debt retirement appropriations 1,391,181	1,279,667	
Reserve for stabilization of rates and contingencies	284,917	250,401
Contributions from the Province of Ontario as assistance for rural construction	126,695	126,695
	1,802,793	1,656,763
	13,162,506	11,385,635

Source: Ontario Hydro. "Annual Report 1978".

Comparative Statistics

	1978	1977	1976	1973	1968
Operating					
Dependable peak capacity (thousands kW)	22,845	21,347	19,677	17,501	10,338
December primary peak demand (millions kW)	15,722	15,677	15,896	13,606	9,994
Primary energy made available (millions kW·h)	95,373	92,855	90,853	78,163	55,789
Customer					
Primary energy sales (millions kW·h)					
Municipalities	61,246a	58,348	57,635	49,340	33,426
Retail	12,901a	13,021	12,436	9,880	6,266
Direct	14,794a	15,187	14,071	14,075	12,252
Total	88,941a	86,556	84,142	73,295	51,944
Secondary energy sales (millions kW·h)	10,393	8,527	4,157	5,564	369
Total Ontario customers (thousands)					
Residential	2,410a	2,358	2,297	2,140	1,941
Farm	115a	118	121	124	131
Commercial and industrial	305a	299	292	273	220
Total	2,830a	2,775	2,710	2,537	2,292
Average annual kW·h per customer					
Residential	9,740a	9,724	9,708	8,620	7,128
Farm	18,068a	17,554	16,955	14,332	10,837
Commercial and industrial	202,000a	201,384	198,722	190,600	162,613
Average revenue per kW·h (c)					
Residential	2.93a	2.80	2.23	1.63	1.29
Farm	3.05a	3.02	2.46	1.87	1.69
Commercial and industrial	2.20a	2.08	1.63	1.13	0.87
Financial					
Bonds and other long-term debt issued ($ millions)	1,847	1,407	1,539	535	240
Gross expenditures on fixed assets ($ millions)	1,694	1,425	1,326	997	329
Revenues ($ millions)					
Primary power and energy	1,849c	1,637b	1,320	794	415
Secondary power and energy	289	210	90	62	2
Assets ($ millions)	13,163	11,386	9,924	6,343	3,749
Staff, average for year	27,850	25,118	24,123	22,962	19,550

Notes:
a) Preliminary.
b) After deducting excess revenues of $122 million.
c) After deducting excess revenues of $130 million.
Source: Ontario Hydro. "Annual Report 1978".

Intervenor Funding Programme of the Royal Commission on Electric Power Planning 1976-80

1976-7 Funding

Beamsville District Secondary School – Mr. K.I. Lee: $150.00
Christian Farmers Federation of Ontario: $2,000.00
Committee on Energy and Economic Planning: $300.00
Conservation Council of Ontario: $11,200.00
Consumers' Association of Canada: $6,500.00
Energy Probe: $23,000.00
Food Land Steering Committee: $18,500.00
Huron Power Plant Committee: $3,500.00
Bruce Mitchell and George Priddle: $1,268.00
Ontario Coalition for Nuclear Responsibility (OCNR): $12,600.00
The Public Interest Coalition for Energy Planning: $40,000.00
Sierra Club of Ontario: $19,000.00
M.P. Sudbury: $1,200.00
Tanfield Enterprises Ltd.: $758.76
University of Windsor: $2,000.00
Whitefish Indian Reserve – Mr. J. McGregor, Chief: $4,000.00

1976-7 Total: $145,976.76

1977-8 Funding

Alternatives magazine: $910.00
ATEED Centre for Environmental Communities: $1,000.00
H. Burkhardt and R. Szmidt: $472.40
Canadian Coalition for Nuclear Responsibility: $24,750.00
Citizens Opposing Radioactive Pollution: $2,500.00
The Conservation Council of Ontario: $2,000.00
Conserver Society Products Co-op: $1,050.00
The Consumers' Association of Canada: $7,000.00
DynamoGenesis Corp.: $700.00
Energy Probe: $15,000.00
Energy Research Group – Ottawa: $5,100.00
Food Land Steering Committee: $17,500.00
National Farmers Union: $1,000.00
Need Committee for the North Channel – Ralph L. Thomas: $7,500.00
Ontario Coalition for Nuclear Responsibility: $13,000.00
Ontario People's Energy Network: $5,700.00
Planetary Association for Clean Energy – A. Michrowski: $500.00
Public Interest Coalition for Energy Planning: $27,500.00
Pollution Probe – Ottawa: $900.00
People Against Nuclear Development Anywhere (PANDA): $1,161.80
Post Kempenfeldt Task Force: $7,000.00
Serpent River Indian Reserve: $4,000.00
The Sierra Club of Ontario: $16,000.00
Funding to other participants who received less than $400 each: $8,000.00

1977-8 Total: $170,323.40

1978-9 Funding

Energy Probe: $12,000.00
Food Land Steering Committee: $17,000.00
Ontario Coalition for Nuclear Responsibility: $3,475.00
Kent Federation of Agriculture: $100.00
Ontario People's Energy Network (OPEN): $125.00
R. Titze: $200.00
DynamoGenesis Corp.: $100.00
Kawartha Engineering: $250.00
North Channel Needs Committee: $1,250.00
G. Wood: $875.00
PANDA: $600.00
Concerned Farmers of United Townships: $132.68

1978-9 Total: $36,107.68

1979-80 Funding

PANDA: $600.00
Food Land Steering Committee: $4,200.00
Kingston Coalition for Nuclear Responsibility: $110.00

1979-80 Total: $4,910.00

Total Intervenor Funding 1976-80: $357,317.84

Staff Members of the Royal Commission 1975-80

Margaret Ann Aboud
Paul Burke
DebbieAnne Chown
Helen Connell
Marc Couse
Sandra Coyne
Neil Cole
Sushil Choudhury
Flora D'Souza
Ann Dyer
Roni Eigles
Penny Evans
Doreen Fenton
Mary-Anne Foster
Betsy Faulkner
Karen Gaynor
Helene Giraud
Linda Hedge
Josephine Hachey
Thelma Hershorn
Catherine Hunt
Fred Hume, Q.C.
Cathy Higgins
Margaret James
Anne Jarvis
Michael Jaffey
Richard Jennings
Keith Lue
Lyse Morisset-Blais
Peter G. Mueller
Madeline Nixon
Helen Noah
Mary Ouchterlony
Gail Randall
Tony Rockingham
Robert Rosehart
Ken Slater
Wayne Stark
Ronald Smith
Robin Scott, Q.C.
Theresa Uszacki
Joan Walsh
Joanna Watts

Statistics of the Hearings of the Royal Commission on Electric Power Planning

List of Hearing Dates and Exhibits

Preliminary Meetings: October 29, 1975-January 22, 1976

Spring Public Information Hearings: March 31-July 27, 1976: Exhibits 1-24

Fall Public Information Hearings: Nov. 2-Dec. 9, 1976; January 12,20, 1977: Exhibits 25-65

Debate Stage Hearings

> Demand Issues: May 17-26, 1977: Exhibits 66-91
> Conventional and Alternate Generation Technology Issues: May 31-June 16, 1977: Exhibits 92-133
>> August 23,24, 1977: Exhibits 202-215
> Nuclear Issue: June 23-July 13, 1977: Exhibits 134-160
>> Sept. 27-Dec. 8, 1977: Exhibits 221-262
>> Jan. 11-April 5, 1978: Exhibits 263-331
> Transmission and Land Use Issues: July 19 - July 28, 1977: Exhibits 160-177
>> August 30,31, 1977: Exhibits 216-220
> Financial and Economic Issues: Aug. 9-18, 1977: Exhibits 178-201
>> March 2, 1978
> Total Systems Issues and Overview: Oct. 2-Dec. 6, 1978: Exhibits 332-375
> Public Participation and Decision-Making Issues: Jan. 9-Feb. 7, 1979: Exhibits 376-401
>> April 30-May 2, 1979: Exhibits 402-410

Southwestern Ontario Regional Hearings: March 6-28, 1979: Exhibits SW 1-72

Eastern Ontario Regional Hearings: April 9-26, 1979: Exhibits SE 1-69

List of Participants in Hearings

Information Hearings

1. Ontario Hydro — Public participation: John Dobson, R. Murray, Art Hill, Mel Bradden, David Robinson, Ken McClymont
2. Ontario Hydro — Generation, technical: Wes James, Bill Morison, Hugh Irvine
3. Ontario Hydro — Generation, environmental: Wes James, Bill Morison, Hugh Irvine
4. Ontario Hydro — Energy utilization: Larry Higgins, Les Skoff, Hal Wright, Harold West
5. Ontario Hydro — Transmission, technical: Neil Thompson, Marino Fraresso, Peter Ralston
6. Ontario Hydro — Transmission, environmental: Jack Winter, Bob Murray, Art Hill, Dr. Dales Black, Bob Walker
7. Ontario Hydro — Socio-economic factors: Dr. David Drinkwalter, Ted Burdette
8. Ontario Hydro — Provincial development and land use: Art Hill, Trevor Johnson
9. Ministry of Agriculture and Food: Norm Watson, Verne Spencer, Claire Rennie, Rob Eaton, Ken Lentz, David George, Charles MacGregor
10. Ministry of Housing: Andrew Zdanowicz, Donald Crosbie, John Burkus, Phil Rennington
11. Ministry of Health: Dr. Max Fitch, Dr. Jean Muller, Dr. Tony Muc, Dr. Doug Harding, J. Keays
12. Ministry of the Environment: Dennis Caplice, Dr. Peter Victor, Mr. Louis Shenfield, Allan F. Johnson, J. Neil Mulvaney (lawyer)
13. Ministry of Energy: Fred Button, Richard Smith, Dennis Timbrell (Minister), Malcolm Rowan (Deputy Minister)
14. Ministry of Natural Resources: Dr. Keith Reynolds (Deputy Minister), Dr. D. Dodge, Ken Lofting, Dr. E. Pye, L. Eckel, D. Drysdale, D. McLean, J. Birch, A Baxter, R. Monzon, A. Elsey, D. Fawcett, K. Fenwick, M. Smith (lawyer)
15. Ministry of Industry and Tourism: W.A. Ledingham, Dick Dillon, F.W. Plumb, W.P. Bratsberg, A. Sandler
16. Ontario Hydro — Project management: Al Jackson, Sam Horton, Ralph Chandler

17. Ontario Hydro – Fuel supply: John Matthew, Stu Hunter, William Skelson
18. Ministry of Treasury, Economics and Intergovernmental Affairs: Rendel Dick (Deputy Minister), Gregory McIntyre, David Redgrave, Brock Smith, Cliff Jutlah, Barry MacFarlane, Bill Milne, Bob Christie, Al Denov
19. Ontario Hydro – Load forecasting: Hal Wright, Bruce MacDonald, Larry Higgins
20. Ontario Hydro – Reliability: Jim Harris, Gerry McIntyre, H.P. Smith, Ken McClymont, Mark Higgins
21. Ontario Hydro – Generation planning: H.P. Smith, Henri Teekman, Don Anderson
22. Ontario Hydro – Transmission planning: Al Watson, Ken McClymont, H.P. Smith, Lou Rubino, Paul Dandeno
23. Ontario Hydro – System interconnections: H.P. Smith, Ken McClymont, Wally Winter, Gerry McIntyre
24. Ontario Hydro – Rate structures: Bruce Maconald, Harold Wright, Dr. David Drinkwalter
25. Canadian Nuclear Association: Donald Douglas, L.R. Haywood, Alan Wyatt
26. Sierra Club of Ontario: Greg Cooper (lawyer)
27. Conservation Council of Ontario: Dr. D. Kendall, Clyde Goodwin, Lyn Macmillan
28. Atomic Energy of Canada Ltd.: Dr. A. Aiken, Dr. G. Pon, Dr. R.G. Hart, Libery Pease
29. Atomic Energy Control Board: Dr. A. Prince, Paul Hamel
30. Ontario Natural Gas Association: William Skewis, Ed Gieruszczak
31. Toronto-Dominion Bank: Dr. Doug Peters, Peter Drake
32. Environment Canada: Dr. Robert Slater, Dr. F. Elder, Dr. R. Durham, D. Foulds
33. Ontario Coalition for Nuclear Responsibility: Ralph Torrie, Paul Carroll, Elaine Marshall, Sharon Belisle, Donna Elliot, George Spangler, Ed Burt, Ken Tilson
34. ARCRAD Ltd.: Philip Coulter, Thomas Wharton
35. Electrical and Electronic Manufacturers Association of Canada: David Armour, W.L. Heatherington, Sundar Raj, W.J. McNicol, W. MacOwen
36. Ontario Institute of Professional Agrologists: Jim McCullough, Charles MacGregor, W.J. McPherson
37. Energy Probe: Bill Peden, Chris Conway, Richard Fine, Barry Spinner
38. Urban Transportation Development Corporation: Kirk Foley, Dr. Bunlee Yang, Dr. Ross Gray, Dr. Jim Parker, Edward Brezani
39. Coal Association of Canada: Garnett Page
40. Association of Major Power Consumers of Ontario: Ken Voss, Alex Munro, John Leighton, D.E. Alderson
41. Food Land Steering Committee: Lloyd Moore, Elbert van Donkersgoed, Mr. MacPherson, Bruce Taylor, Dr. Norman Pearson
42. Huron Power Plant Committee: Adrian Vos, Dr. D.P. Ormrod, Lorne Luther, Phil Durand
43. Ontario Federation of Agriculture: Gordon Hill, Bruce Taylor, David Brown, Ray Cunningham
44. Urban Development Institute: John Boddy, George Dupuis
45. Housing and Urban Development Association of Canada: E. Locke
46. McClintock Homes: Mr. McClintock
47. Ontario Soil and Crop Improvement Association: G. Gardhouse
48. Science Council of Canada: Dr. Angus Bruneau, Dr. Roger Voyer
49. Consumers' Association of Canada: Barbara Shand, J. Wilson, Andrew Kerekes (lawyer)
50. Zero Population Growth Inc.: Chris Taylor
51. Public Interest Coalition for Energy Planning: Rose McMillan, Gail Randall
52. Stelco: Alec Fisher, Jerry Uvira, Stan Paget
53. Algoma Steel: Ron MacColl, Jim Cameron (lawyer)
54. Canadian Steel Industries Construction Council: Alec Carrick, Mr. Johnson
55. Dofasco: Paul Phoenix, Barry Young, Don Eastman, Ross Branston, Dennis Jones
56. Ontario Chamber of Commerce: Jack Cooper, Mr. Carnegie
57. Metro Toronto and Region Conservation Authority: Mr. Henderson, Mrs. Deans
58. Office of Energy Conservation: Ian Efford, Mo Reinberg
59. Ontario Hydro – Corporate Office: Robert Taylor (Chairman of the Board), Bill Rainey, Doug Gordon, Pat Campbell, Henry Sissons, Milan Nastich, Harold Smith

Debate Stage Hearings

Demand

1. Grey County Federation of Agriculture: Grant Preston, Bill Hodges, Fred Karrow
2. Fisheries and Environment Canada: Dr. R. Slater, Bryan Cook, Brian Emmett
3. EEMAC: David Armour, William Blundell, Colin Stairs, Carl Rath, Ed von Arb, Clayton Lemal
4. Martin Mostert
5. E. Bowers
6. D Martyiuk
7. South Grenville District High School: B. Raycroft, G. Foster, R. Throop, (students), Bill Borger (teacher)
8. Canadian Nuclear Association: Alan Wyatt
9. Ontario Ministry of the Environment: Dennis Caplice, Dr. Peter Victor
10. W. Bennett Lewis
11. Baythorn Public School, Grade 4
12. Mrs. L.E. Ayres
13. Walk for Life: Father Vince McGrath
14. Richard Hiner
15. Ontario Hydro – Demand: Larry Higgins, Hal Wright, Dr. David Drinkwalter
16. Ontario People's Energy Network: Michael Bein
17. Aubrey Bernstein
18. Zero Population Growth Inc.: Chris Taylor
19. Mission for Mankind: George Monteith

Conventional and Alternative Technologies

20. Robert K. Swartman
21. DynamoGenesis Corp.: Nick Teekman
22. ARCRAD Ltd.: Thomas Wharton
23. Thomas Markowitz
24. P. Wallheimer
25. John Shewchun
26. Acres Consulting Services: I.W. McCaig, J.G. Warnock, L.H. Anderson
27. Canadian Nuclear Association: Alan Wyatt
28. Minnesota Pollution Control Agency: Aaron Katz
29. Ontario Hydro – Generation, non-nuclear: Bill Morison, Ray Effer, John Matthew
30. Professor J. McNamee
31. Ontario Ministry of the Environment: Dennis Caplice, Bruce Martin, Al Johnson, Rob MacDonald, David Birnbaum
32. EEMAC: William Blundell, Dr. Frank Snape, Norm Williams, William Kostychyn, Robert Sproule
33. Mrs. R.I. Beatty
34. Mike P. Sudbury
35. Fisheries and Environment Canada: David Robinson, Martin Rivers, Tony Bunbrough, Arthur Burgess, Clement Cheng, Floyd Elder, Doug Gillespie, Sandy Lewis, Jim Marshall, Hans Mooij, Ralph Moulton, Victor Shantore, Dr. Robert Slater
36. Ontario People's Energy Network: Michael Bein
37. John Hix
38. Mission for Mankind: George Monteith
39. James Dooley
40. Energy Research Group: Ian Hornby
41. D.K. Sherry
42. Canadian Coalition for Nuclear Responsibility: Ian Connerty (Demand)
43. Dr. Pamela Stokes
44. CCNR (C&A): Ian Connerty

Nuclear

45. National Farmers' Union – Crysler: Joe O'Neill, Gerry Arts
46. Fisheries and Environment Canada: Dr. Robert Slater, Dr. Ray Durham
47. Save the Environment from Radioactive Pollution: Mrs. Pat Lawson, John Veldhuis
48. W. Bennett Lewis
49. EEMAC: W.M. Brown, Gordon Brown, J. Pawliw
50. Institute for Aerospace Studies: Prof. J.H. deLeeuw, Prof. P.C. Stangeby
51. Ontario People's Energy Network: Michael Bein
52. CCNR: Ian Connerty, Gordon Edwards
53. Atomic Energy of Canada Ltd.: Dr. Robert Brewer, Ross Campbell, Dr. A. Mooradian, J.A.L. Robertson, Dr. R.G. Hart, Dr. A.M. Marko, Dr. John Boulton, Liberty Pease, Keith Weaver, Dr. John Scott, Alex Mayman, W.R. Cooper, G. Yaremy, E. Hinchley, Ted Thexton, S. Hatcher, V.G. Snell

Transmission and Land Use

54. Thunder Bay Field Naturalists: Keith Denis
55. Christian Farmers Federation: Elbert van Donkersgoed
56. Paul Tremblay
57. Dr. P. Ormrod, Prof. G. Hofstra, Miss A.P. Humphreys
58. Ontario Institute of Agrologists: J.W. Kennedy, K.A. McGregor
59. Concerned Farmers of the United Townships: Lloyd Moore, Allan McKay, Wayne Johnson, Carl Ditweiler, Mattiah Martin, Pat Daunt, Bruce Schiek, Alvin Vines, George Adams
60. Fisheries and Environment Canada: David Robinson, Dr. Robert Slater, Joseph Bryant, Dell Coleman, Dr. Gordon Rosenblatt, Dr. Hazen Thompson
61. Grey County Federation of Agriculture: William Hodges, Grant Preston, Fred Karrow
62. National Farmers' Union – Crysler: Joe O'Neill
63. National Farmers' Union – Guelph: Blake Sanford, Edward Morton, Joseph Casey
64. Ontario Hydro – Transmission: Art Hill, Peter Ralston, Ken McClymont, Dr. Dales Black

Financial and Economic

65. Huron County Federation of Agriculture: Alan Wolper
66. EEMAC: Ben Ball, J. Ricketts, Mr. Briggs
67. W. Bennett Lewis
68. Ontario Hydro – Financing and Economics: Dr. D. Drinkwalter, Milan Nastich
69. Anneke Killian
70. Canadian Nuclear Association: Leo J. Schofield, Alan Wyatt
71. Aubrey Bernstein
72. Dr. S. Banerjee and Dr. L. Waverman
73. Town of Newcastle: Phil Levine, Mayor Garnett Rickard, Counsellor Bruce Taylor, John Williams (lawyer)
74. SJT Consultants Ltd.: Stan Townsend
75. John Braden
76. Ontario Ministry of Industry and Tourism: W.A. Ledingham, Ken Slater, William Moroz

Conventional and Alternative (cont'd)

77. CANTDU: Paul Carroll, Chris Springer, Tony McQuail
78. Conserver Society Products: Elizabeth Buchan-Kimberly
79. Barber Hydraulics: Arthur Margison
80. Fusion Energy Foundation: Charles Stevens, Joe Brewda
81. SJT Consultants Ltd.: Stan Townsend

Transmission and Land Use (cont'd)

82. EEMAC: Donald Lamont, Neil Bryan, Walter Shakotko, Dr. Andy Schwalm, Colin Stairs
83. Henry A. McKay
84. Moira River Conservation Authority: John Johnston
85. TESLA: Scott Foster

86. Citizens Opposing Radioactive Pollution: Ron Vastokas, Dr. Sudesh Singh, Monte Dennis, Norm Braden
87. L.C. Secord
88. DynamoGenesis Corp.: Nick Teekman, André Martineau
89. Walk for Life: Father Vince McGrath, Margaret Hancock, Deborah Mealia
90. R.G. Bramfitt
91. M. Mostert
92. Norm Braden
93. People Against Nuclear Development Anywhere: Bill Borger, Peter Onstein, Sally Schmidt, Dr. Irving Calder
94. Fusion Energy Foundation: Joe Brewda, Jon Gilbertson
95. Professor J. McNamee
96. Ontario Hydro Employees' Union: Don McMaster, Colin King, Dave Burrows, Mark Breckon
97. Amory Lovins
98. Pollution Probe – Ottawa: Ian Hornby
99. Peggy Walsh Sullivan
100. Dr. Robert Paehkle
101. Dr. H. Burkhardt and R. Szmidt
102. Ontario Coalition for Nuclear Responsibility: Ralph Torrie, Paul Carroll, Elaine Marshall, Donna Elliot, Tony McQuail, Dr. Paul Eisenbarth, Glen Wood, Elaine Hall, Fran McQuail, Sharon Belisle, Chris Springer, Ed Burt
103. Ontario Ministry of the Environment: Dennis Caplice, David Birnbaum, Beverley Thorpe
104. Chemical Institute of Canada: W.F. Brown
105. Canadian Nuclear Association: Don Douglas, Alan Wyatt
106. Whitefish Indian Reserve: Chief Dan McGregor, A.D. Wilkins (lawyer)
107. Hare Report: Dr. Kenneth Hare, Dr. A. Aikin, J.M. Harrison
108. P. Whittaker
109. Canadian Manufacturers' Association: W. Douglas Porter, Douglas Keen
110. Greenpeace London: Richard Curry
111. Serpent River Indian Reserve: James Riordan, Frank Meawasige
112. Prof. Ernest Best
113. Preservation of Agricultural Lands Society: Deborah Kehler, David Serafino
114. Sierra Club of Ontario: Rick Symmes, Don DeWees, Michael Wills, Greg Cooper (lawyer)
115. S.T. Hunnisett
116. Conservation Council of Ontario: Dr. Arthur Timms
117. Federation of Engineering and Scientific Associations: C.M. Bailey, Grant Severeight, Claudette Lassonde
118. Rev. Sam Mo
119. University Women's Club of North York: Verda Young
120. Ontario Mining Association: W. Turner, J. Hester, Mr. Ridout
121. W.G. Artiss
122. Confederation College: Richard Staples, Brian Larson, High McEwan
123. Voice of Women: Donna Elliot, Ursula Franklin, Kay Macpherson, Win Hall, Elaine Hall, Sylvia Porter, Tammy Davis, Marilyn Calmain, Nancy Pocock
124. Prof. Robert Uffen
125. Atomic Energy Control Board: Dr. A. Prince, Paul Hamel, Jon Jennekins, Robert Blackburn, John MacIssac (lawyer)
126. Ontario Hydro – Nuclear: Bill Morison, Hugh Irvine, Bob Wilson, John Deans, John Matthews, Dr. David Robertson, Ed Fenton, David Beattie

Total Systems and Overview

127. D.K. Sherry
128. Fisheries and Environment Canada: Dr. Robert Slater, Dr. G. MacKay, Dr. J. Cooley, R. Lawford, T. Won
129. *Alternatives* Magazine: Ted Schrecker
130. York Energy Group: Samuel Madras

131. Joe Umanetz
132. K.A. Innanen
133. EEMAC: S. Colms, Colin Stairs, R.A. Blanc, D.A. Lamont, George Vaughan
134. C.T. Rose
135. Conservation Council of Ontario: Dr. Arthur Timms
136. Underwood McLellan (1975) Ltd.: Carl Reichart, David Reid
137. W. Bennett Lewis
138. Guido Merkli
139. London Chamber of Commerce: Fred Wieseger
140. Township of March: Reeve Marianne Wilkinson
141. Frances Robinson
142. M.O. Neilson
143. Tymura Solardesigns: Ed Tymura
144. Philip S. Martin
145. Mennonite Central Committee: I. Williams, B. Huntsberg, David Cressman, K.R. Bender
146. Ontario Natural Gas Association: William Skewis, R. Glen Kahey, Robin Rhodes, Simon Wakup
147. Planetary Association for Clean Energy: Andrew Michrowski
148. Rev. Sam Mo
149. M. Zudel
150. Ontario Ministry of Natural Resources: Dr. Keith Reynolds (Deputy Minister), Peter Anderson, Ed Markus, John Birch, Kumr Rashamalla, Dr. Doug Dodge, Mack Williams
151. Kent County Federation of Agriculture: William Taves
152. Ontario Arts Council: Louis Applebaum, Ron Evans
153. University Women's Club of North York: Verda Young
154. Post-Kempenfeldt Task Force: Ying Hope, Elbert van Donkersgoed, Prof. John Sullivan, Prof. Ed Pleva
155. Consulting Engineers of Ontario: Neil A. McGrath, Donald H. MacDonald, Gordon A. Alexander
156. Metal Recovery Industries: Arthur Child
157. SJT Consultants Ltd.: Stan Townsend
158. Kawartha Design and Engineering Co.: Guido Merkli
159. Society of Ontario Hydro Management and Professional Staff: Trevor Faulk, Chris Charlton, Bob Taylor
160. Canadian National Railways: Ross Walker, Richard Michaliszyn
161. Institute of Chartered Engineers of Ontario: Reginald J. Laws
162. Watts from Wafers: Richard Staples, Wes Werbowy
163. Sierra Club of Ontario: Rick Symmes, Donald DeWees, Michael Wills, Sandy Constable, Dr. Melvin Fuss, Greg Cooper (lawyer)
164. Huron Power Plant Committee: Adrian Vos, Philip Durand, Dr. D. Ormrod, Dr. G. Hofstra
165. Ontario Federation of Agriculture: Peter Hannam, Adrian Vos, Otto Crone, Bill Langstaff
166. ATEED Centre for Environmental Communities: J.H. Barrington Nevitt, Arthur Margison, Prof. John Senders, Dan Shatil, Fred Scott
167. Ontario Hydro – Total systems: Art Hill, David Drinkwalter, Doug Wilson, Gerry McIntyre, Ken McClymont, Harold West, Wally Winter

Public Participation and Decision-Making

168. North Channel Needs Committee: Elaine Marshall, Glen Wood, Ed Burt, Sharon Belisle, Judy Smith
169. Elaine Marshall
170. H.I.H. Saravanamuttoo
171. Ontario Colleges of Applied Arts and Technology: C. Flacks
172. Geridex Research Foundation: George Monteith
173. DynamoGenesis Inc.: Nick Teekman
174. Prof. John Cartwright
175. Ron Titze
176. James Holder
177. Robert B. Gibson

178. Consumers' Association of Canada (Ontario): Barbara Shand, Mariam Kramer, Andrew Kerekes (lawyer)

179. Ontario People's Energy Network: Jake Brooks

180. Canadian Nuclear Association: Alan Wyatt, Michael Lewis, James Weller

181. Save the Environment from Atomic Pollution – Darlington: Peggy Clark, Dorothy Bouden, Mavis Carlton

182. Ralph Torrie

183. Electrical and Electronic Manufacturers Association of Canada (Ontario): David Armour, Ian W.W. Hendry (lawyer), Ivan Feltham (lawyer), Rolland Champagne (lawyer), Paul Simon (lawyer)

184. Ontario Municipal Electrical Association: Edwin C. Noakes, A.J. Bowker, Arthur K. Meen (lawyer)

185. Energy Probe: Chris Conway, Robert Crow, Peter Szegedy-Maszak, Brian Marshall

186. Ontario Ministry of the Environment: Dennis Caplice, Vic Rudik

187. Canadian Electrical Association: Wallace S. Reid, Don Campbell, Pat Candy, Peter Mayers

188. Grey County Federation of Agriculture: Grant Preston, William Hodges

189. *Alternatives* Magazine: Robert Paehlke

190. Christian Farmers Federation: Elbert van Donkersgoed

190a. Ontario Hydro – Public participation: Tom Reynolds, Art Hill, Robert Taylor, Trevor Johnson, Sandy McDermid, Brian Potts, David Patriquin, Peter Spratt, Cathy Williams, Doug Bannister, Fred Speer, Ron Weeks

190b. Joan Kurisko

190c. Elaine Marshall

190d. Norman Sloane

Southwestern Ontario Hearings

191. Ontario Hydro panel on load forcasting: Alex Dobronyi, Assistant Commissioner, London Public Utilities Commission; Paul Verosco, Consumer Service Representative, Western Service Region; Larry Higgins; Jim Heller, Economic Analyst, Utilization Division

192. Ontario Hydro panel on systems analysis: Ken McClymont, Tom Rusnov

193. David Peterson, Liberal MPP

194. 3M Canada Ltd.: Fred Weisegger

195. London Public Utilities Commission: Tony Ferrano, Alex Dobranyi

196. London University Hospital: John Agnew

197. London City Council: Howard Culver, M.C. Engels

198. Middlesex Federation of Agriculuture: Bob Baker, Hugh MacKellar

199. Woodstock Public Utilities Commission: John Rousom

200. Tillsonburg Public Utilities Commission: John A. Middel

201. John Labatt Ltd.: G.E. Wilson

202. National Farmers' Union & Gravel Pit Association: Andrew Kittman

203. London Chamber of Commerce: Fred Weisegger, Dr. Ed Pleva

204. Township of London: Albert Bannister

205. Simcoe and District Labour Council: David Fitkowski

206. Ontario Ministry of the Environment, Southwestern Region: E.A.N. Ladbrooke

207. Ontario Hydro panel on load forecasting: Al Clark, Service Representative, Niagara Region, Ontario Hydro; Ivan Bradley, Manager, Waterloo North PUC; Rudy Senyshen, Manager, Kitchener-Wilmont Hydro; Gordon Stacey, Manager, Guelph PUC; Bill Boyle, General Manager, Cambridge and North Dumfries PUC

208. Ontario Hydro panel on systems planning: Ken McClymont, Tom Rusnov, Al Watson

209. Professor E.J. Farkas, University of Western Ontario

210. Hydro Electric Commission of Cambridge and North Dumfries: Bill Boyle

211. Bill Thomson, Regional Municipality of Waterloo

212. Glen Wood

213. Karl Dietrich

214. Wellington Federation of Agriculture: Jim Walker, Brian Crawley, Ivan Suggitt

215. Brantford PUC

216. Ontario Hydro panel on load forecasting: Larry Higgins, Jim Heller

217. Ontario Hydro panel on load forecasting: Larry Higgins; Jim Heller; Walter J. Palmer, Area Manager, Clinton Area, Western Region, Ontario Hydro

218. Huron County Federation of Agriculture: Tony McQuail, André Durand, Bill Fear
219. CANTDU: Paul Carroll, Fran McQuail, Ron Christie, Darryl Carpenter
220. Concerned Farmers of the United Townships: Lloyd Moore, Pat Daunt, George Adams, Lorne Murray, Elmer Lesker, Jim Robinson
221. Bruce County: Art Speer, Ian Jamieson, Garry Harrow, Cathy Cook, Peggy Nulton
222. Turnberry-Howick Hydro Corridor Committee: George Adams, Jim Robinson, Harold Gibson, Wayne Clark
223. National Farmers' Union: Hilda Echland, Elmer Echland, Blake Sanford
224. Food Land Steering Committee: Elbert van Donkersgoed, Lloyd Moore, Blake Sanford, George Adams
225. Kincardine and District Chamber of Commerce: Wayne Peachman, Roger Thomson, Clarence Ackert, John Slade, Larry Steinman
226. Goderich Public Utility Commission: Dr. Jim Peters, Mr. Murphy, Albert Shore
227. Don Haycock, Ian McNaughton, Sam McGregor
228. Wallaceburg Public Utilities Commission: Tom Schuurman
229. Kent Federation of Agriculture: William Taves
230. Union Gas Ltd.: Dr. Simon Wauchop, Arthur Skillington
231. Ontario Hydro panel on load forecasting: Larry Higgins; Clayton Leach, General Manager, Chatham Hydro Electric System; Graham Yates, Consumer Service Supervisor, Western Region, Ontario Hydro
232. Food Land Steering Committee research report: Gary Davidson, Roman Dzus
233. Sarnia Hydro Commission: Mr. Turnbull, Mr. Reynolds
234. Windsor PUC: David Pope, Bill Best
235. Association of Major Power Consumers of Ontario: John P. Bouchard
236. Dow Chemical of Canada Ltd. – Sarnia Division: Joseph P. Zanyk

Eastern Ontario Hearings

237. Ontario Hydro panel on load forecasting: Larry Higgins, Ron Audet, Consumer Service Supervisor, Ontario Hydro; David L. Venutti, Area Manager, Vankleek Hill Area, Ontario Hydro
238. North Country Defence Committee: Joel Ray, Margaret Weitzman, Clyde Morse
239. Dominion Textiles Inc.: Charles Panzer, Manager, Energy and Environment; Bernard Hamel, Plant Manager, Long Sault Plant
240. Canadian Industries Ltd. – Cornwall Works: Roy Rumple, Works Engineer
241. United Townships of Glengarry, Stormont and Dundas: Larry Cotton, County Planner
242. Ontario Ministry of Natural Resources: Jeff Higham, Management Forester
243. Ontario Hydro panel on system facilities: Al Watson, Lew Rubino, Bev Pearson
244. Thyme and Sage Machinery Ltd.: Paul Neelands, President
245. Ontario Non-Nuclear Network: Gary Glover
246. Ministry of Natural Resources, Lanark District: A.S. Corlett, Management Forester
247. Town of Smiths Falls Economic Development Commission: Lucien Lalonde, Director; Smiths Falls Chamber of Commerce: Ross Hill, President
248. Township of South Sherbrook: Stewart Munroe
249. Prof. Gerald Hodge, School of Urban and Regional Planning, Queen's University
250. Kingston Public Utilities Commission: Dr. R.H. Hay, Commissioner
251. Alcan Products Ltd. – Kingston Works: Jim Latimer, Electrical Works Engineer; Sandy Little
252. Cataraqui, Frontenac and St. Lawrence Wards Ratepayers' Association: Mrs. Irene Mooney
253. Helen Henrikson
254. Queen's University Physical Plant: Gerry MacCahill, Director
255. Millhaven Fibres Ltd. (Celanese Canada Inc.): W.M. Campbell, Electrical Engineer
256. Mark Bunting and Michael C. Turcot
257. Kingston Coalition for Nuclear Responsibility: Dr. S.L. Segel
258. South Grenville District High School: Mildred Chang, Cheryl Tuteckey, Lora Graham, Donna Fox
259. Sandra Robertson, Concerned Ratepayers of Edwardsburg
260. People Against Nuclear Development Anywhere: Bill Borger, Sally Schmidt, Emily Finn, Peter Onstein
261. Town of Prescott and Prescott PUC: Mayor Sandra Lawn; Peter Martin, Engineer; Foch Healey, General Manager of PUC

262. Genstar Chemical Ltd.: Ted Bjerkelund, Director, Corporate Services; M. Champagne
263. Grenville Federation of Agriculture: Keith Matthie; Richard Denison, President
264. Blue Church Area Concerned Citizens and Augusta Township Council: William Keith, Andrea Luard, Rod Luard
265. Association of Concerned Ratepayers: Harry Pietsma, John Watts (withdrew)
266. Dundas County Federation of Agriculture: Anna Smail, Ross Dulmidge, John Dalrymple
267. Mrs. Winnifred Veitch, Barbara Dole, William Bartlett of N.Y. State
268. Evelyn Gigantes
269. Department of National Defense: J.V. Johnson, Director, Utilities and Municipal Services
270. Ottawa City Council: Controller Ralph Sutherland
271. Marey Gregory
272. Gloucester Hydro: A.J. Bowker, Commissioner
273. Ottawa Board of Trade and Commercial & Industrial Development Corporation of Ottawa-Carleton: Spencer Ballantine, Past President CIDC; George Pearley, Ottawa Board of Trade; Sean Marque, CIDC
274. Association of Major Power Consumers of Ontario: Alex Munroe, Executive Director
275. R.L. Crain Ltd.: J. Holowka, Plant Engineer
276. Ottawa Hydro Electric Commission: Lloyd Askwith, General Manager
277. Housing and Urban Development Corporation – Ottawa: Mr. Russell and Marian Seymour
278. IVACO Rolling Mills and Eastern Steelcasting: John Griffiths, David Goldsmith, Gordon Silverman
279. University of Ottawa: Ed Butterworth, Energy Manager
280. Nepean Hydro: Wayne Phillips, Vice-Chairman; Doug Bell, Assistant General Manager
281. Milt Maybee and Michael Bein
282. United Counties of Prescott and Russell: M. Laframboise, John Kirby, Councillor; Hydro Hawkesbury: Philip McNeely, Manager
283. Ottawa Health Sciences Centre: Bill Flud
284. Ottawa Civic Hospital: R. Stuart Haslett, Georges Corbeil, George Stencel
285. Consumers' Gas Co.: R.H. Townsend, General Manager, Eastern Region
286. National and Provincial Parks Association: C.W. Woodleigh, President, Ottawa-Hull Chapter
287. Building Owners' and Managers' Association of Ottawa (written only)
288. Badische Canada Ltd.: L.A.M. Marshall, W.R. Lennox
289. Huyck Canada Ltd.: C.G. Hill, Manager of Engineering
290. Playtex Ltd. (written only)
291. North Lanark Energy Conservation Centre: Earl Hansen
292. Town of Renfrew: R. Thomson, Industrial Commissioner
293. Arnprior Business Association: Ben Sauve
294. Northern Municipalities of Almonte, Ramsay and Pakenham: Mr. Swierenga, Reeve of Almonte
295. City of Kanata: Mayor Marianne Wilkinson; Don Farmer, Chairman of Hydro Commission
296. Dr. W.E. Gordon
297. Milt Maybee

Notes to Chapters

Note to Preface
1. C.C. Burwell, M.J. Ohanian, and A.M. Weinber, "Siting Policy for an Acceptable Nuclear Future", *Science*, vol. 204, no. 4397, June 8, 1979, p. 1043.

Notes to Chapter One
1. *Long Range Planning of the Electric Power System*, Ontario Hydro Report 556 SP, February 1974. This report precipitated a major debate between the utility and public interest groups, especially concerning the projected load growths upon which the planning was based.

2. *Interim Report on Nuclear Power in Ontario, A Race Against Time*, Royal Commission on Electric Power Planning, September 1978. This will be referred to throughout this report as the *Interim Report*. The RCEPP's Terms of Reference were amended in December 1977 — in addition to the original requirements, the Commission was required to prepare an interim report on nuclear power.

3. *The Need for Additional Bulk Power Facilities in Southwestern Ontario*, Royal Commission on Electric Power Planning, June 1979.

The Need for Additional Bulk Power Facilities in Eastern Ontario, Royal Commission on Electric Power Planning, July 1979.

4. We have interpreted "environment" in the broad sense embodied in the Ontario Environmental Assessment Act, 1975, Part I, Section 1, subsection (c):

> *Environment* means,
>> (i) air, land or water
>> (ii) plant and animal life, including man,
>> (iii) the social, economic and cultural conditions that influence the life of man or a community
>> (iv) any building, structure, machine or other device or thing made by man,
>> (v) any solid, liquid, gas, odour, heat, sound, vibration or radiation resulting directly or indirectly from the activities of man, or
>> (vi) any part or combination of the foregoing and the interrelationships between any two or more of them, in or of Ontario.

However, from time to time in the Report, especially in Volumes 6 and 7, we interpret "environment" in a limited sense. Furthermore, in Chapter 9 of this volume, "Environmental Concerns — the Conflict with Nature", we interpret environment in the sense of the natural environment.

5. *Shaping the Future*, Report No. 1, Royal Commission on Electric Power Planning, 1976.

6. The existing stock of energy-consuming technology, e.g., buildings and industrial machinery, can be made more efficient by retrofitting, or replacement, but this takes time and is usually costly.

7. The term "end use" refers to the ultimate use of a source of energy. For example, some major end uses of electric energy are metallurgical furnaces, pulp and paper manufacturing, space heating, etc. The concept of "end-use pattern" relates to "patterns of utilization of energy" in the industrial, commercial, and residential sectors.

8. Both avoidable and unavoidable losses (wastes) can, *ipso facto*, be reduced by energy-conservation practices. In converting one form of energy into another (coal or uranium to electricity), as will be shown in Chapter 4, there are unavoidable losses of energy (i.e., a concomitant increase in entropy).

9. We have noted, during the inquiry, that many participants apparently held the view that the major source of atmospheric, soil, and water pollution in the province is Ontario Hydro. The Mississauga train derailment (November 1979) and its potentially dangerous health and environmental implications may cause a change in perceptions. The potential hazards inherent in the manufacture and transportation of highly toxic chemicals may constitute a greater threat to health and the environment than the operations of Ontario Hydro.

10. The Commission's Issue Paper No. 8, *The Decision-Making Framework and Public Participation*, presents the existing structure for decision-making in some detail and identifies the key issues.

11. H. Brooks and R. Bowers, "The Assessment of Technology", *Scientific American*, vol. 222, no. 2, February 1970.

E. Mesthene, *Technological Change*, Cambridge, Mass.: Harvard University Press, 1970.

Notes to Chapter Two

1. See Appendix B for Ontario Hydro's vital statistics. The Tennessee Valley Authority, with a peak generation capacity of over 26,000 MW, is the largest electricity utility in North America. The peak generation capacity of Ontario Hydro, as of July 1, 1979, was 23,600 MW.

2. The Advisory Committee on Energy was appointed by an Order-in-Council in August 1971 and reported to the Cabinet in March 1973. Task Force Hydro was appointed by the Committee on Government Productivity in September 1971 to review the function, structure, operation, financing, and objectives of the Hydro-Electric Power Commission of Ontario. Five reports of Task Force Hydro were published between August 1972 and June 1973. The mandate of Ontario Hydro as set out in the Power Corporation Act, and amended following these reports is:

> The purposes and business of the corporation include the generation, transmission, distribution, supply, sale and use of power and, except with respect to the exercise of powers requiring the prior authority of the Lieutenant Governor in Council under this Act, the Corporation has power and authority to do all such things as in its opinion are necessary, usual or incidental to the furtherance of such purposes and to the carrying on of its business.

3. In general, throughout this Report, reference to "Ontario's electric power system" relates to Ontario Hydro's total system, i.e., the East System and the West System (unless otherwise specified), but excluding the private utilities.

4. The power transmitting capability of 500 kV lines is about four times as great as that of 230 kV lines for short distances, and it is about seven times as great for distances of 300 km.

5. The Power Commission Act specifies the nature of the relationship between Ontario Hydro and the Government of Ontario. It was passed in 1907 and was subsequently amended in response to the recommendations of Task Force Hydro in 1973, when the Power Commission Amendment Act came into force, establishing the Power Corporation Act.

6. Royal Commission on Electric Power Planning, *Report on the Need for Additional Bulk Power Facilities in Southwestern Ontario* and *Report on the Need for Additional Bulk Power Facilities in Eastern Ontario*, Toronto, June 1979 and July 1979, respectively. (See Chapter V in both reports.)

7. "Economics of the Industrial Co-generation of Electricity". Proceedings of a seminar co-sponsored by Ontario Ministry of Energy and Ontario Hydro, December 1978.

Co-generation plants are usually based on gas turbines. Electricity and process-steam for industry can be generated with an overall efficiency of up to 80 per cent. District heating plants, usually coal-fired but with the potential for burning combustible solid municipal waste, produce thermal energy for residential space heating and can be designed to generate electricity as well. These plants, like co-generation plants, are very efficient from the standpoint of the optimum utilization of primary fuels and solid wastes.

See Volume 2 for an in-depth discussion of co-generation.

8. The concept of "added value" is basic in the assessment of the economic growth of a nation or a province. For example, standard of living depends on two things: first, on the natural resources available per capita, and, second, on how these natural resources are converted into goods and services, and in particular how much value is added to the basic raw materials through the processes of production and service. The farmer, for instance, creates wealth by growing crops and breeding animals, and by selling the produce for more than the cost of the seeds, fertilizers, foodstuffs and other purchases.

9. Ontario Hydro submission to the RCEPP Public Information Hearings, "Socio-economic Factors". 1976, p. 4.1-1.

10. Task Force Hydro, Report No. 1, *Hydro in Ontario — A Future Role and Place*. 1972, p. 26.

11. Ontario Hydro submission to the RCEPP Public Information Hearings, "Socio-economic Factors". 1976, p. 4.1-28.

12. See Volume 5 of this Report and Chapter 11 of this volume.

13. This problem is dealt with in Volume 8, and referred to also in Chapter 12 of this volume.

14. "Design for Development". A white paper on provincial and regional development. Ontario Government, April 1966.

15. Ontario's electric power system is interconnected, in effect, with a large power network involving six Canadian provinces and many states of the eastern and central United States.

16. In 1975, Ontario Hydro established a special division, the Route and Site Selection Division, whose duties include the co-ordination of citizen participation in the corporation's planning activities.

Notes to Chapter Three

1. See the Commission's *Interim Report*, Chapters 2 and 13.

2. See the Commission's *Report on the Need for Additional Bulk Power Facilities in Southwestern Ontario*. RCEPP, July 1979.

3. The potential of conservation and load management with respect to each end use, coupled with possible substitution of electricity for fossil fuels in certain cases, will materially affect load shapes in the future. It is essentially for this reason that end-use forecasts should prove to be a significant input to the forecasting process.

4. During the period under consideration, coal was the fuel of the major incremental generation stations.

5. Kenneth E. Boulding, "Science and Uncertain Futures", *Technology Review*, June-July 1979, p. 9.

6. Primary peak demand is defined as the maximum 20-minute demand for electricity generating capacity that occurs during the year as measured at the point of generation.

7. See Volume 3, Chapter 3, and Chapter 11 of this volume for an introduction to the applicability of econometric models to load forecasting.

8. This coefficient corresponds to the "cross-price-elasticity" of demand between oil and electricity, which in turn gives a measure of the change in demand for electricity as oil prices increase.

9. Prior to 1973 there was insufficient variability in oil or electricity prices to provide estimates of how electricity demand would be affected by a significant change in the price of oil.

10. Projections of the effect of saturation and structural change, using the model, may be applied to a given potential economic growth rate, and we can determine the secondary energy demand that might be expected to occur assuming no increase in real energy prices.

11. The Stanford Research Institute study commissioned by the Canadian Electrical Association and Ontario Hydro ("Long Range Electricity Forecast for Canada – A Methodology", November 1978) has provided an excellent basis for econometric modelling of the kind we envisage. The study noted, in particular, "insufficient information . . . to develop a highly reliable forecasting methodology for the commercial sector".

12. See, for example, "Report of Member Electric Systems of the New York Power Pool – Long-Range Plan – 1978", Exhibit 7 – Load Forecasting Methodology, April 1978.

13. Mans Lonnroth and William Walker, "The Viability of the Civil Nuclear Industry". Report prepared for the International Consultative Group on Nuclear Energy, London, England. December 1979.

14. "Report of Institute for Energy Analysis", 1976. See also the "Report on Energy Conservation", Energy, Mines and Resources Canada, 1977.

15. This range is derived simply by subtracting the 1.55 per cent from the projected range for secondary energy growth (assuming no energy-efficiency improvements) of 2.6 to 3.3 per cent.

16. The load forecast is needed for planning capacity expansion and should not be confused with "demand management", which refers to the use of pricing, load control, and other techniques to improve load factors. Such techniques will not alter the need for peak capacity, which depends principally on the weather.

17. Maximum benefits occur when the electricity component of the hybrid system is subject to control by the utility. In effect, this converts a firm load to an interruptible one that can be curtailed in times of system emergency.

Notes to Chapter Four

1. Deuterium and lithium are the basic elements that are being used in nuclear fusion experiments. No self-sustaining nuclear fusion reaction has yet been established, even in the laboratory. If nuclear fusion eventually becomes a commercial reality, both deuterium (an isotope of hydrogen) and lithium (needed for the production of tritium – another heavier isotope of hydrogen) will be required. Although there is enough deuterium in the world's oceans to fuel a massive nuclear fusion programme (on a world-wide basis) for hundreds of thousands of years, lithium is by no means so abundant.

2. It is worth noting also that all physical and biological processes involve energy conversion in one form or another. Indeed, some of the most complex energy conversion processes known to man are involved in the conversion of food energy into "muscle power" and hence mechanical energy, and into "brain power" based on electric and chemical energy. Note also that the conversion of a very small percentage of the energy stored in the nucleus of an atom of uranium-235 into thermal energy, and then into electric energy, is a highly significant example of man's scientific and engineering skills.

3. There are, however, methods available for extracting energy from the oceans at very low efficiency, some of which have been demonstrated experimentally. For example, the temperature difference

between surface layers of water and water at very deep levels, especially in tropical latitudes, is appreciable, and could theoretically be utilized in a "thermodynamic engine". Further, experiments are well advanced in connection with the conversion of ocean wave energy into electricity. But to date neither of these approaches has been shown to be cost-effective.

4. The efficiency of conversion in this stage is particularly important. The maximum efficiency depends solely on the input temperature of the steam entering the turbine and the output temperature of the condensed water. (The condensation process normally involves a large quantity of lake or river water which, through a heat exchanger, condenses the turbine steam by removing its latent heat.) The heat of condensation is subsequently discarded as "waste thermal energy" and the condensate is recycled. The thermodynamic efficiency of an ideal, or "reversible", heat engine is given by the expression $(T_1 - T_2)/T_1$, where T_1 is the absolute temperature, expressed in degrees Kelvin of the input steam, and T_2 is the absolute temperature of the condensate. The theoretical upper limit on conversion efficiency for this stage, assuming a steam temperature of $500°C$ and a condensate temperature of $30°C$, is about 60 per cent.

5. Not only is the "entropy law" one of the most fundamental in the physical sciences, but it has important implications for economic theory. See, for example, Nicolas Georgescu-Roegn, *The Entropy Law and the Economic Process*, Harvard University Press, 1970.

6. See also R.K. Swartman, "Alternative Generation Technologies", in *Our Energy Options*, Royal Commission on Electric Power Planning, Toronto, 1978.

7. Amory B. Lovins, "Soft Energy Paths: Toward a Durable Peace", Ballinger, 1977.

"Alternative Long-Range Energy Strategies", Joint Hearings before the Select Committee on Small Business and the Committee on Interior Affairs, U.S. Senate, 94th Congress, Washington, D.C. U.S. Government Printing Office, Interior Committee Serial No. (94-47) (92-137), December 1976.

The Commission is particularly indebted to Amory Lovins for his appearances in both its public and its *in camera* hearings. Not only did he provide us with an eloquent discourse on the distinction between hard- and soft-technology paths, but he brought new insights to bear on energy policy issues. We believe his perceptions have advanced the energy debate materially, especially by putting into focus a new frame of reference. While we do not endorse all of his claims, we nevertheless strongly support his view that the position of the adherents to hard-energy technologies should be challenged.

8. The term "decentralized system" has various meanings. Indeed, the existing electric power system is decentralized in so far as the public utility commissions are concerned. Wood-fuelled stoves, for example, can be regarded as decentralized energy units.

9. Active solar energy systems, as opposed to passive systems, incorporate collector panels, etc. The sources of information are the National Research Council (NRC) Building Research Division; the NRC Renewable Energy Task Force; the Saskatchewan Research Council in connection with the Saskatchewan Conservation House; and studies by the architectural firms of Raymond Moriyama and John Hix. However, a major difficulty at present is that inadequate scientifically monitored data is available for a realistic assessment.

10. M.J. Helferty and R.G. Lawford, "Meteorological Information for Use in Assessing the Auxiliary Energy Requirements of Solar and Wind Energy Systems in Five Ontario Locations", Internal Report SSU-78-9, Ontario Region, Atmospheric Environment Service, Environment Canada, Downsview, Ontario.

A complementary study in the United States, "Solar Availability for Winter Space Heating: An Analysis of SOLMET Data, 1953 to 1975", J.G. Ashbury *et al. Science*, vol. 206, November 9, 1979, p. 681, concludes that solar space-heating systems in many regions of the U.S. require large back-up power on peak days even in the case of solar heating systems with comparatively large thermal storage.

11. *Ibid.*

12. The term "co-generation" normally applies to the generation of electric power and industrial process-steam. However, as well as "dual-mode generation", "co-generation" is used for describing all systems in which electric and thermal energy are generated. Accordingly, the term is frequently used in connection with district heating plants that generate electricity.

13. In Chapter 7 we discuss the concept of "system responsiveness", or "fine tuning of the system". During the 1990s, if the load growth in Ontario increases more than expected, there may be an important role for co-generation technologies, operating in a base-load mode, that can be put in place comparatively quickly (e.g., two to three years). Especially if operated as dual-purpose stations, such systems may prove very attractive.

14. Lovins. *Op. cit.* See also, P.H. & S. Streiter, "Multiple Paths for Energy Policies: A Critique of Lovins' Energy Strategy", *Energy Communications*, vol. 4, no. 4, 1978.

15. RCEPP transcript, vol. 163, p. 23194.

16. "Roads to Energy Self-Reliance", Science Council of Canada, Report No. 30, Ottawa, June 1979.

17. These are particularly undesirable atmospheric pollutants that give rise to a range of environmental threats including "acid rain". The topic is dealt with in depth in Volume 6, and our general conclusions relating to it are given in Chapter 9.

18. "Hearst Wood Waste Energy Study: Summary Report", Acres Shawinigan Ltd., Toronto, January 1979. Prepared for the Ontario Ministries of the Environment, Energy, Natural Resources, Industry and Tourism, and Northern Affairs, Ontario Hydro, the Town of Hearst, and the Hearst Lumbermen's Association.

19. The manufacture of solar panels, thermal storage systems, and ancillary equipment necessitates the use of non-renewable energy resources and materials.

20. J.P. Holdren, "Fusion Energy in Context: Its Fitness for the Long Term", *Science*, vol. 200, April 14, 1978, p. 168.

21. Ontario Hydro, Design and Development Division "Residential District Heating with Moderator Heat from CANDU Power Plants", Report No. 78201, September 1978.

Notes to Chapter Five

1. H.J. Otway, "Review of Research on Identification of Factors Influencing Social Response to Technological Risks." Report CN-36/4, International Atomic Energy Agency, Vienna, Austria.

2. For a detailed discussion of radiation exposure risks, see the RCEPP *Interim Report* and Volume 6 of this Report.

3. L. Sagan, "Radiation and Human Health". *Electric Power Research Institute Journal*, September 1979, p. 9.

4. E. Radford, Chairman of the Committee on the Biological Effects of Ionizing Radiation (BEIR III), established by the United States National Academy of Sciences, has recommended that there should be an appreciable tightening up of these standards. But this view is not accepted by all members of the committee.

5. See, in particular, evidence of Dr. E. Radford presented at hearings of the Ontario Legislature Select Committee on Ontario Hydro Affairs, July 10-11, 1979.

6. The most comprehensive study of the probability of occurrence and the consequences of a major nuclear accident was undertaken by Professor Norman Rasmussen of MIT.
Reactor Safety Study: An Assessment of Power Plants. U.S. Nuclear Regulatory Commission. Washington, D.C., October 1975. WASH-1400 (NUREG-75/014). See RCEPP *Interim Report*, pp. 77-9.

7. *Risk Assessment Review Group Report to the U.S. Nuclear Regulatory Commission*, Chairman H.W. Lewis, September 1978. (NUREG/CR-0400).
This *ad hoc* review group prepared a critical review of the Rasmussen Report (WASH-1400).

8. *Report of the Committee on Biological Effects of Ionizing Radiation.* BEIR III, National Science Foundation, Washington, D.C. May 1979.

9. H.T. Peterson, letter to the editor of *Science*, August 1979.

10. We reiterate that, although understanding of radiation-induced disease is obviously inadequate, it far transcends our understanding of the carcinogenic physical and chemical agents that occur in the environment and induce cancer by damaging the DNA. While the cancer-inducing effect of ionizing radiation has been known for many years, and while methods for detecting ionizing radiation have been in use for at least 60 years, with a massive increase in sensitivity of detection, the risk of man's exposure through the environment to a plethora of toxic chemicals is just beginning to be recognized. Nor are there simple means of detection or measurement. Nor, as in the case of ionizing radiation, does any international committee of experts, or indeed, national committee, exist to establish protective guidelines that might lead, for example, to the banning of all unnecessary potential carcinogens.

11. We refer in particular to *Risk Assessment Review Group Report to the U.S. Nuclear Regulatory Commission*, Chairman, H.W. Lewis, Washington, D.C. September 1978; and a selection of "significant event reports" relating to the operation of the Pickering and Bruce nuclar generating stations. The latter were made available through the proceedings of the Select Committee on Hydro Affairs, April 1979.
Report of the Inter-Organizational Working Group established by the AECB. This report reviews current general safety principles and criteria (relating to the CANDU reactor) and presents recommendations that relate to modified guidelines. As far as we are aware, these recommendations have not yet been adopted by the AECB.

12. The recent decision of the AECB (January 1980) not to endorse the recommendations relating to

relaxed public standards of radiation exposure in the event of a major nuclear accident, as well as the rejection of the "escape clause", should be noted. However, the Board accepted the recommendations relating to licensing criteria with minor modifications.

Note that the "Board has not accepted the more controversial aspects of the IOWG Report".

13. The design of the Three Mile Island reactors is quite different from that of CANDU. The reactor is a pressurized light water reactor (the moderator and coolant are light water rather than heavy water), and the reactor core is contained in a large pressurized "reactor vessel" through which the coolant flows, extracting heat from the reactor and delivering it to the boilers. In CANDU, a large number of pressure tubes containing pressurized heavy water are used instead of the vessel configuration.

There have been many reports and articles relating to the Three Mile Island nuclear generating station accident of March 28, 1979. We draw attention to the following:

"Three Mile Island nuclear generation accident of March 28, 1979". Report of the scientific assessment team to the Ontario Inter-Ministerial-Agency co-ordinating committee, April 20, 1979.

E. Marshall, "Three Mile Island — News and Comment", *Science*, vol. 199, April 20, 1979, p. 281.

Correspondence relating to the Three Mile Island accident: *Science*, vol. 204, May 25, 1979, p. 794.

14. *Report of the Presidential Commission on the Accident at Three Mile Island.* Washington, D.C. October 31, 1979.

15. K.S. Pitzer. Letter to the editor of *Science*, June 22, 1979.

16. H.T. Peterson. Letter to the editor of *Science*, August 1979.

17. The "deaerator" and storage tank in the CANDU condensate recycling circuit (i.e., the feed train) contain a large amount of water that would continue to feed the boilers for long enough either to restore operation of the condensate recovery pump or to establish a stable reactor system condition.

18. Ontario Hydro. Presentation to the Select Committee of the Ontario Legislature. Exhibit E-1. April 25, 1979.

19. See the RCEPP *Interim Report*, p. 80.

20. It is on the record of the Select Committee hearings that these significant event reports accompanied by a letter to Mr. E. Sargent, MPP, were from Mr. "M. Schultz" (a pseudonym). Subsequently the letter and the reports were tabled at a Select Committee hearing.

21. The Significant Event Reports in question are Bruce Generating Station, SER 77-47, SER 77-50, SER 77-53, SER 78-1, SER 78-4, SER 78-22, SER 79-9. These reports are available at the Ontario Hydro library and at the Legislative Library, Queen's Park, Toronto.

22. See the Compendium of Major Findings in the RCEPP *Interim Report*. There appear to be an average of seven significant event reports at the Bruce Generating Station per month, and consequently a total of several hundred to date. The selected reports exemplify certain weaknesses in operational procedures. The official reports on the TMI accident were those of the Ontario Government Task Force, Ontario Hydro's presentation to the Select Committee on Hydro Affairs, and a summary of the United States Presidential Report on the accident. In addition, we have read several definitive articles on the accident that have appeared in *Science* (the journal of the American Association for the Advancement of Science) during the last eight months.

23. A similar proposal has been put forward by Peter A. Morris, see "Nuclear Reactor Operation", *Science*, July 13, 1979. p. 148.

24. This recommendation was made by the Select Committee on Ontario Hydro Affairs. Subsequently, the Ontario Legislature resolved, in December 1979:

> That, in the opinion of this House, the Government of Ontario should request the Atomic Energy Control Board to commission a study to analyze the likelihood and consequences of a catastrophic accident, such as a nuclear meltdown in a nuclear reactor, or radiation escape, but if, within six months of the date of the request, the Atomic Energy Control Board has not commissioned such a study, the Government of Ontario should undertake the study on its own initiative.

25. The Rasmussen Study was headed by the Chairman of the Department of Nuclear Engineering at MIT, and included at least 20 professionals with expertise in nuclear engineering, probability and statistical theory, etc. The cost of the study was in the order of $5 million.

26. Transactions of Ontario Legislature Select Committee on Hydro Affairs, July 10-11, 1979.

27. An important issue is — when all the tailings areas in the vicinity of the mill operations have been filled, what next? The sheer bulk of the tailings militate against long-distance transportation — the problem deserves serious study.

28. *A Public Hearing by the Environmental Assessment Board into the Expansion of the Uranium Mines in the Elliot Lake Area*, D.S. Caverly, Chairman. Final Report May 1979. Summary and Recommendations, p.x.

29. The high level radioactive wastes from nuclear weapon production are necessarily liquid wastes, and, in general, are more difficult to handle than solid wastes. On the other hand, because reprocessing of spent nuclear fuel on a large scale is restricted to one or two countries (Canada is not one of them), the spent fuel from nuclear power plants is at present in its original solid form.

30. The concrete silos at present under development at the Whiteshell Nuclear Research Establishment are 5 metres high, and each is designed to store about 5 tonnes of spent fuel in retrievable form. The cost of such interim storage is likely to add no more than 0.2-0.3 mills/kW·h to the cost of electricity.

31. Because of Ontario Hydro's reduced load forecast (1978-9) there has been an inevitable stretching out of the nuclear power programme envisaged in the early 1970s. For this reason, there is no urgency with respect to a decision relating to the reprocessing of spent fuel. While we advocate that the option be kept open, we also advocate that the decision be delayed as long as possible.

32. F.K. Hare, *et al.*, *The Management of Canada's Nuclear Wastes*. Ottawa: Energy, Mines and Resources Canada. Report EP 77-6. August 31, 1977.
The members of the study group were Dr. A.M. Aikin, Dr. J.M. Harrison and Dr. F.K. Hare (Chairman).
R.J. Uffen, "The Disposal of Ontario's Used Nuclear Fuel", Submission to RCEPP. Exhibit 316.

33. By the year 2010 there may be a strong case for reprocessing the CANDU spent fuel because of world shortages of uranium. In anticipation of this potential situation, it would be desirable to store the spent fuel, on site, in bays or large concrete silos, until such time as reprocessing is required and deemed to be acceptable to the public.

34. Atomic Energy of Canada Limited has prime responsibility for spent fuel management research and development programmes, and for research in spent fuel reprocessing and the disposal of high-level radioactive liquid wastes. It is important to note that one of the major energy demonstration programmes, recently recommended by the Science Council (Science Council Report No. 30), is concerned with the disposal of spent nuclear fuel.

35. The demonstration of an "Acceptable Irradiated Fuel Management and Disposal System" recommended by the Science Council of Canada (Report No. 30) is identifiable as the Whiteshell programme.

36. This committee would in every sense parallel the Committee on the Technical Aspects of the Waste Disposal Problem.

37. The subject is treated fully in Volume 5 of this Report, Appendices A and B. See also Chapter 11 of this volume.

38. This assumes a 25 per cent excess capacity to ensure adequate reliability. It is also assumed that the annual capacity factor of nuclear power stations is 75 per cent.

39. "Cost Comparison of 4 × 750 MW Fossil-fuelled and 4 × 850 MW CANDU Nuclear Generating Stations". Ontario Hydro Report No. 584 SP, January 1979.
The Commission's conclusions relating to coal versus nuclear costs were largely based on:
S. Banerjee and L. Waverman. "Life Cycle Costs of Coal and Nuclear Generating Stations". Draft Report to RCEPP, April 1978. Exhibit 194. See RCEPP *Interim Report*, Chapter 7.

40. An assessment of the comparative environmental impacts of coal and nuclear power is provided in Volume 6, Chapters 2 and 4, and in summary in Chapter 9 of this volume.

41. See *Roads to Energy Self-Reliance*, Science Council of Canada. Report No. 30. Ottawa, 1979. p. 46; and *Energy Futures for Canadians*, EMR Report EP 78-1, June 1978, p. 132.

42. Present nuclear capacity in the rest of Canada consists of the Gentilly II and the Point Lepreau 600 MW reactors, under construction in Trois Rivières, Quebec, and Point Lepreau, New Brunswick, respectively. The 250 MW CANDU reactor in Trois Rivières, the Gentilly I (cooled by boiling light water rather than pressurized heavy water as in other CANDUs), has been mothballed by Hydro-Québec following a long history of operating problems. Further plans for another 600 MW CANDU in both Quebec and New Brunswick appear to be uncertain due to Quebec's large remaining hydro potential and New Brunswick's small system.

43. Science Council of Canada, *op. cit.*

44. These issues are considered in some detail in Chapters 7, 8, and 10 of the *Interim Report*.

45. R.L. Meehan. "Editorial", *Science*, vol. 204, no. 4393. May 11, 1979, p. 571.

46. Kemeny, *op. cit.*, p. 8.

47. See transcript of a hearing of Select Committee on Hydro Affairs, August 1, 1979, and August 2, 1979.

48. R. Chalk and F. von Hippel. "Due Process for Dissenting Whistle Blowers", *Technology Review*, June/July 1979, p. 53.

49. *Ibid.* The concepts for the procedures detailed in Recommendation 5.22 are based on this article.

50. The majority of models of nuclear reactor behaviour in normal and abnormal circumstances are necessarily of a probabilistic nature.

Notes to Chapter Six

1. The power transmitted through a conductor is the product of the voltage and the amperage. However, since the thermal losses are approximately proportional to the square of the amperage, it is clear that, to minimize the thermal losses (and hence to maximize the power transmitted), it is desirable to utilize as high a voltage as practical. The limits to the voltage level are set by the efficacy of the insulators and by the height and width requirements of the transmission towers.

2. A major disadvantage of the underground cable has to do with the installation operation, which is highly energy-intensive. The work involved in installing overhead lines and underground cables of equal capacity is compared in the following table, which gives the requirements for one mile of a two-circuit 500 kV transmission line (data provided by Ontario Hydro):
Material installed (tonnes): 1,225 overhead; 17,879 underground
Material removed (tonnes): 667 overhead; 13,787 underground
Material excavated (tonnes): 667 overhead; 30,000 underground

3. An excellent example of the potential impact of severe weather conditions on the bulk power transmission system occurred during the "Woodstock Tornado", which occurred in August 1979. As a direct consequence of the tornado, a 230 kV double circuit was put out of service. However, in order to remedy the situation without risking serious damage at generating stations and at transformer and switching stations, five 230 kV and four 115 kV circuits were lost before full service was restored. Ontario Hydro is undertaking a comprehensive series of computer simulations in which specific lines are "put out of service" and appropriate limits are set on other lines to warn operators that emergency steps have to be taken to minimize power outages. However, in the case of the Woodstock incident, the line that was forced out of service had not been subjected to simulation tests, and, in consequence, no limits were set in the system. Had limits been set, the number of circuit outages would have been appreciably reduced. This is not, however, a criticism of the utility; we appreciate that, in order to cover every contingency, many thousands of simulations would be required.

4. In the case of AC power, in addition to the "real" power, there is the so-called "reactive power", which, in effect, is needed to produce the magnetic flux necessary for the operation of transformers and induction motors. To ensure high efficiency of operation of AC motors the amount of reactive power drawn from the supply should be minimized in order to achieve a high "power factor". This problem does not arise with DC motors.

5. In Canada, such a link exists between the Hydro-Québec and New Brunswick Electric Power Commission systems at Eel River in New Brunswick.

6. Ontario Hydro has developed, with input from outside experts, an extensive computer-aided approach to help the planner in identifying broad transmission bands in a given region. In this approach, the relevant region of Ontario is divided into a 2 km × 2 km grid, and each cell is characterized with respect to nine environmental factors – human settlement, agricultural production, timber production, mineral extraction, wildlife game resources, recreation, aquatic communities, terrestrial communities, and the appearance of the landscape. For each factor, several objectives are identified and all are placed in order of priority as items to be avoided when routing the line. Existing urban and non-urban areas of human settlement, for instance, is the category at the head of the list. A composite factor map can then be produced that permits a global look at the situation and shows the planner how to avoid routes of significant undesirable impact.

7. High-voltage transmission lines, because they carry alternating electric currents, give rise to alternating electric and magnetic fields in their vicinity. The intensity of a field diminishes as the distance from the conductor increases. Under a typical 500 kV line, at its closest proximity to the ground, the field intensity is in the order of 5 kV per metre at a point 1.5 m above the ground, and the induced current flowing through a person standing at this point would be about 100 microamperes. This should be compared with the standards set for the leakage currents associated with certain household appliances, which range up to 500 microamperes for portable appliances.

8. References are given in Volume 9. Note, also, that because the U.S.S.R. is contemplating the use of ultra-high-voltage transmission (up to 1.5 million volts) the early studies are apparently not being taken seriously.

9. *Report on a Study of Biological Effects of Electric Fields*, Department of Preventive Medicine and Biostatistics, University of Toronto, 1979. Copies of a summary of the main findings are available; the research will eventually be reported in the medical literature.

10. To date, the most comprehensive review of the subject is available in: A.R. Sheppard and M. Eisenbud, *Biological Effects of Electric and Magnetic Fields of Extremely Low Frequency*, New York: New York University Press, 1977. See Also: A. A. Marino *et al.*, "Electric Field Effects in Selected Biologic Systems", *Annals of the New York Academy of Science*, vol. 238, 1974, p. 435, and R.O. Becket, prepared testimony before the State of New York Public Service Commission, Case 26559, 1974.

11. See the University of Waterloo study prepared for the RCEPP by J.C. Boyer *et al.*, "The Socio-Economic Impacts of Electric Transmission Corridors – A Comparative Analysis", April 1978, and comments from Ontario Hydro. B. Mitchell, G. Priddle, *et al.*, "The Long-Term Socio-Economic Impact of an Electrical Transmission Corridor on the Rural Environment: Perception and Reality". A submission to the RCEPP by the University of Waterloo, April 1976.

12. For a detailed discussion of the need for additional bulk power facilities in southwestern Ontario and eastern Ontario, together with a technical discussion of the capabilities of the bulk power facilities in these regions, reference should be made to the Commission's two regional reports.

13. P. Daunt. RCEPP transcript vol. 260, p. 40725, March 13, 1979.

14. RCEPP. *Report on the Need for Additional Bulk Power Facilities in Eastern Ontario*, July 1979, p.v.

Notes to Chapter Seven

1. However, see Chapter 4 for explanation.

2. See Chapter 11 for a detailed discussion.

3. *Report on the Need for Additional Bulk Power Facilities in Eastern Ontario*, RCEPP, July 1979, pp. 90-91.

4. During emergency situations when some load-shedding is essential, it would be desirable to shed load on a decreasing priority basis, e.g., electric water heaters and electric washers and dryers would probably be assigned lower priority than, for example, electric stoves. The digital (chip) meter, together with associated storage and local control facilities, would not only enhance the load-management process but would provide more flexibility and hence resilience in the operation of the system as a whole.

5. See Chapter 4, Volume 2, for a detailed discussion of reliability.

6. It is important to distinguish between "installed reserve" and "actual reserve". The former relates to the peak capability of all the installed generation in the system minus the peak load, while the actual reserve corresponds to the peak capability of the generating facilities that can be operated at a given time, minus the peak load. In discussing the generating capability of an electric power system, it is convenient to refer to "dependable peak resources" over a specific period, which have a 98 per cent chance of being exceeded. To facilitate determination of optimum reserve capacity, Ontario Hydro maintains detailed operating records that are used to determine the "availability" of various types of generating facilities, that is, the percentage of time that these facilities are available to supply electricity. The large thermal units of Ontario Hydro have availability in the order of 77 per cent while the hydraulic units have availability of about 95 per cent.

7. The complementary planning concepts relating to bulk power transmission are considered in the Commission's report on southwestern Ontario. They are not repeated here.

8. "Planning of the Ontario Hydro East System", Ontario Hydro Report No. 573 SP, part 1, vol. 1, June 1976, Section 6.0.

9. With respect to the potential of co-generation, it is interesting to note that a utility-industrial energy centre of 61 MW capacity was put in service in Eugene, Oregon, in December 1976 after a total construction period of 18 months. The capital cost per installed kilowatt was $166. Noteworthy also is the wood-burning Willamette steam electric plant of 32 MW capacity, which also services the Eugene area.

10. Bibliography on pumped storage to 1975: Hydroelectric Power Subcommittee of the IEEE Power Generation Committee Paper F76 007-5, Winter Meeting, New York, 1976, O.W. Bruton and R.L. Mittelstadt, "Planning for Pumped Storage in a Hydro-Thermal System", American Society of Civil Engineering, 22nd Annual Hydraulics Division Conference, Knoxville, Tennessee, July 1974; and R.H. Resch and D. Predpall, "Pumped Storage Site Selection: Engineering and Environmental Considerations", ASCE Engineering Foundation Conference on Pumped Storage, August 1974, pp. 39-70.

11. *The Need for Additional Bulk Power Facilities in Southwestern Ontario*, RCEPP, June 1979. *The Need for Additional Bulk Power Facilities in Eastern Ontario*, RCEPP, July 1979.

12. "Planning of the Ontario Hydro East System", Ontario Hydro Report No. 573 SP, part 1, vol. 1, June 1976, pp. 1-7.

13. Such an arrangement, or arrangements, might delay by several years the construction of a Darlington-type nuclear power station, or at least a major part thereof, while enhancing the adaptability of

Ontario's electric power system, for reasons given previously. It would, we believe, be a welcome co-operative venture between provinces that would receive public approval, whereas we predict lengthy and even bitter opposition to each new nuclear power plant.

14. Interconnection between Ontario Hydro and New York and Michigan utilities in effect means that Ontario Hydro is interconnected with a vast power grid — indeed the utility contributes only about 5 per cent to the "total mechanical inertia" of the interconnected system as a whole.

15. Because of the nature of the Hydro-Québec system, with generation located at great distances from major load centres, it cannot operate in parallel with the Ontario Hydro system. The only means of exchanging power between the two systems is by isolating load or generation, as is done now, or by installing an asynchronous link in the form of a DC tie-line.

16. The first interconnection for power exchange between Ontario and Quebec was established more than 50 years ago at Paugan Falls. Several other interconnections have been established during the intervening years, essentially for importing power to Ontario from Quebec. The most notable interconnection is at Beauharnois, where there are tie-lines with a nominal winter capacity of 1,060 MVA. (The megavolt ampère (MVA) unit of power takes into account both "real" and "reactive" power.)

17. A proper economic analysis of hydro purchase versus the nuclear alternative would require that each be analysed within a complete simulation of the Ontario Hydro system over at least 30 years. The foregoing observations suggest, however, that the least-cost strategy for Ontario Hydro may not be an either/or decision, but rather the purchase of Manitoba hydroelectric power for a fixed term and its replacement by nuclear in Ontario thereafter. Much would depend on the rate of price escalation in the purchase agreement. The lower the rate, the longer would be the optimum contract period. Also critical is the portion of the investment required to strengthen the Winnipeg-Thunder Bay system.

18. Canada/United States Electricity Exchanges, May 1979, U.S. Department of Energy, Washington, D.C., and Energy, Mines and Resources Canada, Ottawa.

19. In Chapter 7 of Volume 2, the effect of a 3 per cent average annual growth in load to the year 2000 on Ontario Hydro's fuel requirements and contracted supply are discussed. It is shown that, if the load were to grow at 3 per cent, substantial reductions in contracted supply of U.S. coal would be required in the early 1980s. In the late 1980s, unless some committed nuclear capacity is deferred, contract cutbacks will be substantial. It is also pointed out that contract cutbacks of about 35 per cent will be required in the early 1980s even under Ontario Hydro's 1979 Load Forecast of about 4.5 per cent per annum growth rate.

Notes to Chapter Eight

1. We have interpreted "land use" as, essentially, "land and water use".

2. Soil classification under the Canada Land Inventory is categorized as follows:

Class 1 — soils in this class have no significant limitations in use for crops.

Class 2 — soils in this class have moderate limitations that restrict the range of crops or require moderate conservation practices.

Class 3 — soils in this class have moderately severe limitations that restrict the range of crops or require special conservation practices.

Class 4 — soils in this class have severe limitations that restrict the range of crops or require special conservation practices, or both.

Classes 5-7 — soils in these classes are in general not suitable for arable culture or permanent pasture — improvement practices are marginally feasible in the case of Class 5.

3. In the RCEPP Issue Paper No. 5, "Land Use", the interesting analogy is drawn between "quality of land" and "quality of energy". The agricultural potential of Class 1 land is readily degradable by the works of man (e.g., when a shopping centre is built on such land). Furthermore, land degradation of this kind is irreversible. Analogously, the quality of energy is also readily degradable, for example, when electricity (i.e., high-quality energy) is degraded into low temperature thermal energy (i.e., low-quality energy). And again, in the case of a conservative system, the degradation of high-quality energy to lower-quality energy is an irreversible process.

4. H. Brooks and J.M. Hollander, "United States Energy Alternatives to 2010 and Beyond — the CONAES Study". *Annual Review of Energy*, 1979, 4:1-70.

5. Ontario Hydro Memorandum with respect to the Public Information Hearings, "Land Use"; also Norman Pearson, "Land Use Implications of Electricity Supply Facilities", a study funded by the RCEPP; and "Food Land and Energy Planning", RCEPP Research Report for the Foodland Steering Committee, September 1976, Exhibit 95-2.

6. The data in Tables 8.2 and 8.3 are extracted from: D. Pimental *et al.*, "Energy and Land Constraints in Food Protein Production", *Science*, November 21, 1975, vol. 190, no. 4216, pp. 754-61.

7. H.J. Evans, ed., "Enhancing Biological Nitrogen-Fixation", National Science Foundation, Washington, D.C., 1975.

8. "A Strategy for Ontario Farmland". Statement by the Ministry of Agriculture and Food. March 1976, pp.4-5.

9. See, in particular,"Food Land Guidelines — A Policy Statement of the Government of Ontario for Agriculture 1978", Ontario Ministry of Agriculture and Food.

10. These were the meetings that culminated in the tragic air crash of September 4, 1976 — see the Commission's report entitled "The Meetings in the North".

11. RCEPP Exhibit 396. "Energy Planning in a Conserver Society: Implementation Strategies". Energy Probe, January 1979.

12. "The Hearst Wood Waste Energy Study, Summary Report". Prepared for the Ministries of the Environment, Energy and Natural Resources, the Town of Hearst, and the Hearst Lumbermen's Association, assisted by the Ontario Ministry of Northern Affairs, the Ontario Ministry of Industry and Tourism, Ontario Hydro, and the Northern Ontario Development Corporation. Prepared by Acres, Shawinigan Ltd., January 1979.

13. "Northern Ontario Water Resources Studies: Summary Report on Engineering Feasibilities and Cost Investigations". Inland Water Resources Directorate, Water Resources Branch, Environment Canada, 1973.

14. Ontario Hydro. Memorandum with respect to Information Hearings entitled "Generation — Technical", 1976, RCEPP Exhibit 2.

15. A mixture of lignite and forest-industry wastes might also be used as fuel — this would stretch out the resource.

16. Onakawana Feasibility Study — Shawinigan-Steag Company, March 1978.

Notes to Chapter Nine

1. Because no energy conversion process is 100 per cent efficient (although the conversion of electricity to thermal energy is nearly so) there are inevitably energy losses. In some processes, such as the conversion of coal to electricity through the steam cycle, the energy losses may be in the order of 60 per cent.

2. It is noteworthy that, during the Three Mile Island incident, the majority of people identified the generating station with the massive cooling towers — not with the actual nuclear reactor building, which was dwarfed by the towers.

3. Norman Morse (with the collaboration of Donald Chant), "An Environmental Ethic — Its Formulation and Implications", Report No. 2 of the Canadian Environmental Advisory Council, January 1975. For a non-technical, ecologically-oriented introduction to the environmental ethic, see D.A. Chant, "Toward an Environmental Ethic", *Ontario Naturalist*, March 1977.

4. D.A. Chant, "Toward an Environmental Ethic", *Ontario Naturalist*, March 1977.

5. It is interesting to note that the ability of vegetation to absorb many common pollutants acts as a form of natural air purifying system. But there is an inevitable impact on plant growth. For example, acute damage to plants occurs when sulphur dioxide concentrations are in the order of $0.0005 \, g/m^3$. But the susceptibility of plants varies greatly. Thus, sulphur dioxide levels of $0.0002 \, g/m^3$, which are common in the vicinity of large industrial towns, can inhibit the growth of certain grasses by 50 per cent.

6. G.M. Woodwell, "The Carbon Dioxide Question", *Scientific American*, vol. 238, no. 1, January 1978, p. 34.

7. *Ibid.*

See also: *Science*, November 23, 1979, vol. 206, pp. 912-13. A report of the United States National Academy of Sciences group established to assess the "critical argument" of recent studies that have concluded that increased carbon dioxide levels will increase the average temperature of the earth, cannot find any fault with the argument. It is concluded that, if the carbon dioxide concentration is doubled, the average temperature rise will probably be in the range of 1.5°C to 4.5°C. On the basis of a 4 per cent per annum growth rate in coal utilization (the growth rate up to a few years ago), the time to doubling would be 50 years, i.e., to the year 2030.

The above conclusion is supported in a lead editorial in *Nature*, no. 279, May 3, 1979, entitled "Costs and benefits of carbon dioxide". It is stated there that the concentration of carbon dioxide in the atmosphere is increasing and that this can be attributed to man's activities. Specific conclusions are:

(i)There will be a higher mean annual temperature, with marked latitudinal asymmetry. The higher the latitude, the more marked the increase.

(ii)There will be increased activity of the hydrological cycle, with an increased level of precipitation and of evaporation.

8. "Carbon Dioxide and Climate: Carbon Budget Still Unbalanced", Research News, *Science*, vol. 197, no. 4311, September 30, 1977, p. 1352.

9. There are many anomalies that relate to the pH levels of the lakes, and research is not sufficiently far advanced to provide firm conclusions regarding the sources of the acid condition. It is not known, for example, why of two lakes in the same area one should be normal and the other should have low pH (i.e., acidic) levels.

See, in particular, "Acid Precipitation in the Western United States", W.M. Lewis and M.C. Grant. Research Report, *Science*, vol. 207, January 11, 1980. This report suggests that acid rain may be caused by plants situated at distances of up to 1,000 km, and, furthermore, that the acidic components may move against prevailing weather patterns.

10. An analogous situation exists in Europe. There is evidence that Norwegian and Swedish lakes and streams are being polluted by acid rain originating in the United Kingdom. Sulphur oxides from coal-burning power stations and other sources may be transported across the North Sea for upwards of 1,000 km.

11. Pickering cooling water is heated by $11°C$ in passing through the condenser. But note that the water flow is about 3 million litres per minute per reactor. When all four reactors are operating at full power, 12 million litres of warm water are being discharged every minute into Lake Ontario.

12. The thermal discharge tends to float on the surface of the lake in a comparatively thin "plume". The size and general configuration of this plume is determined by the design of the outlet, lake currents, lake temperature gradients, and the prevailing wind. For example, if the air is warmer than the discharge water, and if the humidity is high, the evaporation and consequent heat loss in the plume will be very small. These interacting factors complicate appreciably the study of the impact of thermal discharges.

13. The Elliot Lake uranium tailings, amounting to approximately 150 million tonnes, already constitute a major environmental problem. Depending on the future level of mining activity, hundreds of millions of tonnes of additional uranium tailings could result.

14. The conclusion that exposure to relatively high concentrations of air pollutants is probably harmful to health does not imply that long-term exposure to low levels may also be harmful until there is evidence to the contrary. More epidemiological research is needed before the risks can be evaluated adequately. The same applies to the risk of damage to aquatic and terrestrial ecosystems resulting from "acid rain".

15. It has been estimated that, in the United States, tripling the level of coal production by the year 2000 (this is the Administration's target) will result in 370 fatalities and 42,000 disabling injuries each year in the coal-miner population. These deaths and injuries do not include those that will be attributable to "black lung" disease.

16. Various techniques have been proposed for the determination of relative weights in environmental assessment – notable is the Delphi technique. See for example:

A.L. Delbecq and A.H. Van de Ven, "The Effectiveness of Nominal Delphi, and Interacting Group Decision Making Processes", *Journal of the Academy Of Management*, December 1974.

But probably the most definitive approach to the problem is that of Otway and Edwards:

H.J. Otway and W. Edwards, "Application of a Simple Multi-attribute Rating Technique to Evaluation of Nuclear Waste Disposal Sites: A Demonstration". An interim report on research being undertaken by the International Institute for Applied Systems Analysis, Vienna, Austria, on behalf of the International Atomic Energy Agency, June 1977.

17. On the one hand deforestation (especially in the tropics) would reduce the earth's CO_2 sink, and on the other hand the combustion of wood, ethanol, or methane would tend to increase CO_2 levels.

Notes to Chapter Ten

1. See Chapter 4. The first law of thermodynamics states that, for a conservative system, energy can neither be created nor destroyed. This law remains consistent with nuclear energy by invoking the Einstein equivalence of energy and mass. The second law of thermodynamics, or the entropy law, relates to energy transformation, energy flow, and the quality of energy. In its simplest form, the law states that, without external intervention, heat will not flow from a cold to a contiguous hot body or environment. Heat only flows from a hot to a cold body or environment. However, it is important to note

that, by using the heat-pump principle and an external source of energy, we can in effect transfer heat from a low-temperature source of energy to a high-temperature source of energy. Indeed, the operational principle involved is analogous in every respect to the operation of a refrigerator.

2. One megajoule is equivalent to about 950 BTU, and 20.278 kW·h. Usually, and unfortunately, because of the need to standardize energy units, the energy content of oil and natural gas is stated in terms of British Thermal Units (BTU) – i.e., the amount of thermal energy required to heat one pound of water at 39.2°F through 1°F.

3. High-quality sources of energy are gravitational energy, electric energy, chemical energy (because these can be converted into other forms of energy and especially into high-temperature thermal energy). Low-quality energy is identified with low-temperature sources of thermal energy. Note that high-quality sources of energy are degraded step by step until they become low-quality energy sources.

4. T.F. Widmer and E.P. Gyftopoulos, "Energy Conservation and a Healthy Economy", *Technology Review*, June 1977, p. 31. The authors point out that about 60 per cent of total energy consumption in the United States is based on comparatively low second-law efficiencies. On the other hand, they note that the energy conversion processes involved in the production of steel are 21 per cent efficient, which, compared with space heating, is a high-efficiency operation. See also "Physics 1975: AIP Conference Proceedings", ed. H.C. Wolfe, No. 25, "Efficient Use of Energy", American Institute of Physics, 1975.

5. Although a bicycle is a much more efficient means of transportation than an automobile, for distances greater than a few miles the bicycle may be unsuitable because of the time factor. On the other hand, if no more than two people are involved, a motor cycle provides much more efficient transportation, as far as the optimization of fuel is concerned, but it is not as comfortable as a car. Hence, even in the application of second-law efficiency criteria, we must bear in mind that time and comfort factors should be given appropriate weight in deciding which alternative is most suitable.

6. Ontario Ministry of Energy, *Ontario Energy Review*, Toronto, 1979, p. 34. Note particularly that, in the 1979-80 fiscal year, the Ministry is spending $7.2 million, and Ontario Hydro $6.2 million, on energy conservation programmes.

7. Report prepared for the Ontario Arenas Association, supported by the Province of Ontario, the Ministry of Energy, and the Ministry of Culture and Recreation, by McIntosh and Moeller, Architects and Engineers, Hamilton, Ontario: "Energy Conservation in Existing Arenas (Three Case Studies)". Available from the Ontario Government Bookstore.

8. Ontario Ministry of Energy, *Ontario Energy Review*, Toronto, 1979, p. 37.

9. "The Economics of Industrial Co-generation of Electricity", proceedings of a seminar co-sponsored by the Ontario Ministry of Energy and Ontario Hydro, December 13-14, 1978.

10. R.W. Besant, R.S. Dumont, and G. Schoenau, "The Saskatchewan Conservation House: Some Preliminary Performance Results", *Energy and Buildings*, 2 (1979), pp. 163-74.

11. *Ibid.*

12. Besant *et al.* have stated:

> Based on the $15,000 cost for an active system (collectors – $6,000; storage tank – $3,000; piping, heat exchangers, controls, etc., and installation – $6,000), the economics do not appear as attractive as the energy-conserving features for residential applications. Perhaps a packaged active system for hot water at a cost of approximately $2,500 [installed] would be a more economically desirable system for residences.

13. "Energy Study of Black Creek Pioneer Village Visitors' Centre", report prepared for the Ontario Ministry of Energy by Raymond Moriyama, Architects and Planners, and Solatherm Engineering Inc. (Ontario), March 1978.

14. Members of the Commission visited the Saskatchewan Conservation House in July 1979 and were given an in-depth briefing. The indoor temperature on a warm summer day was 23°C.

15. K.V. McDowell and N.A. Kischuk, "Housing Satisfaction", a report on research funded by Saskatchewan Research Council Grant No. 3-385-048, 1978.

16. An excellent survey of energy conservation potential in industry is given in the July 1979 issue of *Modern Power and Engineering*, "MP&E's Energy Conservation Casebook". The casebook presents reviews of a multiplicity of energy conservation practices currently in operation. For example, it itemizes low-cost, energy-conserving ideas for the commercial and industrial sectors under the headings: boiler and steam efficiency, steam lines and traps, fans, pumps and motors, compressed air, and lighting. In all, 47 specific steps whereby energy consumption in a plant or building could be reduced are listed.

17. For technical discussion of the importance of nitrogen in the agricultural production system, see:

R.L. Halstead, "Nitrogen Research in Agriculture Canada", proceedings of a symposium "The Nitrogen Cycle in Canada" held at the Institute for Environmental Studies, University of Toronto, May 1976, EE-3.

18. M. Rawitscher and J. Mayer, "Energy, Food and the Consumer", *Technology Review*, August/September 1979, pp. 45-52. The corresponding data for Ontario (i.e., food-energy equivalents) were computed on a simple per capita basis.

19. Throughout man's evolution, the technologies of communication (beginning with speech) and of the harnessing of energy have proceeded closely in parallel, but communication has always been ahead. For example, the Gutenberg press (the major landmark in societal communication) and the steam engine (the major landmark in harnessing energy) were 300 years apart. Similarly, man's latest achievement in communications technology — the micro-electronic computer, processor, and communicator — will probably be at least 50 years ahead of nuclear fusion energy technology.

20. A. Porter, Inaugural Lectures, Imperial College of Science and Technology, University of London, 1955-6, published by Imperial College, London, England.

21. The "R value" is an index of insulation efficiency, that is used in establishing insulation codes. It is predicated on the heat flow through an insulation layer for a given temperature differential — the larger the R value the lower the rate of heat flow across the insulation.

22. The Canadian Standards Association has recently published basic material and related standards for active solar energy systems — see "Solar Collectors", CSA Preliminary Standard F378-M 1979, August 1979, Rexdale, Ontario. These standards will appreciably facilitate the introduction of solar energy as a conservation technology because they will ensure the enhanced reliability of such systems. It is noteworthy that the Royal Commission on Electric Power Planning provided the initial stimulus to the Canadian Standards Association to undertake a standards investigation and to promulgate recommended standards.

Stanford Research Institute, "Residential Hot Water Solar Energy Storage Sub-systems", Technical Report NSF RA-N-75-095, January 1976.

Additional references:

H.D. Foster and W.R.D. Sewell, "Solar Home Heating in Canada: Problems and Prospects", Office of the Science Advisor, Report No. 16, Dept. of Fisheries and Environment, Ottawa, 1977.

United States Council on Environmental Quality, "Solar Energy — Progress and Promise", Washington, D.C., April 1978.

"Capturing Sunlight: A Revolution in Collector Design", Research News, *Science*, vol. 201, July 7, 1978, pp. 36-9.

23. A. Juchymenko, "Co-generation Potential in Ontario and the Joint Venture Approach", proceedings of the Economics of Industrial Co-generation of Electricity Seminar, December 1978.

24. With thermodynamic efficiencies in the order of 80 per cent, the rationale for burning natural gas in a co-generation mode is not unreasonable — indeed, the efficiency of utilization of the natural gas is greater than that achieved in conventional domestic gas-fired furnaces.

As an example, the Great Lakes Forest Products Ltd. co-generation installation at Thunder Bay comprises two 13.8 MW units. Each unit operates in conjunction with a separate kraft mill. One of the boilers is bark burning, and it is estimated that at least 50 per cent of the energy used in the mill complex is supplied by wastes (i.e., draft liquors and wood waste).

Several examples of wood-burning co-generation installations in New England and Oregon have been brought to the Commission's attention. Some of these have been in place for several years and our understanding is that they have provided a reliable source of electricity and steam.

25. D.A. Drinkwalter, "Relationship of Industrial Generation and Ontario Hydro's Expansion Program", proceedings of the Industrial Co-generation Seminar, December 1978, p. 25.

26. D.J. Ben Daniel and E.E. David, "Semiconductor Alternating-Current Motor Drives and Energy Conservation", *Science*, November 16, 1979, vol. 206, no. 4420, pp. 773-6.

27. Modern Power and Engineering — Energy Conservation Casebook, 1979, p. 41.

28. The term "energy intensity" has been used as a measure of the energy required to manufacture a specific product. This is an important first step in the assessment of industrial energy needs, based on end uses, on a province-wide basis. We anticipate the institution of energy audits, that are analogous to financial audits, especially in the case of large users.

29. We understand that the silicon chip meter and associated microelectronic equipment is now available, on a limited scale, in both the United States and the U.K. Furthermore, we understand that demonstration projects in both countries are at present being planned. We are grateful to Robert A.

Peddie, Chairman of the South Eastern Electricity Board, U.K., for providing the Commission with details of the prototype meter, the ancillary equipment, and the proposed demonstration project.

30. While the energy/GNP ratio gives a rough indication of the energy needed to sustain standard of living, we believe that its use in a comparative sense (e.g., the comparison of Canada's ratio with that of, say, Sweden) is not particularly meaningful and indeed can be misleading, as pointed out by Schipper and Darmstadter, "The Logic of Energy Conservation", *Technology Review*, January 1978.

31. Analogously, we can drive from point A to point B at an average speed of 100 km/h and consume x litres of gasoline, or reduce our average speed for the journey to 80 km/h and, as a result, consume less than x litres of fuel. Consequently, we save some available energy but increase the time for the journey. (Note that the additional fuel is used up in performing work to overcome the extra air resistance.) A second example is that of the supersonic jet, the future of which is problematical on environmental, economic, and thermodynamic grounds.

32. In the electricity sector, marginal cost pricing, as the term implies, is based on the marginal cost of electricity (see Chapter 11). At present, the marginal cost is appreciably higher (essentially because of inflation) than costs averaged over the last decade.

33. "The Role of Electricity in the National Energy Policy", an advisory brief to the Government of Canada, January 1978.

34. *Engineering*, vol. 13, no. 9, September 17, 1979, p. 8.

35. The concept of daylight saving time, and how it could be put into effect, was due to William Willett, a Fellow of the Royal Astronomical Society – the date was 1907.

The proposal was submitted at a public hearing of the Commission by a research group from York University (Exhibit 337, K.A. Innanen and S.E.H. Innanen, Physics Dept., York University). See also: I.R. Bartky and E. Harrison, "Standard and Daylight-Saving Time", *Scientific American*, vol. 240, no. 5, May 1979, pp. 46-53.

Notes to Chapter Eleven

1. Ontario Hydro's current and evolving load-forecasting techniques are considered in detail in Volume 3, Chapter 3. See also the Commission's regional reports – *The Need for Additional Bulk Power Facilities in Southwestern Ontario* and *The Need for Additional Bulk Power Facilities in Eastern Ontario*, June and July 1979, respectively.

2. The implication is that "energy gaps" and "energy surpluses" could indeed occur – in fact, in the real world, the total electric power being generated will always balance the load. But the load, as already intimated, may have to be seriously curtailed to avoid a major breakdown of the total system.

3. It is indisputable that the real price of electricity affects the demand for electricity – the key question is, by how much? There is evidence as well that the demand for electric energy also responds to the increasing price of oil and natural gas. Hydro's model suggests, counter-intuitively, that the short-run demand for electric energy actually decreases as the price of oil and natural gas increase (because of the depressant effect on the economy).

4. *Energy – Global Prospects 1985-2000*. Report of the Workshop on Alternative Energy Strategies, McGraw Hill, New York, 1977, pp. 3-5.

5. This was evident during the Commission's southwestern Ontario hearings.

6. Ontario Ministry of Energy. *Energy Security for the Eighties*. September 1979.

7. However, increasing government support can be expected for electrically powered public transport systems, notably the subway, streetcar, and trolley bus, as part of a programme to reduce oil consumption. On the other hand, electric vehicles based on electric battery power are unlikely to make a major impact before the end of the century (we have been told that a 2-3 per cent penetration is in the right order). Nevertheless, some battery systems are in advanced stages of research and development; they are essentially non-polluting, and have overall efficiencies equal to that of the diesel engine. Consequently, the potential of mass-produced electric cars, probably powered by batteries of the sodium-sulphur, nickel-zinc, etc., types appears to be very promising for the period 1990-2010.

8. See Volume 5 – national estimates of the appropriate "output multiplier" have been used because macro-economic models of the Ontario economy are still in an early stage of development. Nevertheless, we believe that the national estimates provide a reasonable approximation.

9. Although this figure is based on expenditure and employment, respectively, in connection with the design, manufacture, and construction of nuclear stations and heavy-water plants in 1977 (study by Leonard & Partners, Consulting Engineers, Ottawa), we believe it provides a reasonable estimate for the programme as a whole because, for the last few years, the nuclear programme has dominated Hydro's capital expenditures.

10. Ontario's industrial power rates are very competitive with most other jurisdictions, but higher than those in, for example, Manitoba and Quebec (see next section).

11. Capacity sales normally occur when the importing utility is facing an emergency during peak power periods. Economy sales, on the other hand, take place when the fuelling costs of the importing utility's available generating capacity are greater than the running costs of the spare capacity in the exporting system.

12. If the load growth is according to Hydro's 1980 forecast of about 3.4 per cent and the Darlington Generating Station is kept on schedule, the surplus above required reserve may remain at about 3,500 MW until the early 1990s. It will include the Lennox Generating Station (2,200 MW oil-fired) and Hearn (1,200 MW gas/coal-fired). Sales would be mainly from coal-fired units during the winter peak with some nuclear energy available during the summers in the late 1980s.

13. Initial studies suggest that pay-back on the costs of the cable would take place within the five-year life of the contract, leaving Hydro with a permanent and paid-for interconnection with a new market area (Pennsylvania) in a new regional reliability council (Mid-Atlantic Area Council).

14. It is important to differentiate between "internal" social costs, which are the costs incurred directly by the utility in undertaking, for example, environmental studies, and the "external costs" (e.g., the environmental effects of burning coal, or discharging warm water into the Great Lakes), which are incurred by society as a whole.

15. Ontario Ministry of Energy. *Ontario Energy Review*, Toronto, June 1979. Figures 28 and 29, p. 26.

16. Amory B. Lovins, *Soft Energy Paths*, 1977. Chapter 8.

17. Ontario Energy Board, "Report to the Minister of Energy on Principles of Electricity Costing and Pricing for Ontario Hydro", HR-5, Toronto. December 20, 1979.

18. The concept of marginal cost is basic to the study of economics. It relates essentially to the capital and operating costs of producing (or importing) an additional unit of a good or service, such as a barrel of oil or a kW·h of electricity. Alternatively, it might be interpreted as the "replacement cost" of a unit of a good or service.

19. In the 1980s, the coal component of the base load will be declining in relative terms but will probably continue to be "on the margin" of base-load capacity for the remainder of the decade.

20. The point here is that rates should not encourage the addition to the system of long-term loads that would be extremely costly to serve once the period of temporary surplus capacity had passed. A developer's decision to heat a subdivison electrically is an example of an investment decision that will affect Hydro's need for peaking capacity long after the temporary surplus of fossil generation has been absorbed.

21. But note that marginal cost pricing need not increase the total electricity bill — especially if the accounting cost revenue requirement is still to be met; it just redistributes the charges. A shift to marginal cost-pricing will increase costs to those customers who continue to consume during peak periods, whereas those that adjust their consumption patterns may actually experience reduced electricity bills.

22. In its December 1979 report (see footnote 18), the Ontario Energy Board stated, however, that time-differentiated rates can equally be based on average or historic costs.

23. We understand that the silicon chip meter and associated micro-electronic equipment is now available on a limited scale in both the United States and the U.K. and that demonstration projects in both countries are being planned. We are grateful to Robert A. Peddie, Chairman of the South Eastern Electricity Board, U.K., for providing the Commission with details of the prototype meter, the ancillary equipment, and the proposed demonstration projects.

24. For a more detailed analysis of the comparative economics of nuclear and coal-fired generating stations, see S. Banerjee and L. Waverman, "Life Cycle Costs of Coal and Nuclear Generating Stations", RCEPP Research Report, July 1978.
See also the Commission's *Interim Report*, September 1978, Chapter 7.

25. It was pointed out in the *Interim Report*, Chapter 8, that unless a minimum order level for CANDU reactors corresponding to about 1,200 to 1,700 MW of nuclear capacity a year can be achieved and maintained, the nuclear industry would be in serious economic trouble. According to Hydro's 1980 load forecast (3.4 per cent per annum), Ontario's requirements will not support this level of production.

26. This compares with Leighton and Kidd's 1976 estimates for the Commission of a 1985 potential of 2,085 MW and Hydro's estimate of a 1,950 MW potential by 2000.

27. IBI Group. "Impact of Solar Heating on Electrical Power Generation in Ontario". RCEPP Research Report. Toronto, 1977.

IBI Group. "Solar Heating: An Estimate of Market Penetration". RCEPP Research Report. Toronto, 1977.

Middleton Associates. "Background Notes on Alternative Energy Sources". Prepared for RCEPP. April 1979.

28. L. Doody, "Solar Domestic Hot Water Heating", *Engineering Digest*, vol. 23, no. 7, August 1977. The author suggests that the pay-back period for residential solar water-heating systems is in the order of eight years.

29. M.J. Helferty and R.G. Lawford, "Meteorological Information for Use in Assessing the Auxiliary Energy Requirements of Solar and Wind Energy Systems in Five Ontario Locations". Internal report SSU-78-9, Ontario Region, Atmospheric Environment Service, Environment Canada, Downsview, Ontario.

J.G. Asbury, C. Maslowski, and R.O. Mueller, "Solar Availability for Winter Space Heating: An Analysis of SOMET Data, 1953 to 1975", *Science*, vol. 206, November 9, 1979.

30. Co-generation has been implemented for many years in Europe, notably in Scandinavia and West Germany, on a comparatively broad scale. Solar heating systems have reached a reasonable level of commercial development in Australia and Israel, and the United States government is providing major funding support to both co-generation and solar space-heating programmes, as well as to solar central power and solar-electric systems.

31. D.K. Foot *et al.*, *The Ontario Economy 1977-1987*, Ontario Economic Council, Section 3.4.

32. D.K. Foot, *ibid.*, pp. 48-9.

33. J.E. Pesando, "Capital Markets", *The Ontario Economy 1977-1987*, Ontario Economic Council report. 1977.

34. The Ontario government has already begun to help in the financing of energy-sector investments in the province(e.g., conservation, solar energy, and public transit) and elsewhere in Canada (e.g., natural gas pipelines from the frontier and synthetic crude oil plants). The amounts involved have not been large relative to Ontario Hydro's capital expenditures and will probably not become so, but there could be some encroachment on Ontario Hydro's potential to incur long-term debt.

Notes to Chapter Twelve

1. D.L. Bazelon, "Risk and Responsibility", *Science*, vol. 205, July 20, 1979. This is a particularly significant article. The author is a Senior Circuit Judge of the United States Court of Appeals. His thesis, essentially, is that although scientists have an important role in energy decision-making in connection with risk regulation, their judgements relating to the social implications are no more significant than those of the non-scientist.

While the physical scientist devises a conceptual system that reflects the regularities inherent in nature, and the technologist uses this knowledge (for example, to design energy conversion systems), the social scientist is concerned with the implications of the resulting changes for society as a whole. Unfortunately, certainly in academia, there has been a lack of co-operation between natural scientists and technologists, on the one hand, and social scientists, on the other.

2. The Association of Professional Engineers of Ontario's prescribed code of ethics begins:
   1. A Professional Engineer owes certain duties to the public, to his employers, to his clients, to other members of his profession, and to himself, and shall act at all times with: a) fairness and loyalty to his associates, employers, clients, subordinates and employees; b) fidelity to public needs; c) devotion to high ideals of personal honour and professional integrity.
   2. A Professional Engineer shall: a) regard his duty to public welfare as paramount . . .

3. A.Porter, "Control, Automation and Computers", I.E.E. Centenary Lecture, May 1971.

4. Compare the resilience concept as applied to decision-making with the resilience concept as applied to electric power system planning, with special reference to the generation mix — see Chapter 7 of this volume.

5. The need to examine the "redundancy issue" is implied in the mandate of the RCEPP. Indeed, at the time of announcing the setting up of the Commission, the Honourable Allan Grossman mentioned specifically that the redundancy issue was of fundamental concern.

6. Since 1972, there has been a marked increase in Ontario Hydro's budget for "public participation". At present, more than 100 employees at the Head Office and in the regions are involved full time in this activity.

7. Task Force Hydro. *Hydro in Ontario — A Future Role and Place*. Report No. 1, 1972. p. 24.

8. The Ontario Cabinet, because of the timing of the Act, ruled that the Darlington Generating Station would be exempt from the environmental assessment procedures laid down in the Act.

9. This can be interpreted as "need" for a specific facility. And this, in turn, relates to the extent to which, in the case of Ontario Hydro, the current load forecast is acceptable as a basis for long-term planning. The Act can be interpreted as requiring the EAB to examine the viability of the load forecast.

10. In creating an all-party Select Committee on Ontario Hydro Affairs, the Ontario Government established a review process at the political level; this has much to commend it, since the procedures have important educational aspects. However, it is important to note that many of the critical issues are of a highly technical nature and involve a depth of background knowledge and practical professional experience that are unlikely to be found in the legislative body. In other words, it is our belief that the government, at present, has no credible mechanism that is both open and independent for the formulation of general energy policies.

11. White Paper on The Planning Act. Government of Ontario, May 1979. p. 45.

12. We draw special attention to the following RCEPP exhibits:
- Exhibit 397 – Ontario Ministry of the Environment. "The Decision-Making Framework and Public Participation".
- Exhibit 400 – Ted Schrecker and W.J. Ramp. "Public Interest and Public Participation: Towards a Democratic Energy Future".
- Exhibit 400-2 – Robert Paehlke. "Comments on Public Participation in Energy Decision-Making".
- Exhibit 382 – DynamoGenesis Corp. (N. Teekman). Submission to the Debate Stage Hearings, January 10, 1979.
- Exhibit 388 – Consumers' Association of Canada (Ontario Branch). Submission to the Debate Stage Hearings, January 16, 1979.

Each of the above submissions advocated in one form or another the "single-window" in contrast with the "multiple-window" approach.

13. As a guide, it should be noted that the New York Public Service Commission has a small staff of system planners (15) and environmental planners (15). The Commission reviews the long-range proposals of seven New York electricity utilities. On this basis, it would appear that the professional staff of the OEC would total about 20.

14. See the section, Major Activities of the OEC, for details of these hearings, and especially the relationship between the OEC, the EAB, and the OMB. Note that the OEC (which is essentially an expanded OEB) would be responsible for hearings of the sort currently carried out under the authority of the OEB.

15. A. Porter, a letter to the people of Ontario from the Chairman of the RCEPP, 1975.

16. The Commission contrived to arrange meetings in the evenings as well as the mornings, to accommodate the public. Noteworthy, too, is the fact that on many occasions eminent scholars and scientists from abroad, and from across Canada, participated in the hearings. However, except for the Commission itself and the official representatives at its hearings, the views of these distinguished people went largely unnoticed.

17. Although the Commission's mandate (time period, 1983-93 and beyond) obviously did not require consideration of the issues surrounding the Halton Hills controversy (concerning the completion of the 500 kV transmission line from the Bruce Generating Station to the Milton Transformer Station), we were surprised that the public interest group responsible for the main opposition to the line did not appear before the RCEPP, or submit a brief. Indeed, this is an excellent example of the "own backyard syndrome". It should be noted that, at the time when opposition to the line was originally voiced, Ontario Hydro's public participation procedures were not fully in place.

18. The adversary approach is patterned essentially on courtroom procedures which usually call for yes or no responses, and for guilty or not guilty decisions. But especially when the concern is with the future, there are obviously no clear-cut answers. Furthermore, decisions must be couched in this way: "... on the one hand A, but on the other hand B". While some level of quantification may be possible, value judgements based on qualitative factors are invariably essential. We interpret the "inquisitorial approach" in the sense of an examination and a defence of a thesis or a brief. It is a more open and less structured process than the "adversary approach".

19. By general consent, the most successful activities of the RCEPP were the nuclear symposium, the seminars that preceded each of the major subjects of the debate stage hearings, the Ashby seminar on public participation in decision-making, and the three major Commission workshops that were held at "retreat" locations and lasted for two or three days each.

20. Executive Director, the Ontario Conservation Council.

21. Dolores Montgomery undertook the task of co-ordinating the Public Coalition office with extraordinary enthusiasm and competence. Her contributions to the inquiry, especially during the public information hearings, were particularly noteworthy. She lost her life in the tragic aircraft disaster at Fraserdale, Ontario, on September 4, 1976.

22. We are particularly impressed with the virtually single-handed efforts of Mrs. L.E. Ayers, at Baythorn Public School, Thornhill and her special classes on energy for grades IV to VII. The presentation of a brief by her Grade IV students on May 24, 1977, was a "highlight" of the Commission's hearings.

The Colleges of Applied Arts and Technology have successfully launched a broad range of energy courses relating to alternative energy sources and energy conservation. Furthermore, imaginative programmes have been organized by the institutes of environmental studies at the universities of Toronto, York, and Waterloo. But, in general, the universities have not risen to the challenge, especially in so far as involving the general public is concerned.

23. In connection with the potential role of nuclear power in Sweden, some 9,000 study circles involving more than 75,000 people were established. The cost to the Swedish government was about $650,000. See D. Nelkin and M. Pollak, *Public Policy*, vol. 25, 1977, p. 333.

24. E. Ashby, *Reconciling Man with the Environment*, p. 85, Stanford, California: Stanford University Press, 1978. 1